FOOD SCIENCE AND TECHNOLOGY

TOMATOES

CULTIVATION, VARIETIES AND NUTRITION

FOOD SCIENCE AND TECHNOLOGY

Additional books in this series can be found on Nova's website under the Series tab.

Additional e-books in this series can be found on Nova's website under the e-book tab.

AGRICULTURE ISSUES AND POLICIES

Additional books in this series can be found on Nova's website under the Series tab.

Additional e-books in this series can be found on Nova's website under the e-book tab.

FOOD SCIENCE AND TECHNOLOGY

TOMATOES

CULTIVATION, VARIETIES AND NUTRITION

TADAHISA HIGASHIDE
EDITOR

nova publishers
New York

For permission to use material from this book please contact us:
Telephone 631-231-7269; Fax 631-231-8175
Web Site: http://www.novapublishers.com

Library of Congress Cataloging-in-Publication Data

Tomatoes : cultivation, varieties and nutrition / editor: Tadahisa Higashide.
p. cm.
Includes index.
ISBN 978-1-62417-915-0 (hardcover)
1. Tomatoes. 2. Tomatoes--Varieties. 3. Tomatoes--Health aspects. I. Higashide, Tadahisa.
SB349.T62 2013
635'.642--dc23

2012051677

Published by Nova Science Publishers, Inc. † New York

PREFACE

Tomatoes attract people such as growers, food and distribution companies, breeders, plant physiologists, molecular biologists, other researchers, and nutrition-minded and health conscious consumers. Recently, tomato industries and researches have been advanced tremendously so that many findings are released. However, because of the rapid progress, each tomato specialist could be ignorant of fields outside her/his specialty. For example, some molecular biologists have little knowledge on climate control in a greenhouse, and some plant physiologists don't know so much how lycopene of tomatoes influences in human body though they must be interested in the knowledge on tomatoes each other. This book must be helpful for them. Although this book is not a comprehensive guide on tomatoes, it is accessible to the ordinary reader. The book is consisted of recent topic written by tomato experts in a wide range of fields. These authors are top-class researchers in their field in the entire world, Asia, Africa, South and North America and Europe.

The book has two parts. First part is consisted of 10 chapters on tomato production. This part describes physiology of tomatoes such as yield components and photosynthate translocation, cultivation techniques such as sensing and control in a greenhouse and in tomato diseases, effects of cultivation methods on product, and conditions of international trade of tomatoes. These chapters in this part target mainly improvements of stability and efficiency of tomato production. Second part is consisted of 7 chapters on nutritional contents of tomato fruits. This part describes nutritional components, especially antioxidant such as lycopene, effects of ripening and processing on these contents, and effects of the contents on human health. These chapters target improvements of nutrition in tomatoes as a food and of human health.

I would like to thank each author for their efforts and providing the latest findings. I will be glad if this book is any help to the people concerned on tomatoes. I would like to hear many readers' opinions, comments and suggestions.

Tadahisa Higashide
Tsukuba, April 2012

I. Production of Tomatoes

In: Tomatoes: Cultivation, Varieties and Nutrition
Editor: Tadahisa Higashide

ISBN: 978-1-62417-915-0

Chapter 1

GREENHOUSE TOMATO YIELD AND SOLAR RADIATION

Tadahisa Higashide
National Agricultural and food Research Organization,
Institute of Vegetable and Tea Science, Japan

ABSTRACT

Tomato (*Solanum lycopersicum*) yield is determined by many processes such as leaf photosynthetic rate, fruit set, light interception, total dry matter production, distribution to fruits. These processes are called yield components. The important point is that the components are arranged in a hierarchy and that upper components are determined by related lower components, e.g., fresh fruit yield is directly determined by dry weight yield and fruit dry matter content. This chapter focused solar radiation and tomato yield components. Yield and yield components are affected directly or indirectly by solar radiation. Leaf photosynthetic rate affects directly on dry matter production per intercepted photosynthetic active radiation; light use efficiency (LUE). Light extinction coefficient; *k*, also affects on LUE. In a crop canopy that has a small *k*, light interception by lower leaves is higher than that in a large *k*. Although light-photosynthesis curve could limits leaf photosynthetic rate in upper leaves, the rate in the lower leaves in small *k* canopy is higher than that in large *k* canopy. Photosynthetic rate per plant is integrated leaf photosynthetic rate from lower to upper leaves. Thus, LUE in a crop canopy that has a small *k* is higher than that in a large *k*. Greenhouse tomato yield in The Netherlands has increased tremendously over the past 50 years. This increase in yield is caused by many factors, e.g., improvements in greenhouse transmissivity and developments in cultivation techniques such as longer cultivation, soilless culture, high wire training systems, CO_2 enrichment, environmental control by computers, and breeding for higher yielding cultivars. Higashide et al. investigated higher yielding cultivars of tomatoes that were released over the past 50 years. They reported that total dry matter production of plants increased with the year of release, dry matter partitioning to fruit of modern cultivars was not increased and fruit dry matter content was not decreased. Thus, an increase in yield over the past 50 years in Dutch tomato was caused by an increase in total dry matter production. Then, the increase in dry matter production was caused by high LUE. Light extinction coefficient in the crop canopy decreased while leaf photosynthetic rate increased significantly with the year of release.

INTRODUCTION

Greenhouse tomato (*Solanum lycopersicum*) yield in The Netherlands has increased tremendously from 30 kg•m^{-2} per year in the 1980s to 60 kg•m^{-2} per year in 2005 [1]. This increase in yield is caused by many factors, e.g., improvements in greenhouse transmissivity and developments in cultivation techniques such as longer cultivation, soilless culture, high wire training systems, CO_2 enrichment and environmental control by computers.

Tomato is also the most important vegetable in Japan. For many years, total production value of tomato occupies first place among vegetables in Japan [2]. However, the yield in Japan has not increased from the 1980s and has remained less than 30 kg•m^{-2} per year. Although cultivation techniques and breeding new cultivars have also been improved in Japan, an increase in the yield still has not been succeeded.

Solar radiation is the most important factor for plant production. In modern greenhouse, many conditions such as temperature, humidity, nutrient solution, diseases and pests are well-controlled. However, controlling solar radiation, except shading, is difficult so that the yield is often limited by solar radiation.

To produce tomatoes efficiently and improve the yield, all tomato growers and researchers should comprehend a relationship between light interception by crop canopy and tomato production. This chapter focused on interception of solar radiation by a plant canopy in a greenhouse and an influence of the intercepted light on tomato yield.

INTERCEPTED LIGHT BY TOMATO LEAVES AND PLANTS

Light Strength within a Plant Canopy and Intercepted Light by the Canopy

Irradiated solar radiation is intercepted by tomato plants. Tomato plants, mainly leaves, perform photosynthesis by the solar radiation. In a field or a greenhouse, tomato plants usually make a plant canopy. Light strength under the plant canopy is lower than that above the plant canopy. We can measure fraction intercepted light by the plant canopy. Intercepted light by the canopy may be derived from subtraction the light strength under the canopy from that above the canopy. When light strength under the canopy is 10% of that above the canopy, intercepted light by the canopy is 90% (100 – 10; Figure 1).

Intercepted light may also be measured within the canopy. Intercepted light may usually increase as a leaf number increases. Leaf number and cumulative leaf area from above the canopy are higher at a lower layer of the canopy than those at an upper layer. Then, light strength at the upper layer is higher than that at the lower layer. Leaf area index (LAI) indicates cumulative leaf area per land area. For example, LAI is 2 when cumulative leaf area of a canopy is 2 m^2 per 1 m^2 land area. LAI at a point in a lower layer of the canopy is higher than that in an upper layer (Figure 1).

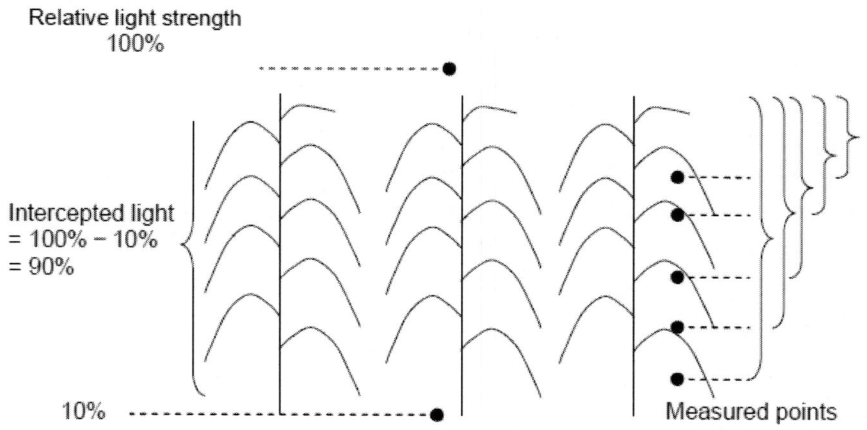

Figure 1. Relative light strength and intercepted light in a plant canopy.

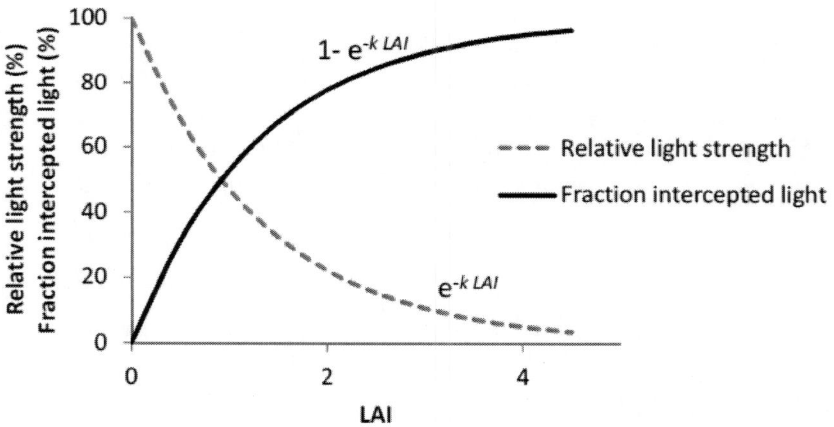

Figure 2. Relative light strength and a intercepted light in a plant canopy as a function of leaf area index (LAI) and light extinction coefficient (k = 0.75).

Monsi and Saeki [3] showed that a relationship between light interception and LAI in a plant canopy conformed to the Beer's rule. Then, relative light strength within a plant canopy can be described as a function of LAI (Figure 2). Light strength decreases exponentially as LAI increases, i.e., lower layer of the canopy.

Does Light Interception Increase Always with an Increase in LAI?

Figure 2 shows that light interception increases from about 30% to 50% when LAI increases from 0.5 to 1.0. Light interception increases from about 50% to 80% when LAI increases from 1 to 2. However, light interception increases from about 90% to 95% when LAI increases from 3 to 4, and in LAI increasing from 4 to 5, only a little increase in light interception. Thus, even if LAI increases in high LAI plants the light interception little increases since leaves are covered each other as plant growth.

Figure 2 suggests two points of crop management in tomato production. First, plants should be grown avoiding too high LAI. In too high LAI plants, the lower leaves may be received light lower than light compensation point. These leaves cannot produce dry matter by photosynthesis but consume it by their respiration. Heuvelink et al. [4] confirmed that tomato yield in a greenhouse increased as an increase in LAI and little increased at LAI higher than 4.0. Second, in low LAI, light is not effective use of greenhouse area. Tomato plants are grown in a short greenhouse in some countries. The plant height is limited by height of the short greenhouse so that LAI is also limited. Even if in a tall greenhouse LAI often may be lower than 3 at young plants or in a low plant density. In these cases, there are still redundant light that are not intercepted by the plants. Then, tomato yield may be increased by control in appropriate LAI such as high plants density and growing an extra stem.

Light Extinction Coefficient; *k*, Shows How Much Light Penetrates in a Plant Canopy

k in Figure 2 is called a light extinction coefficient. When irradiated light penetrates much to lower leaves, a value of *k* may decrease. When many part of the light intercept by upper leaves and not reach to the lower leaves, the value of *k* may increase. A value of the coefficient in a greenhouse tomato is about 0.6–0.9 [5]. The coefficients may differ according to species and cultivars and generally may not differ depending on plant density and plant height.

LIGHT INTERCEPTION AND PHOTOSYNTHESIS

Photosynthesis in an Individual Leaf

Photosynthesis is one of the most important process in plant growth and fruit production. Figure 3 shows photosynthetic rate in an individual leaf as a function of light strength [6]. The photosynthetic rate increases linearly as light strength increases at low light level. However, the increase in the rate is getting slowly and saturates. This curve is called a light-photosynthesis curve. We can accordingly estimate photosynthetic rate of an individual leaf by the light strength. Photosynthetic rate at saturated is called maximum photosynthetic rate, P_{max}, that vary with the species and cultivars. Since clear-sky insolation may be higher than P_{max}, higher P_{max} plants may have an advantage in growth and yield.

Photosynthesis in a Plant Canopy

Light strength in a plant canopy is higher at upper leaves but lower at lower leaves. Even if upper leaves reached P_{max}, photosynthetic rate of lower leaves may still not reach P_{max} accordingly. Then, in contrast with individual leaf photosynthetic rate, photosynthetic rate of a plant canopy with sufficient LAI may not draw a saturation curve. As described previously, light strength within a plant canopy may be determined as a function of LAI and light extinction coefficient (Figure 2). Photosynthetic rate of each leaves may be estimated with the light strength and light-photosynthesis curve (Figure 3). We may accordingly estimate light strength on each leaves from lower to upper layer. Since photosynthetic rate of a plant canopy is an integration of photosynthetic rate of each leaves, we can also estimate photosynthetic rate of the plant canopy consequently.

Under supplied enough solar radiation, the photosynthetic rate and dry matter production of the leaves may be determined by the maximum photosynthetic rate. However, even if no difference in maximum photosynthetic rate, light intensity and photosynthetic rate of lower leaves differ depending on light permeability in the plant canopy (Figure 4).

For example, as long as the same maximum photosynthetic rate and LAI, if light intensity on lower leaves is higher, the photosynthetic rate of the leaves and of the whole plant are also higher. Thus, canopy photosynthetic rate is determined by not only maximum photosynthetic rate but also light permeability in the canopy, i.e. light extinction coefficient. Light permeability in a plant canopy is determined by the canopy structure such as a leaf shape, angle and distribution. Light extinction coefficient; *k*, indicates light permeability in a plant canopy. In a large *k* canopy, most part of light is intercepted by upper leaves. In contrast, in a small *k* canopy, light penetrates to even if lower leaves. Photosynthetic rate by lower leaves is therefore higher in a small *k* canopy than that in a large *k* canopy, and the rate of the whole plant is also higher in small *k* canopy accordingly.

As described above, photosynthetic rate of a plant canopy is determined by not only maximum photosynthetic rate but also light extinction coefficient and light-photosynthetic curve.

Individual leaf photosynthetic rate may be influenced by previous environmental conditions [7; 8]. Maximum photosynthetic rate of a leaf grown under high light intensity may be higher than that grown under low light. To be more precise, photosynthetic rate of a plant canopy might be affected these differences in the maximum rate.

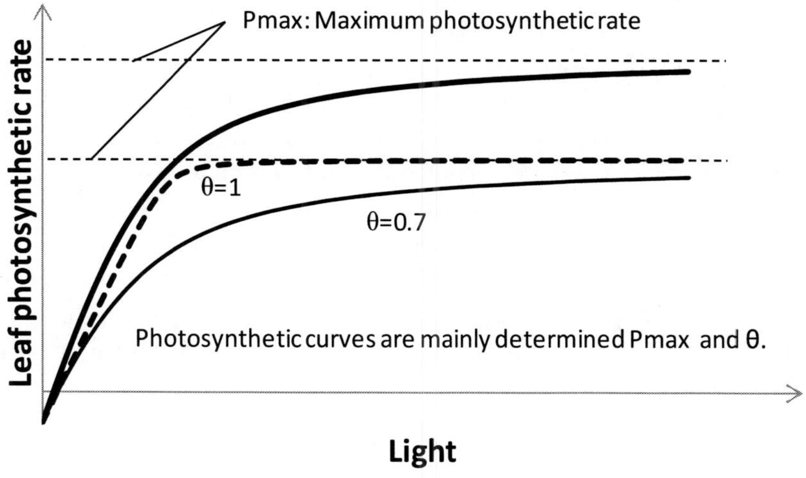

Figure 3. Light-photosynthetic curve in an individual leaf. θ determines shape of the curve. $0 \leq \theta \leq 1$.

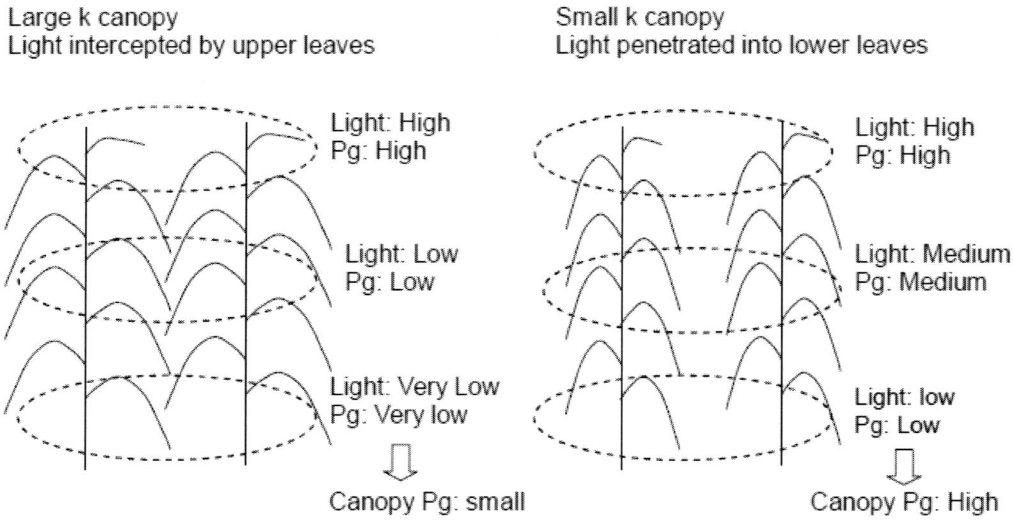

Figure 4. Light extinction coefficient affects on canopy photosynthetic rate. Pg; photosynthetic rate.

However, since leaves grown low light level are usually positioned lower layer in the canopy, received light and photosynthetic rate of the leaves may not reach levels of the maximum photosynthetic rate. There is therefore little problem to estimate photosynthetic rate of a whole plant canopy without accounting the difference in each leaf photosynthetic rate derived from previous environmental conditions.

CO_2 Level

Leaf photosynthetic rate as a function of CO_2 concentration also shows a similar curve like a light-photosynthesis curve. Photosynthetic rate increases linearly as CO_2 concentration increases at low CO_2 level but the increase in the rate is getting slowly and saturates. To improve tomato yield, CO_2 enrichment is often used in a greenhouse. An effect of CO_2 enrichment is higher under higher light intensity but still effective under low light level.

CO_2 enrichment may often result abundant leaves and increase in their thickness. When growers see these tendencies, they often assume that distribution to leaves increases instead of distribution to fruits.

However, there might have been no report to decrease in fruit distribution by CO_2 enrichment. Conversely, distribution to fruits may increase by CO_2 enrichment because of an improvement of fruit set. CO_2 enrichment is effective against plant growth for an initial a few months but the effect may decrease for the following term because of acclimatization to high CO_2 concentration [9; 10].

Sink/Source Affects Photosynthetic Rate

Fruit number, leaf area and their balance may affect leaf photosynthetic rate. A tomato plant with growing fruits allocated the produced photosynthate to fruits and utilizes it for the fruit growth and development. When the plant has a few fruits because of low fruit set or heavy pruning, a plant might not use up photosynthate. When a lot of photosynthate is produced under high light intensity and high CO_2 level, the photosynthate is also abundant. In these cases, supply of photosynthate is higher than demand of it.

Accordingly, residual photosynthate in leaves might restrict the leaf photosynthetic rate. The restriction mechanism might be related metabolism and translocation of photosynthate. Parts of producing and consuming photosynthate are called "source" and "sink", respectively. Then prominent source compared to sink may restrict photosynthesis. However, tomato fruit yield may be mainly restricted by source strength. This can be understand based on that weight per fruit decreases as an increase in fruit number per truss [4; 11].

MEASUREMENT ON LIGHT INTERCEPTION BY TOMATO PLANTS IN A GREENHOUSE

Light interception may contribute directly dry matter production of tomato plants. Light interception could be employed on cultivation management. For example; an adequate amount of nutrient element supply per period could be determined based on light interception and a crop growth model [12].

Measurement on light interception and obtaining a light extinction coefficient are fundamental process to investigate and promote crop productivity.

As previously mentioned, light attenuation in the crop canopy can be described by the light extinction coefficient (k), cumulative LAI, and the equation: $I = I_0 \, e^{-k \, \mathrm{LAI}}$, in which I = horizontal light strength in the canopy, I_0 = horizontal light strength above the canopy (Figure 2); [3]. For a tomato plants canopy, light extinction coefficients differed between tomato cultivars [5]. Incidentally, it was reported that light attenuation in the tomato canopy on a slope was smaller than that on a flat even if in the same cultivar [13].

Measurement Light Interception by a Light Sensor

To obtain a light extinction coefficient (Figure 1), photosynthetic photon flux (PPF) is measured by a long PPF sensor (e.g. LI-191SA; LI-COR, Lincoln, US) at different heights in a closed plant canopy. The measured PPF level by the sensor is not light level on surface of the leaves but is horizontal light level. PPF above the plant canopy is also measured by a PPF sensor (e.g. LI-190SA; LI-COR) and recorded by data recorder.

Based on these PPF data, relative light strength at the different height against PPF above the canopy are calculated. Individual leaf area of the plants is required.

Then, cumulative LAI from the top at different heights is calculated based on the individual leaf area and leaf number at that height. Light extinction coefficient is obtained as a slope in logarithmic regression of relative light strength against cumulative LAI at the different heights.

Physiological and morphological changes were reported in tomato cultivars released over the past 50 years [5].According this report, light extinction coefficients differed significantly between cultivars (0.57–0.91), and each of the coefficients was negatively correlated with the year of release.

Measurement Light Interception by Solar Radiation Recording Films

The light interception by plants is also able to measure by using simple recording films for measuring cumulative solar radiation (Opt leaf, Taisei E and L, Tokyo, Japan). The solarimeter film can measure radiation by the remaining percentage dye after exposure to radiation directly at the same time on many points such as individual leaves in a plant canopy [14, 15].

The film is attached on each point and measures cumulative solar radiation. The cumulative solar radiation as light interception by each leaf is calculated by using a calibration line. Individual leaf area of the plants is also required in this measurement to obtain light extinction coefficient. By using the films, light interception by tomato plants grown on a slope and a flat was measured [13].

Although the data was spread widely, light interception was significantly correlated with the cumulative LAI in both the slope and the flat. The light attenuation in the plant canopy on the slope was smaller than that on the flat and then light interception by lower leaves of tomato plants on the slope was higher than that on the flat.

Simulation for Light Interception by Modeling Three-Dimensional Plants

Light interception by plants which are planted in an orchard may be affected by the distance between plants and by the orientation of rows, compared to other crops such as a wheat and rice. The relation between light interception in an orchard and the arrangement and orientation of rows were investigated [16; 17; 18; 19]. Digitized plants may make it possible to efficiently use data in crop production such as in the calculation of light interception [20; 21; 22; 23; 24; 25].Another approach to represent plant architecture was based on a growth algorithm such as the L-system [26; 27; 28; 29; 30].

A ray-tracing simulator was developed for an analysis by modeling three-dimensional (3D) plants (Figure 5), and simulated light interception by individual leaves in similar conditions as the previous slope and flat [13]. Decrease rate of light interception as an increase in LAI was simulated an also expressed in the exponential regression equation.The coefficient k of the model plants on the slope was significantly lower than that on the flat indicating that light interception by leaves at lower positions on the slope was higher than that in plants on the flat. The simulation showed light interception in details that had been difficult to measure by actual measurement, e.g., diurnal changes in light interception by the upper, middle and lower leaves.

Wiechers et al. also [31] developed 3D plants approach on light distribution in a cucumber canopy. They showed that simulations of leaf level PAR based on combinations of detailed 3D surfaced-based plant and light distribution models. These models are suitable to derive light-induced physiological responses on organ level.

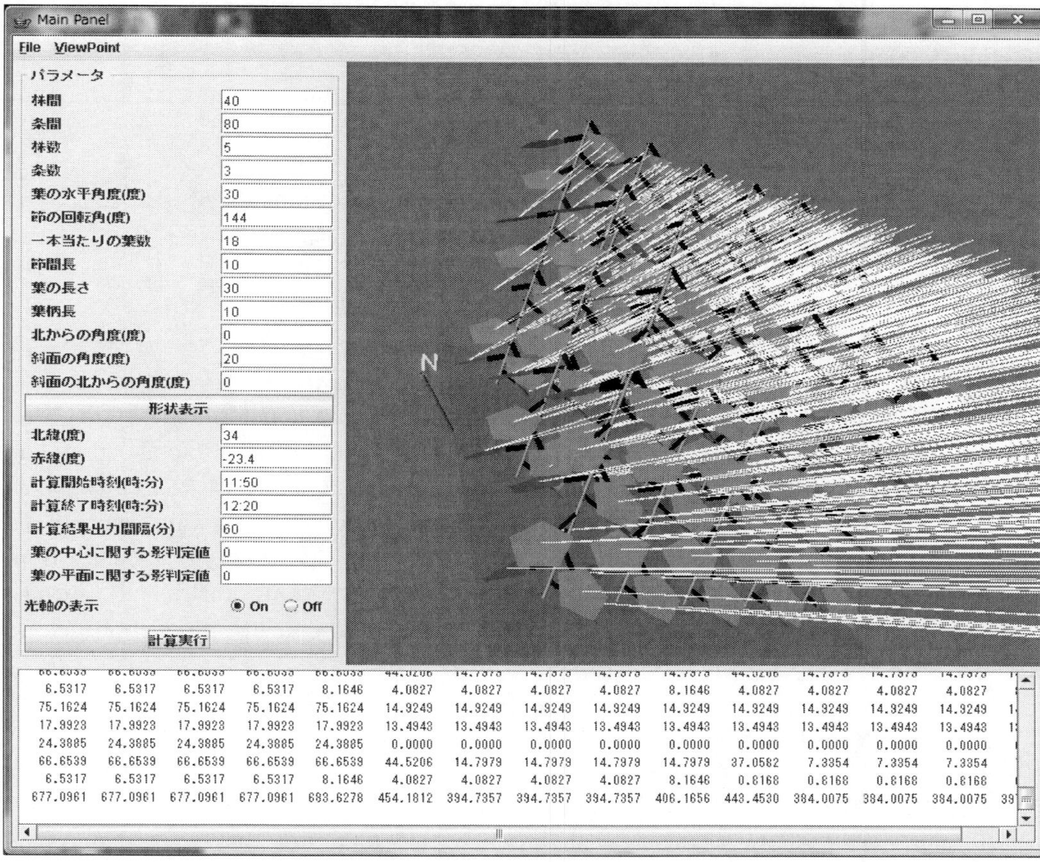

Figure 5. A ray-tracing simulator for an analysis by modeling three-dimensional plants. Simulated rays and results of calculation of light interception [32].

YIELD COMPONENTS

Yield Components Hierarchy

A grower often says that high yield is obtained in this season because of fine weather. However, it is not easy to identify a cause of the high yield because yield is determined by many processes such as photosynthetic rate, fruit set, light interception, dry matter distribution and so on. These processes that determine crop yield are called yield components. Yield components affect tomato yield hierarchically (Figure 6); [5]. For example, possible steps are as follows: fine weather = high solar radiation → high photosynthetic rate in an individual leaf → high dry matter production , and distribution to fruits increase or not change → high dry fruit yield, and fruit dry matter contents decrease or not change → high fruit yield. The followings are typical yield components and interpretation on them.

Dry matter production; Sum of the dry weight produced by crops. Fruit including harvested and not harvested + leaves including removed + stem + roots dry weight (g m^{-2}).

Fruit dry matter contents; dry fruit weight divided by fresh fruit weight (g g^{-1}), 3.5−7.5 in a usual tomato fruit.

Light use efficiency; dry matter production per light interception (g MJ^{-1}). Total dry matter production = light use efficiency × intercepted light.

Fraction of light intercepted; percentage of light intercepted to irradiated light. Light interception by plants per area (MJ m^{-2} (MJ $m^{-2})^{-1}$)

Individual leaf photosynthesis; see Figure 3.

Light extinction coefficient; light attenuation rate in a plant canopy. See Figure 2.

Sink strength; an attraction ability of assimilate produced in leaves.

Yield Components in Tomatoes

Figure 6 shows as follows. First, fresh fruit yield of tomato is determined by dry fruit yield and dry matter content in fruits. An increase in dry fruit yield conduces to an increase in fresh fruit yield. A decrease in dry matter content in fruits also conduces to an increase in fresh yield. However, a decrease in dry matter content may indicate high water content and consequently low quality of fruits. Secondly, an increase in dry fruit yield results from increases in total dry matter production and/or in fraction of dry matter distributed to fruits. The fraction is calculated dividing dry fruit weight by total dry matter production, and is called harvest index.

A distribution of assimilate to fruits is affected by sink strength per fruit and fruit number per plant. Sink strength imply an attraction ability of assimilate produced mainly in leaves. A distribution of assimilate in a plant is determined by sink strength. In a tomato plant, fruits are the largest sink organ and many part of assimilate is allocated to fruits.

The other determining element in dry fruit yield; total dry matter production, is determined by light use efficiency and fraction of intercepted light. Theoretically, an increase in fraction of intercepted; light interception by plants per land area, is not effective in increase in a yield. As described previously, when plants have enough LAI, an increase in LAI has an only small effect on light interception.

An increase in light use efficiency is effective in an increase in a yield more than an increase in fraction of light intercepted accordingly. Individual leaf photosynthesis and light extinction coefficient may determine light use efficiency, i.e., the efficiency may increase by an increase in individual leaf photosynthesis and/or decrease in light extinction coefficient.These factors are mainly determined by characteristics of cultivars when there is no restriction of cultivation condition such as temperature, nutrition and water supply. CO_2 level also affects yield components. High CO_2 level enhances leaf photosynthesis without increasing light and then may increase light use efficiency.

Improvement High-Yielding Cultivars in Crops

Many crops have increased in yield by breeding of high-yielding cultivars over the long time. Plant breeding of high-yielding cultivars has much to contribute to increase in crop yield, especially, for crops grown on open field without controlling environmental conditions.

To find out causes of yield improvement, the genetic components have been investigated in several crops. It was shown that increases in yield of wheat (*Triticum aestivum*) and barely (*Hordeum vulgae*) over the past century was caused by increased in harvest indices [33].

Harvest indices of modern rice (*Oriza sativa*) cultivars were also higher than those of old cultivars [34]. In soybean cultivars released in Canada, the yield increased 0.5% per year from 1930's and the increase was related to increase in harvest index, photosynthetic rate and stomatal conductance and decrease in LAI. In tomatoes, partitioning dry matter to fruits determines the yield of different types such as beef, cherry and round tomatoes [34], i.e., the partitioning dry matter to fruits in large round tomatoes is higher than that in cherry tomatoes.

These show that yield improvement by breeding in these crops is primarily related to increase in harvest index rather than total dry matter production. However, an increase in maize (*Zea mays*) yield by breeding has not been related to harvest index but to total dry matter production [33]. Light use efficiency of modern maize cultivars was higher than that of old cultivars despite no difference in light extinction coefficient and in light absorption by crops [36]. Among the same type of tomatoes, partitioning to fruits of modern tomato cultivars was also not higher than that of old cultivars [37].

HIGH-YIELDING CULTIVARS IN TOMATOES

Tomato Yields of Modern Cultivars Are Higher Than Those of Old Cultivars

As mentioned above, greenhouse tomato yield in The Netherlands has doubled from 30 kg m^{-2} $year^{-1}$ in the 1980's to 60 kg m^{-2} $year^{-1}$ in 2005 [1]. This increase in yield is caused by many factors such as improvements greenhouse transmissivity and developments cultivation techniques.

Breeding for high-yielding cultivars have also played a role in this increase in yield. van der Ploeg et al. [37] reported that yield of modern cultivars under current conditions was 40% higher than that of cultivars in the same greenhouse conditions.

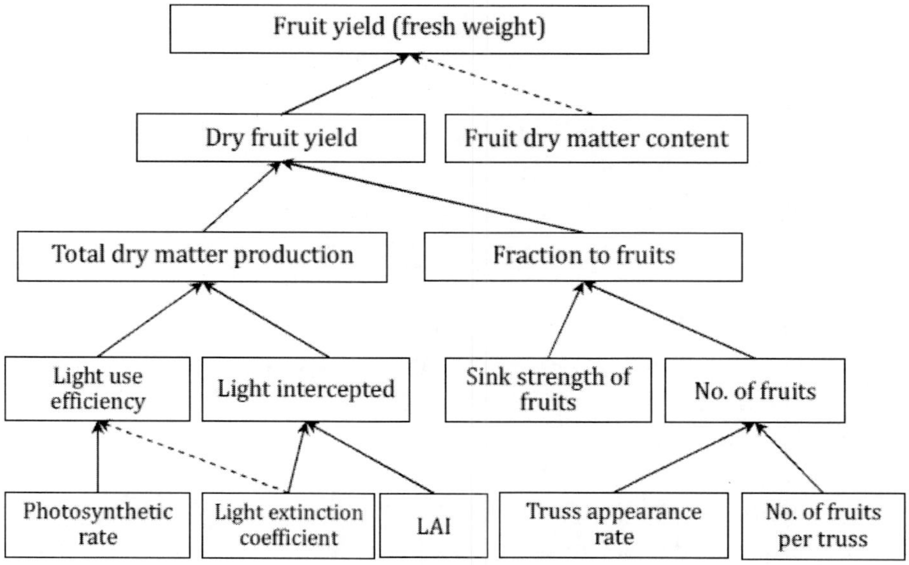

Figure 6. Hierarchy of yield components in tomatoes; arrowed line: positive relation; dashed-arrowed line: negative relation.

Higashide and Heuvelink [5] also reported that higher yield of modern tomato cultivars was caused by increase in total dry matter production of plants, not by an increase in dry matter partitioning to fruits. They compared nine tomato cultivars released from 1950 to 2002 under the current conditions. As a result, fresh tomato yield increased about 0.9% per year from 1950 to 2002. Dry fruit yield and total dry matter production of plants increased with the year of release, whereas the fruit dry matter content and dry matter partitioning to fruits were not correlated with the year of released. These results indicate that the increase in tomato yield is caused by an increase in total dry matter production of plants.

High Light Use Efficiency Resulted in an Increase in Total Dry Matter Production

As mentioned before, total dry matter production is determined by light use efficiency and fraction of intercepted light. Light use efficiency was correlated with the release year of tomato cultivars [5]. However, there was no correlation between the year of release and fraction of light intercepted and LAI. The increase in total dry matter production may thus be caused by an increase in light use efficiency and not by an increase in intercepted light. Light extinction coefficient was negatively-correlated with the release year and leaf photosynthetic rate was correlated with the release year.Thus, an increase in photosynthetic rate and a decrease in light extinction coefficient may result in the increase in light use efficiency.

In photosynthetic rate of an individual leaf of rice, wheat and cotton, there were no difference between old and modern cultivars [38; 39; 40; 41; 34). These report showed that an increase in yield of these crops was mainly caused by change in morphological factors such as light extinction coefficient. However, in tomatoes, leaf photosynthetic rate of modern cultivars were higher than that of old cultivars [5].

These results on tomatoes were also supported by correlations between each yield components [5], i.e., fresh fruit yield were strongly correlated with dry fruit yield and total dry matter production; total dry matter production was correlated strongly with light use efficiency; there was negative correlation between light use efficiency and light extinction coefficient.

The report also mentioned on a Japanese tomato cultivar [5].The yield of the Japanese cultivar was lower than those of the Dutch modern cultivars though the dry matter content of the Japanese cultivar was higher than those of the Dutch modern cultivars.

Improvement in Cultivation Techniques for High-Yield in Tomatoes

Greenhouse Transmissivity

Greenhouse transmissivity shows how much radiation penetrates into a greenhouse. For example, when solar radiation decreases 30% by frames and covering materials compared to outside solar radiation, transmissivity of the greenhouse is 70%.

In The Netherlands, 1% rule in greenhouse crops is well-known by growers and researchers. The simple rule is that change in 1% in light occur change in 1% in yield, e.g., 2% crop yield may decrease by decrease in 2% of total solar radiation. To increase in solar radiation in a greenhouse and then in yield, greenhouse transmissivity have improved and reached about 80% in state-of the-art greenhouse in The Netherlands.

Rockwool System

To produce dry matter by photosynthesis, plants absorb CO_2 through stomata on a leaf. When a plant is not supplied enough water, stomata may close to decrease transpiration and to keep water within the plant body. CO_2 absorption may also decrease by closing stomata accordingly. Plants grown in a rockwool system are supplied enough water and nutrients constantly. The unconstrained plants absorb CO_2 and produce dry matter efficiently.

High Wiring System and LAI Control

A high wiring system in a greenhouse combines control in LAI and workability by growers. As described above, to improve tomato yield, a grower has to control and to keep an adequate LAI. Growers could control LAI by modifications of plant density.

However, without a high wiring system, high plant density by low height plants may decrease significantly in workability because of shortages of working space. A high wiring system make easily to control stem density depend on seasons. The stem density may be kept low and then be kept low LAI in the winter season because of low light. However, stem density may be increased toward summer and be kept high LAI during the summer.

CO_2 Enrichment and Environmental Control by a Computer

CO_2 is substrate of photosynthesis. Since high CO_2 level makes an increase in absorption of CO_2 by leaves and then increases in photosynthesis, CO_2 enrichment is often installed in a greenhouse producing tomato. CO_2 enrichment is more effective in higher light condition. However, even if in low light condition, net-photosynthetic rate may increase because of restriction in photorespiration.

As mentioned above, CO_2 enrichment makes thick and broad leaves. However, there may be no reports that decrease in dry matter partitioning to tomato fruits by CO_2 enrichment. Fruit set and number may be improved by CO_2 enrichment and then increase in dry matter partitioning to fruits.

In a greenhouse, environmental conditions are controlled by several devices. For example, air temperature is controlled by ventilation, shading and heating system. Humidity is also affected based on these devices consequently. Without computers, it is difficult to control in levels of these elements. To maximize dry matter production and the yield, we need a computer for a greenhouse.Then, photosynthesis and dry matter production are expressed as a function of several elements such as light, CO_2, humidity and temperature.

FLUCTUATION IN WEEKLY YIELD IN TOMATOES

Even if in a controlled greenhouse, weekly tomato yield is often changed. Yield fluctuations cause price fluctuations, a decrease in reliance on the growers, a shift by purchasers to other suppliers, and an inadequate distribution of labour. High temperature might cause fluctuations in tomato yield, since temperature affects fruit set and development [42].

Adams, Adams and Valdes [43; 44] recorded increased tomato yields after raising air temperature for 1 week, but subsequently recorded lower yield, with no effect on cumulative yield. De Koning [45] showed that high temperature enhanced early fruit growth at the expense of vegetative growth. The increase in assimilate partitioning to fruits resulted in reduced leaf growth, increased flower abortion and subsequently delayed fruit growth. Bertin [46] reported that strong competition for assimilates (i.e., strong sink strength) negatively affected fruit set in tomatoes. These results suggest that much of the photoassimilate would have been partitioned into fruits rather than into flowers and thus flower abortion increases. Change in solar radiation might also cause yield fluctuations.

Lower light levels (i.e., lower source activity) caused by shading reduce fruit set and yield [47; 48; 49; 50]. Higashide [51] showed that solar radiation before anthesis was one of the important factors in prediction of tomato yield under warm greenhouse conditions.In his report, it was investigated that effects of solar radiation and temperature in different periods before and after anthesis on yield of tomato in summer greenhouse cultivation.As a result, fluctuations in yield were caused mainly by the variation in the fruit number rather than the fruit weight, and the number of harvested fruit and the yield of plants grown in summer and autumn were significantly and positively correlated with solar radiation during 10–4 days before anthesis. Then, it was succeeded to predict fluctuations in fruit number and yield by a model based on the solar radiation before anthesis.

CONCLUSION

Solar radiation is the most important factor on tomato production but is generally unable to control except for reducing light by a screen in a greenhouse. Irradiated solar radiation toward ground is absorbed by crops and affects dry matter production and then fruit yield in tomatoes. The absorbed radiation, i.e., light interception, is determined by LAI and light extinction coefficient in the plant canopy.

Efficiency of dry matter production per intercepted light, i.e., light use efficiency is determined by not only leaf photosynthetic rate but also light extinction coefficient. Measurement on the light interception requires considerable efforts due to measure on leaf area and obtain light extinction coefficient in the canopy.

However, to study and improve crop yield by breeding or cultivation techniques, yield components such as LAI and light interception must be analyzed. An increase in yield in modern greenhouse tomato was caused by an increase in light use efficiency. Solar radiation before anthesis affected fluctuations in tomato yield in warm conditions.

REFERENCES

[1] Kwantitatieve Informatie voor de Glastuinbouw. Proefstation voor de bloemistrij en glasgroente (PGB). *Kwantitatieve Informatie voor de Glastuinbouw, 2005–2006*, Wageningen, The Netherlands (2005).

[2] Ministry of Agriculture Forestry and Fisheries of Japan. *Survey of vegetables production and shipments, Crop survey,* Ministry of Agriculture Forestry and Fisheries of Japan, Tokyo, Japan (2009).

[3] M. Monsi and T. Saeki, *Jpn. J. Bot.*, 14, 22 (1953).

[4] E. Heuvelink, *Ann. Bot.,* 75, 369 (1995).

[5] T. Higashide and E. Heuvelink, *J. Am. Soc. Hort. Sci.,* 134, 460 (2009).

[6] J. H. M. Thornley, *Mathematical models in plant physiology*, Academic Press, London, UK (1976).

[7] I. Terashima, S. Miyazawa and Y. T. Hanba, *J. Plant Research* 114, 93 (2001).

[8] G. Trouwborst, S. W. Hogewoning, J. Harbinson and W. V. Ieperen, *J. Hort. Sci. Biotechnol.*, 86, 403 (2011).

[9] J. D. Cure and B. Acock, *Agr. For. Meteorol.,* 38, 127 (1986).

[10] R. T. Bestford, L. J. Ludwig and A. C. Withers, *J. Exp.Bot,* 41, 925 (1990).

[11] R. Matsuda, K. Suzuki, A. Nakano, T. Higashide and M. Takaichi, *Sci. Hort.,* 127, 520 (2011).

[12] T. Higashide, R. Matsuda and M. Takaichi, *Acta Hort.*, 907, 193 (2011).

[13] T. Higashide, *Agr. For. Meteorol.,* 149, 756 (2009).

[14] A. Isoda, T. Yoshimura, T. Ishikawa, Y. Nakamura, H. Nojima and Y. Takasaki, *Tech. Bul. Fac. Hort. Chiba Univ.*, 43, 39 (1990).

[15] S. Watanabe, Y. Nakano and K. Okano, *Environ. Cont. Biol.,* 39, 121 (2001).

[16] J. Goudriaan, *Agr. For. Meteorol.,* 43, 155 (1988).

[17] J. W. Palmer, *J. Appl. Ecol.,* 14, 601 (1977).

[18] R. S. Johnson and A. N. Lakso, *HortScience*, 26, 1002–1004 (1991).

[19] A. P. Papadopoulos, S. Pararajasingham, *Sci. Hort.*, 69, 1 (1997).

[20] H. Sinoquet, B. Moulia and R. Bonhomme, *Agr. For. Meteorol.*, 55, 233 (1991).

[21] B. Andrieu, N. Ivanov and P. Boissard, *Agr. For. Meteorol.*, 75, 103 (1995).

[22] R. W. Percy and W. Yang, *Oecologia*, 108, 1 (1996).

[23] H. Shono and S. Sase, *Environ. Cont. Biol.*, 34, 75 (1996).

[24] A. Takenaka, Y. Inui and A. Osawa, *Func. Ecol.* 12, 158 (1998).

[25] T. Honjo and H. Shono, *J. Agr. Meteorol.*, 57, 101 (2001).

[26] P. Prunsinkiewicz and A. Lindenmayer, *The algorithmic beauty of plants*, Springer-Verlag, New York (1990).

[27] T. Higashide, H. Shimaji and M. Takaichi, *Acta Hort.*, 519, 43 (2000).

[28] T. Takeda, T. Hirano, S. Urano and I. Horiguchi, *J. Agr. Meteorol.*, 57, 145 (2001).

[29] T. Watanabe, J. S. Hanan, P. M. Room, T. Hasegawa, H. Nakagawa and W. Takahashi, *Ann. Bot.*, 95, 1131 (2005).

[30] K. Kahlen, *Acta Hort.*, 718, 51 (2006).

[31] D. Wiechers, K. Kahlen and H. Stützel, *Agr. For. Meteorol.*, 151, 906 (2011).

[32] T. Higashide, *Bull. Natl. Agric. Res. Cntr .West. Reg.*, 9, 37 (2010).

[33] R. K. M. Hay, *Ann. Appl. Biol.*, 126, 197 (1995).

[34] K. Saitoh, H. Shimoda and K. Ishihara, *Jpn. J. Crop Sci.*, 62, 509 (1993).
[35] L. Ho, *J. Expt. Bot.*, 47, 1239 (1996).
[36] M. Tollenaar and A. Aguilera, *Agron. J.*, 84, 536 (1992).
[37] A. van der Ploeg, M. van der Meer and E. Heuvelink, *Euphytica*, 158, 129 (2007).
[38] P. E. Abbate, F. H. Andrade, L. Lazaro, J. H. Bariffi, H .G. Berardocco, V. H. Inza and F. Marturano, *Crop Sci.*, 38, 1203 (1998).
[39] E. Kuroda and A. Kumura, *Jpn. J. Crop Sci.*, 59, 283 (1990).
[40] E. Kuroda and A. Kumura, *Jpn. J. Crop Sci.*, 59, 298 (1990).
[41] K. Saitoh, H. Shimoda, and K. Ishihara, *Jpn. J. Crop Sci.*, 59, 130 (1990).
[42] S. Sato, M. Kamiyama, T. Iwata, N. Makita, H. Furukawa and H. Ikeda, *Ann. Bot.*, 97, 731 (2006).
[43] S. R. Adams, *Acta Hort.*, 593, 19 (2002).
[44] S. R. Adams and V. Valdes. *J. Hort. Sci. Biotechnol.*, 77, 461 (2002).
[45] A. N. M. De Koning, *Acta Hort.*, 248, 329 (1989).
[46] N. Bertin, *Ann. Bot.*, 75, 55 (1995).
[47] K. E. Cockshull, C. J. Graves and C. R. J. Cave, *J. Hort. Sci.*, 67, 11 (1992).
[48] M. P. N. Gent, *HortScience*, 42, 514 (2007).
[49] V. M. Russo, *HortScience*, 28, 1133 (1993).
[50] T. Wada, H. Ikeda, K. Matsushita, A. Kambara, H. Hirai and K. Abe, *J. Jpn. Soc. Hort. Sci.*, 75, 51 (2006).
[51] T. Higashide, *HortScience*, 44, 1874 (2009).

In: Tomatoes: Cultivation, Varieties and Nutrition
Editor: Tadahisa Higashide

ISBN: 978-1-62417-915-0

Chapter 2

TRANSLOCATION AND DISTRIBUTION OF PHOTOSYNTHATES IN TOMATO PLANTS

Takashi Nishizawa * ***and Yoshihiro Shishido***
Faculty of Agriculture, Yamagata University

ABSTRACT

Carbon budgets of photosynthates in tomato plants as influenced by environmental factors were studied using radioisotopes as tracers. In $^{14}CO_2$-pulse feeding studies, both translocation and distribution of the ^{14}C-photosynthates were affected not only by the relative sink strength of each organ, but also by the phyllotaxis. Among the environmental factors, temperature was the most effective factor for altering the distribution rate of ^{14}C-photosynthates among sink organs but not so much for the export rate from the source leaf. Because the $^{14}CO_2$-pulse feeding method can detect only the destination of temporary assimilated carbon, the $^{14}CO_2$-steady state feeding (SSF) method was developed for clarifying the carbon budget for a whole day including the respiratory loss. As a result, translocation and distribution patterns of the ^{14}C-photosynthates by both $^{14}CO_2$-feeding methods were similar to each other, indicating that they were comparable techniques for detecting the destination of the assimilated carbon. In addition, the carbon budget of young tomato seedlings as affected by different environmental conditions, such as root temperatures, photoperiods and light intensities, were also investigated using the SSF method. Lastly, the positron-emitting tracer imaging system (PETIS) was introduced as a new tracing technology for visualizing the real-time movement of ^{11}C-photosynthates from the source leaf to sink organs.

INTRODUCTION

Both photosynthesis and translocation are two of the most important physiological phenomena for tomato production. In a broad sense, mass production by plants is defined as the difference between photosynthesis and respiration [1, 2]. In commercial tomato

* E-mail address: nisizawa@tds1.tr.yamagata-u.ac.jp.

production, however, the situation is more complicated because the plant growth is often regulated by pinching, plant growth regulators, root restriction etc. for promoting the growth of fruit. In general, the growth of plant organs is regulated not only by photosynthesis of the leaves, but also by translocation and distribution of the photosynthates which are also influenced largely by environmental factors as well as mechanical and chemical growth regulations [3]. Thus, translocation and distribution of photosynthates show the efficiency of the leaf to the productivity of the edible parts. The relation between supply and demand of photosynthates is called a source-sink relationship [4]. Non-photosynthetic organs such as fruit, grains and roots play the role of main sink organ which lead to competition for the demand of photosynthates among sink organs. Although green leaves usually play the role of main source organ, they also often require the supply of photosynthates, especially at an immature stage [5, 6].

^{11}C, ^{13}C and ^{14}C have been used commonly as tracers for the study of source-sink relationships. Because ^{13}C is a stable isotope and is detected by mass spectrometry, there is no strict limitation on the use of both the place and method, while it is often unsuitable for high-precision analysis [7]. ^{11}C is easy to detect because of its strong radioactivity but difficult to launch a long-term search of the assimilated carbon due to its shortened half-life (approximately 20 min.). On the other hand, the half-life of ^{14}C is very long (approximately 5,700 years) and it has often been used for investigating the partitioning of photosynthates, regardless of its strict usage limitations.

When ^{14}C is used as a tracer, leaves are usually closed in the assimilation chambers made of acrylic resin and $^{14}CO_2$ is generated in the chamber by gently dropping a lactic acid solution. Thus, a mix of $^{12}CO_2$ and $^{14}CO_2$ is supplied for 30 to 60 minutes under light conditions. This short-term $^{14}CO_2$-application is called the $^{14}CO_2$-pulse feeding method. The plants are then placed under light or dark conditions for a certain period, and harvested plants are immediately separated into each component part, dried, weighed, and powdered. A 10 to 20 mg dried sample is oxidized by combustion using a sample oxidizer, generated $^{14}CO_2$ is absorbed into a known amount of sintillator, and the ^{14}C radioactivity is counted using liquid scintillation spectrometry. In this Chapter, at $^{14}CO_2$ pulse feeding [8], the technical terms used for understanding the translocation and distribution of photosynthates were defined as;

1) Source leaf: $^{14}CO_2$ fed leaf. $^{14}CO_2$ was applied to single leaf in all the experiments.
2) ^{14}C-export rate: Percentage total export of ^{14}C-photosynthates from the fed leaf = (^{14}C-recovered in a whole plant except the source leaf / Total ^{14}C-recovered) x 100
3) ^{14}C-distiribution rate: Percentage distribution of the ^{14}C-phtoynthates to each part = (^{14}C-photoynthates recovered in each part / ^{14}C-recovered in a whole plant except the source leaf) x 100

1. Effects of Feeding Time on Export and Distribution of the ^{14}C-Photosynthates in Non-Fruiting Tomato Plants

Tomato seedlings were grown at 24/17°C (day/night) under 16-hour photo periods until the 6th leaf counting from the bottom unfolded (non-fruiting stage). When $^{14}CO_2$ was applied to the 3rd leaf (source leaf) at 24 °C for 30 minutes, the time course export of the ^{14}C-photosynthates from the source leaf increased rapidly during the first 2 to 4 hours, irrespective of feeding time and light conditions after feeding indicating that the ^{14}C-photosynthates exported rapidly soon after the feeding (Figure 1) [1].

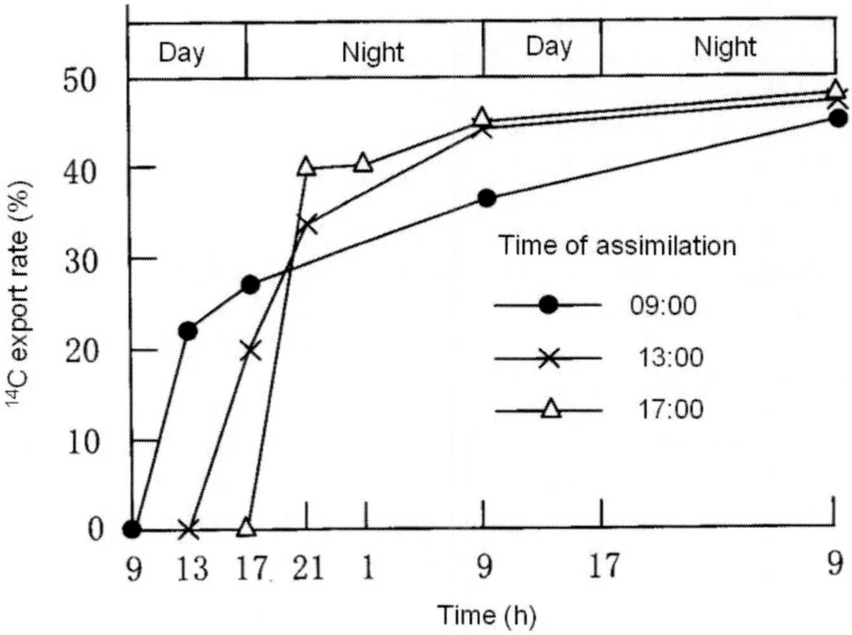

Figure 1. Time course of total export of ^{14}C-photosynthates as affected by feeding time. $^{14}CO_2$ was applied to the 3rd leaf of non-fruiting tomato plants with 6 unfolded leaves and temperatures were kept at 24/17°C (day/night) (n=3).

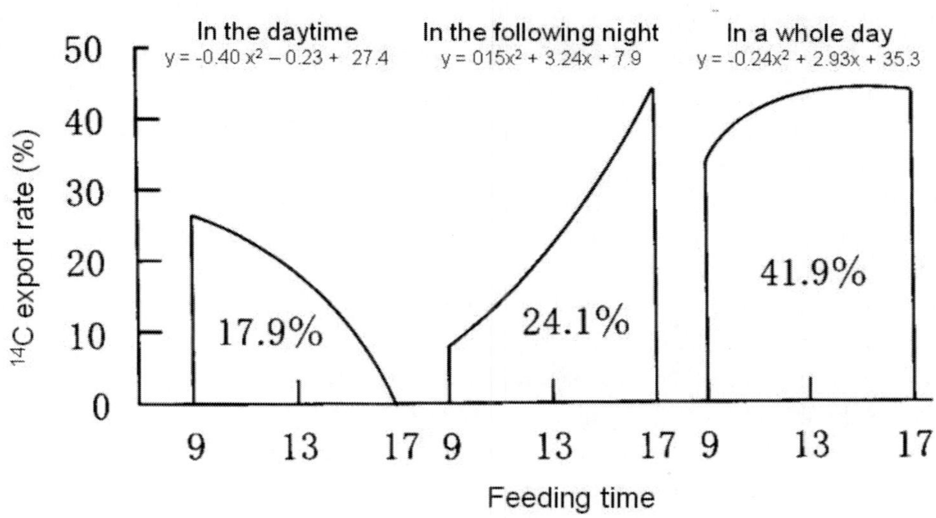

Figure 2. Total export of ^{14}C-photosynthetic products in the daytime, in the following night, and in a whole day as affected by feeding time (n=3). y = total ^{14}C-export rate (%), x = feeding time.

Among the different feeding times, increase in the export was more rapid in latter feeding, reaching 23% after 4 hours when fed at 9:00 compared to 40% after 2 hours when fed at 17:00. Thus, the proportion of remained ^{14}C-photosynthates in the source leaf was higher in the morning than in the evening. Such a comparatively slow export in the morning would be due to the incorporation of a part of ^{14}C-photosynthates into storage reserves following successive photosynthesis. When $^{14}CO_2$ was applied just before darkening, however, most ^{14}C-photosynthates would be in a soluble pool such as sucrose and exported rapidly as shown in potatoes [9].

In other words, a part of photosynthates is utilized preferentially in the source leaf when a shortage of the carbon pool occurs during night, but they also start to export to sink organs immediately after the photosynthesis even under such conditions.

Differing from the rapid export of ^{14}C-photosynthates during the first 4 hours, the additional ^{14}C-export thereafter was small, irrespective of feeding time, especially after 12 hours. Although Khan and Sagar recognized the export even after one week of CO_2 assimilation [10], the re-translocation of photosynthates from temporarily stored organs such as roots and stems should also be considered.

Based on Figure 1, the total ^{14}C-export rate in the light period (9:00-17:00), in the following dark period (17:00-9:00), and in a whole day was diagrammed and formulated (Figure 2). By integrating these equations, it was indicated that the total ^{14}C-export rate for a whole day was 41.9% and the ^{14}C-export rate during the light period during the following dark period was approximately 3:4.

2. Effects of Night Temperatures on Export and Distribution of the ^{14}C-Photosynthates in Non-Fruiting and Fruiting Tomato Plants

It is well known that temperatures directly affect the growth of plants [2, 5], with sifts in temperature often resulting in changes in the developmental patterns of plants. Since plant growth is mainly defined by the accumulation of energy, in other words, the difference of photosynthesis and respiration, both export and distribution of the photosynthates will be also influenced by temperature.

$^{14}CO_2$ was applied to the 3rd leaf of non-fruiting tomato seedlings as described above, then the seedlings were immediately kept at different temperatures from 17 to 24°C under dark conditions [1].

The total ^{14}C-export rate from the leaves during a 16-hour night period was 30 to 43%, irrespective of night temperature and no consistency was found (Figure 3).

However, the ^{14}C-distribution patterns showed a clear trend: the lower the night temperature, the higher the ^{14}C-distribution rate to the parts below the source leaf, and *vice versa*. Thus, rapid temperature change immediately affected distribution of photosynthates in the plant but not exports from the source leaf.

Among the parts above the source leaf, ^{14}C was greatly distributed to leaves especially in the uppermost unfolded leaf (data not shown). Sink strength of each plant organ is often shown as the ^{14}C-radioactivity per unit weight, which is called specific activity (SA).

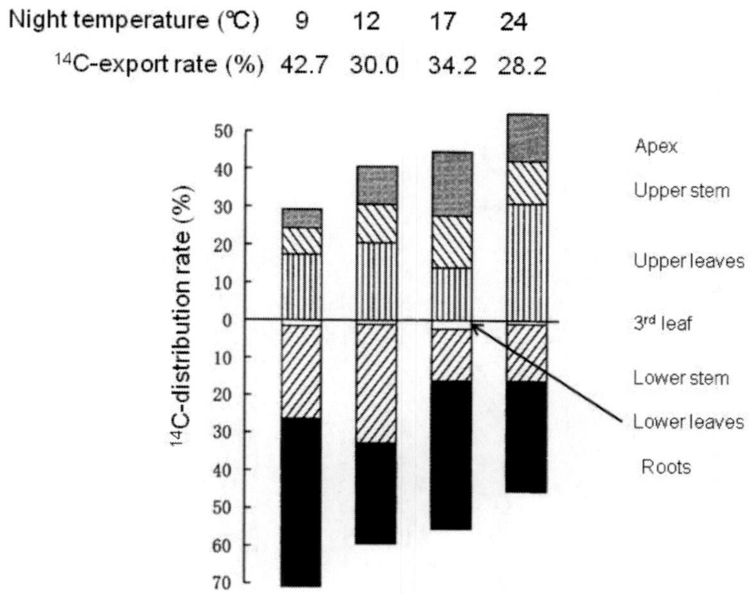

Figure 3. Export and distribution of ^{14}C-photoasynthates in non-fruiting tomato plants with 6 unfolded leaves as affected by night temperatures. $^{14}CO_2$ was applied to the 3rd leaf (n=3).

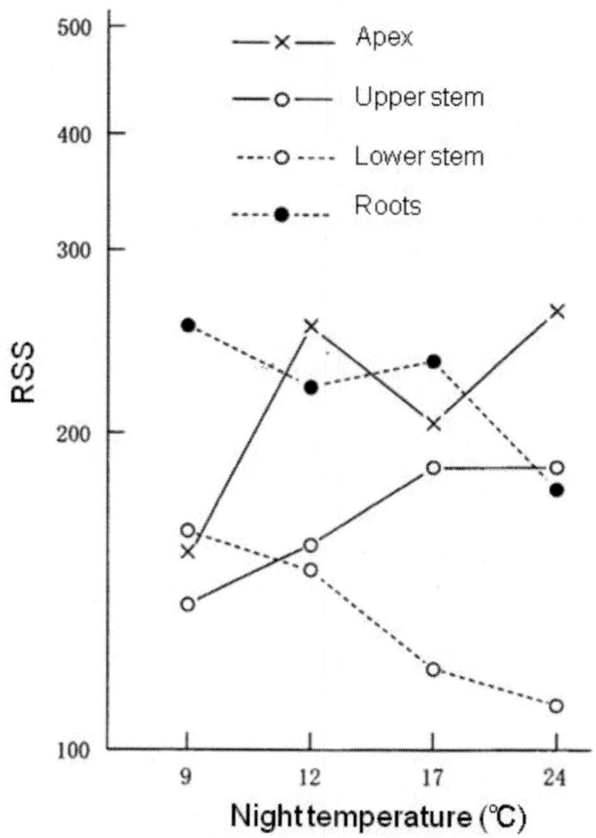

Figure 4. Relative sink activities of each plant part in non-fruiting tomato plants as affected by night temperatures (n=3).

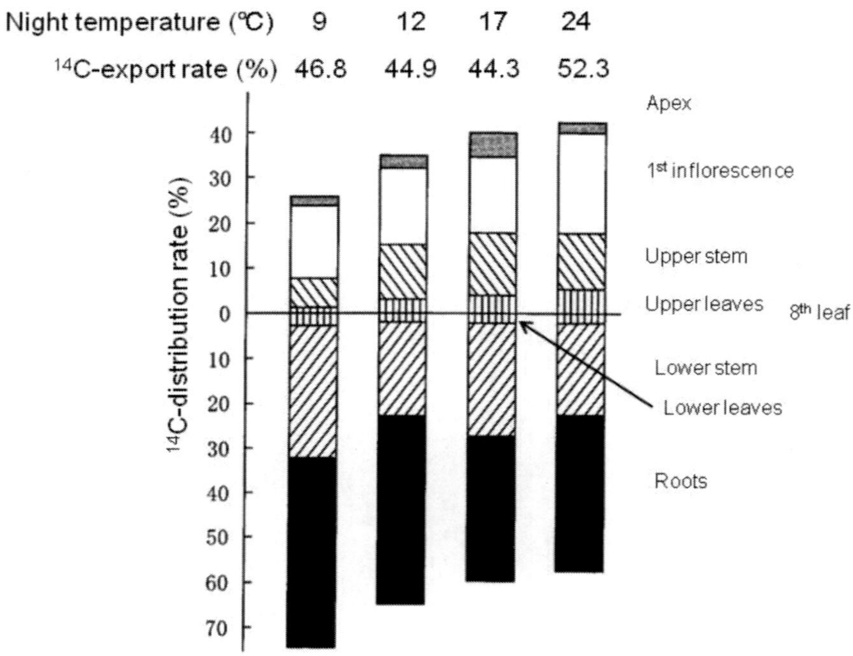

Figure 5. Export and distribution of ^{14}C-photosynthates in fruiting tomato plants with 15 unfolded leaves as affected by night temperatures. Tomato plants were acclimatized at different night temperatures for one week before $^{14}CO_2$ was applied to the 8th leaf (n=3).

For example, young plant organs which are dividing cells such as the plant apex commonly show high SA but the ^{14}C-distribution rate is relatively low because of its small dry mass. On the other hand, stems show low SA, regardless of the high ^{14}C-distribution rate because of its large dry mass. The concept of relative strength of each plant part as a sink organ (RSS) was introduced as follows [1]:

RSS = (SA of each sink / SA of a whole plant) x 100 = (^{14}C-distribution rate in each sink / Distribution rate of the dry matter: the fed leaf was excluded) x 100

Changes in RSS values as affected by night temperatures also corresponded well to the ^{14}C-distribution patterns, and the plant apex (includes a young unfolding leaf) was the relatively strongest sink organ when the night temperature was kept at 24°C (Figure 4).

To investigate the effect of fruiting on export and distribution of the ^{14}C-photosynthate, tomato seedlings were grown until the 15th leaf unfolded. At this stage, the 1st inflorescence had grown for two weeks after anthesis.

The seedlings were subsequently acclimatized for one week at different night temperatures from 9 to 24°C, and then $^{14}CO_2$ was applied to the 8th leaf at 24°C.

Total ^{14}C-export rate during a 16-hour dark period was between 47 and 52%, irrespective of night temperature, while the ^{14}C-distribution rate to the lower stem and roots increased as night temperature lowered (74% at 9°C vs. 53% at 24°C) (Figure 5).

On the other hand, the ^{14}C-distribution patterns among the seedlings were affected by night temperatures: 74% of the ^{14}C was found in the roots and lower stem at 9°C, while 53% was found at 24°C. In addition, the ^{14}C-distribution rate to the parts above the source leaf including the 1st inflorescence increased as the night temperature rose. This also corresponded to the RSS (i.e., the values in both the plant apex and 1st inflorescence increased as night temperature rose) (Figure 6).

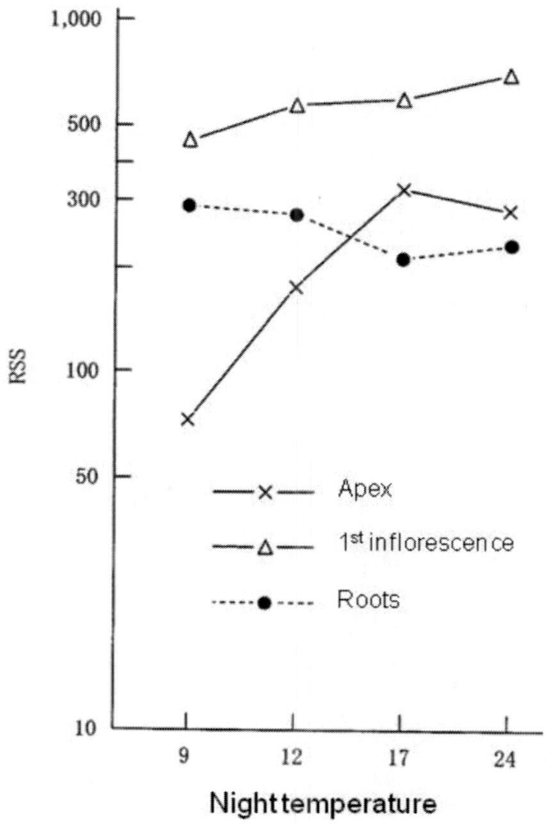

Figure 6. Relative sink activities of each plant part in fruiting tomato plants as affected by night temperatures (n=3).

3. Effects of Day/Night Temperature on Export and Distribution of the ^{14}C-Photosynthates in Fruiting Tomato Plants

The results mentioned above showed that the ^{14}C-distribution rate rather than the ^{14}C-export rate was more strongly influenced by the changes in night temperatures. In commercial tomato production, however, day/night temperatures change frequently and will affect both the export and distribution of photosynthates.

Therefore, tomato seedlings were grown at 24/17°C (day/night) until the 9th leaf unfolded and the 1st inflorescence just bloomed, acclimatized at different day/night temperatures (i.e., 30/24, 24/17 and 17/12°C), respectively for two weeks, and then $^{14}CO_2$ was applied to the 8th leaf. Although the ^{14}C-export rate during the 16-hour dark period increased 12% from 17/12°C to 24/17°C, it did not correspond well to the changes in day/night temperatures as shown in the ^{14}C-export rate that was affected by night temperatures (Figure 7).

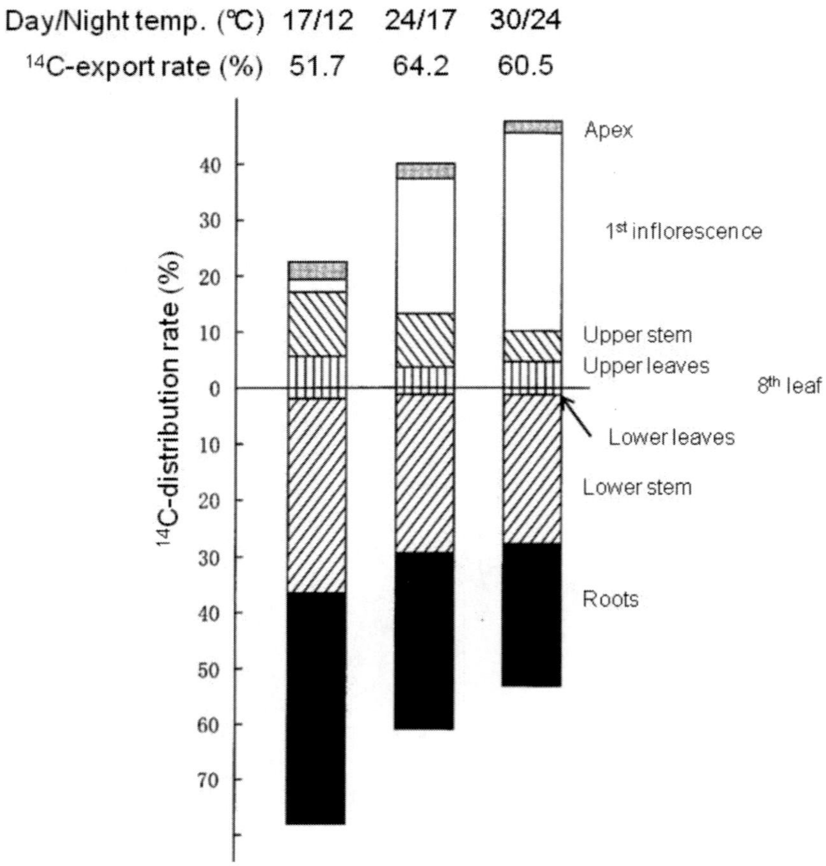

Figure 7. Export and distribution of ^{14}C-photosynthates in fruiting tomato plants as affected by night temperatures (n=3). $^{14}CO_2$ was applied to the 8th leaf.

On the other hand, changes in the ^{14}C-distribution rate corresponded well to the changes in day/night temperatures (i.e., the ^{14}C-distribution rate to the 1st inflorescence increased from only 2% at 17/12°C to higher than 30% at 30/24°C, while that to the roots decreased).

4. Effect of Different Air and Root Temperatures on Export and Distribution of the ^{14}C-Photosynthates in Fruiting Tomato Plants

As shown in Figure 5, the ^{14}C-distribution rate to the parts below the source leaf increased as night temperature lowered. Recently, hydroponic systems have been installed widely in commercial tomato production and not only air but also soil temperature is often controlled in the system, resulting in the difference in temperatures between air and soil [11]. To investigate the effect of different root and air temperatures on export and distribution of the ^{14}C-photosynthates, therefore, tomato seedlings were grown at 24/17 or 17/12°C (day/night) until the 4th leaf unfolded, transferred to different root temperatures (i.e., 12, 17 and 24°C) until the 9th leaf unfolded, and then $^{14}CO_2$ was applied to the 4th leaf.

In this experiment, the ^{14}C-export rate from the source leaf during the 16-hour dark period increased as night temperature rose from 12 to 24°C, irrespective of day temperature (Figure 8) [12].

Moreover, the ^{14}C-export rate during the light period was higher at 24°C than at 17°C. Unlike the result that was shown in Figures 3 and 5, the dry mass of each plant part differed among different day/night temperatures because of a long acclimation period (Figure 9), suggesting that the ^{14}C-export rate was mainly affected by the difference in total sink demands for the photosynthates.

The ^{14}C-distribution rate to the parts below the source leaf, such as roots, lower stem and lower leaves, was higher at 17/12°C than at 24/17°C of air temperature, while the rate increased with raising root temperatures. Moreover, a decrease in the ^{14}C-distribution rate to the parts below the source leaf at lower root temperatures was higher at 17/12°C than at 24/17°C of air temperature. These results indicate that the distribution rate of photosynthates increases as the root temperature rises when the plants are well acclimatized at different night temperatures due to the increase in the dry mass; nevertheless, the ^{14}C-distribution rate to the parts above the source leaf increases when the air temperature is independently kept at higher levels as shown in Figures 3 and 5.

In conclusion, the following two points are noted: Firstly, distribution of the photosynthates but not the export is more strongly influenced by both day and night temperatures, irrespective of thermal acclimation [13].

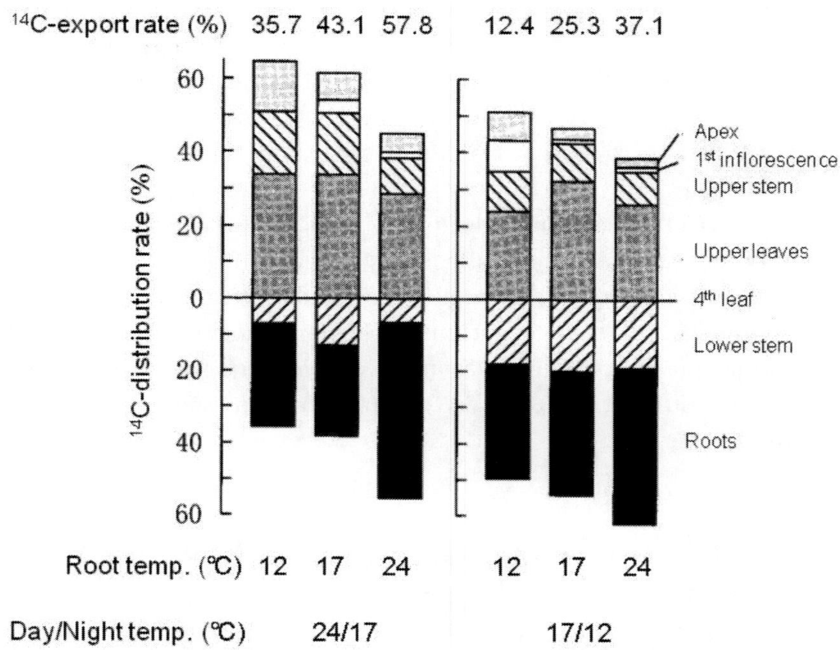

Figure 8. Export and distribution of ^{14}C-photosynthates as affected by independently controlled air and root temperatures (n=3). The plants were grown at 24/17°C or 17/12°C (day/night) until the 4th leaf unfolded, then grown at different night temperatures of 12, 17 or 24°C until the 9th leaf unfolded. $^{14}CO_2$ was applied to the 4th leaf.

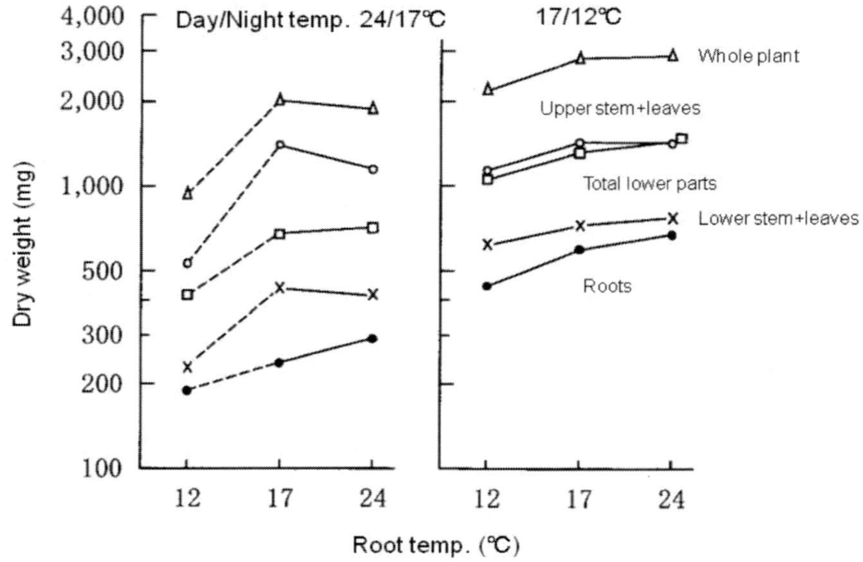

Figure 9. Changes in dry mass as affected by independently controlled air and root temperatures (n=3).

Secondly, the difference in the ^{14}C-respiration rate in each plant part has not been taken into consideration in the $^{14}CO_2$ pulse feeding method [14].

It is well known that the respiration rate is also greatly influenced by temperature [15].

For instance, when night temperature is kept too high, plant growth is also promoted temporarily, but soon it hastens to early deterioration due to the consumption of photosynthates, caused by the high respiration rate [16]. Accordingly, higher export and distribution rates of the ^{14}C-photosynthates do not necessarily indicate higher accumulation amounts of photosynthates.

In this sense, the $^{14}CO_2$-pulse feeding method is not an adequate technique to measure the actual accumulation of the photosynthates that was exported from source leaves since this method is used mainly for detecting the recovered ^{14}C-radioactivities from each plant organ after a $^{14}CO_2$ feeding [17]. Although the quantity of a photosynthetic product accumulated in the entire growing period can be roughly estimated by measuring the dry matter [18], this method is also very difficult to estimate the translocation and accumulation of the photosynthates for a whole day.

5. Effect of Leaf Position on Export and Distribution of the ^{14}C-Photosynthates in Non-Fruiting Tomato Plants

Distribution patterns of photosynthates are affected not only by environmental factors such as air and soil temperatures, but also by the phyllotaxis or the vasculature as reported by the soybean [19], tomato [20], cucumber [21], melon [22] etc. Tomato plants are known to show sympodial branching: their stem is made up of a series of superposed branches developed from the axillary buds, and each of the superposed branches terminates in an inflorescence [23]. In general, tomato plants appear to have eight leaves below the 1st inflorescence and differentiate inflorescences successively after developing every three leaves on it. In such a plant, the 9th leaf seems to be at the position higher than the 1st inflorescence (Figure 10) [24].

However, the nine leaves alternate on the main stem with four orthostichies and their divergence is a combination of 180 and 90°. The axially bud of the 9th leaf grows vigorously and forms the 1st sympodial branch. The 9th leaf coalesces on its stalk with the base of the 1st sympodial branch (recoalescence but not concoalescence) [25].

At the same time, the 1st inflorescence is pressed to the opposite side to the 9th leaf, and as the result, appears as if it differentiated prior to the 9th leaf and developed as a lateral branch. Following to the 9th leaf, the 10th, 11th and 12th leaves develop with divergence of 270, 180 and 270°, respectively, as the 1st sympodial branch and terminated in the 2nd inflorescence.

The axially bud of the 12th leaf grows to form the 2nd sympodial branch and the 13th, 14th and 15th leaves differentiate with a divergence of 90, 180 and 90°, respectively, and it terminates again in the 3rd inflorescence. The same growing processes are repeated on the 3rd sympodial branch and upward. Thus, one of the four orthostichies is occupied only by inflorescences, except the 2nd and 5th leaves being situated below the 1st inflorescence.

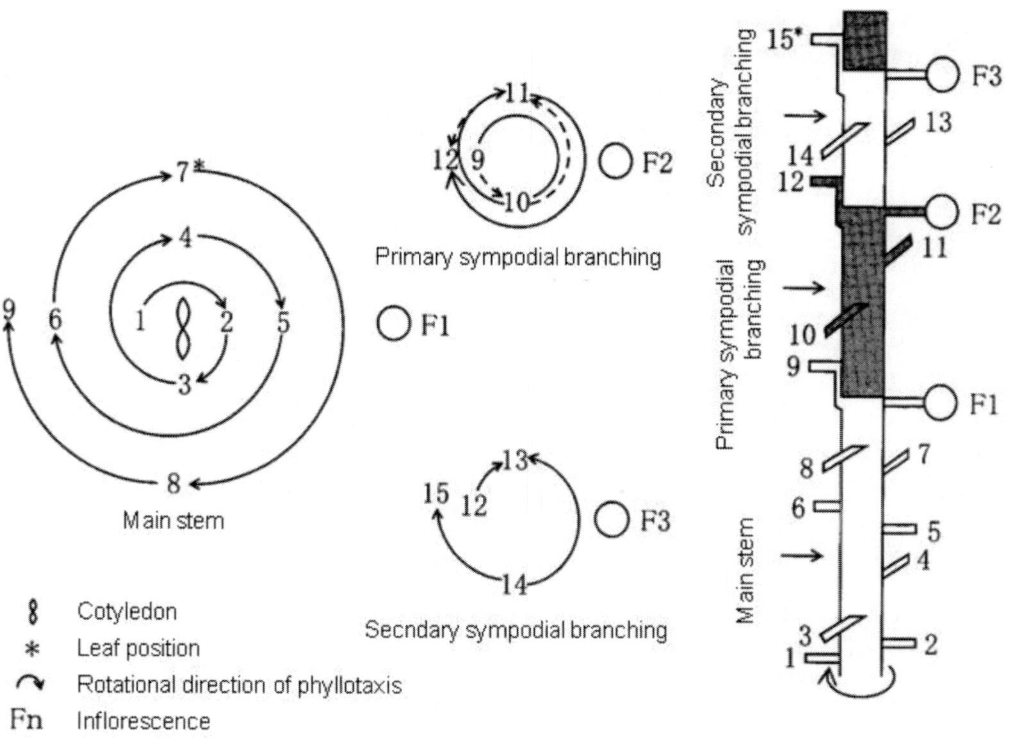

Figure 10. Arrangement of leaves and inflorescences on the stem in tomato plants. 1) On the primary stem, leaves differentiate successively with divergence of a combination of 180° and 90° until the stem terminates in the 1st inflorescence. 2) There can be found clockwise as well as counterclockwise rotation in the ratio of 1:1, and the 3rd leaf is considered to be the determinat of the rotative direction. 3) If the rotative direction may be considered to be reversed on the sympodial branches having odd numbers, divergence of a combination of 90, 180 and 90°, instead of 270, 180 and 270°, will be enough for explanation. 4) Terms "primary stem" and "Nth sympodial branch" are used tentatively.

The leaves just above the inflorescences are situated on the opposite orthostichy to the inflorescences, and the other leaves above the 1st inflorescence are on the side orthostichies of the inflorescences. Furthermore, it may be a modification of the orixate type in phyllotaxis, in which the leaves alternate with four orthostichies and their divergence is a combination of 180, 270, 180 and 90° [26].

Figure 11 shows export and distribution patterns of the ^{14}C-photosynthates from the 2nd, 3rd and 4th leaves when the 6th, 7th and 8th leaves unfolded, respectively [24]. At the 6th and 7th unfolded leaf stages, the ^{14}C-export rate from the 2nd to 4th leaves during 20 hours of a dark period was lower as $^{14}CO_2$ was applied to the upper leaves, irrespective of the plant age.

At the 8th unfolded leaf stage, the 3rd leaf exported the highest ^{14}C-photosynthates among the three source leaves but the difference became smaller. This would be due to the changes in the relative strength of each leaf as a source of carbon which was related to maturity.

The ^{14}C-distribution rate to the parts below the source leaf was higher when $^{14}CO_2$ was applied to the upper leaves. This was partly due to the fact that the volume of the lower stem increased as $^{14}CO_2$ was applied to the upper leaves. As far as the roots, however, a higher ^{14}C-distribution rate was obtained when $^{14}CO_2$ was applied to the upper leaves. The distribution pattern of the ^{14}C-photosynthates also changed with the plant growth (i.e., the ^{14}C-distribution rate to the upper parts increased, regardless of leaf position, probably due to the enhanced growth). Among the upper leaves, the shoot apex (60 to 80 mg in dry mass) and two unfolded leaves just below the apex were found to have high activity as sinks. The ^{14}C-distribution rate to the uppermost leaf was higher, with one exception, than that of the next uppermost one, when they were on the same orthostichy with, or on the side orthostichies of the source leaf (Refer Nos. 2, 3 and 6, but No. 7 was exceptional.) (Table 1).

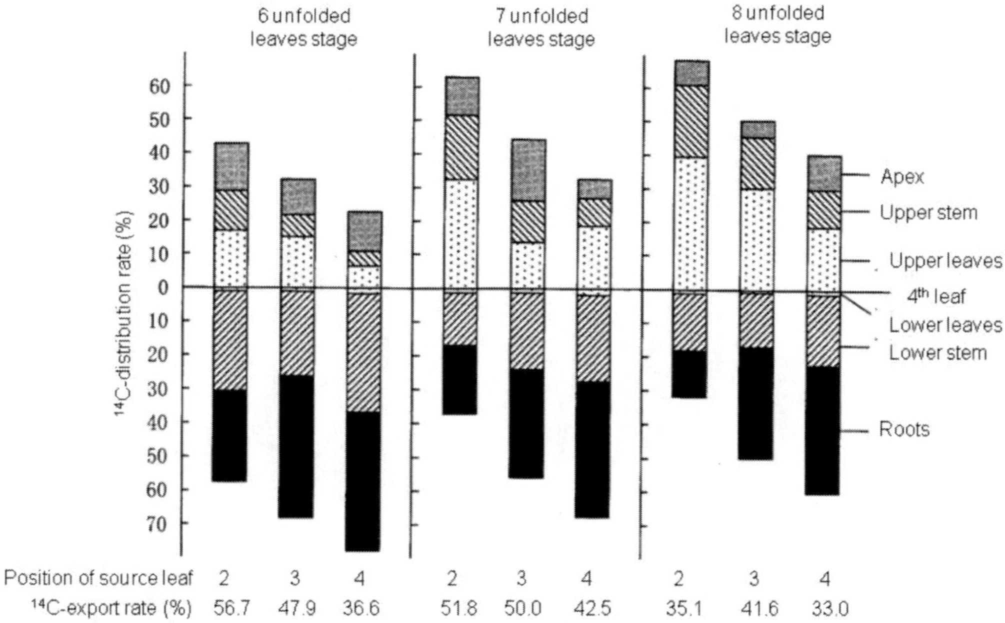

Figure 11. Export and distribution of ^{14}C-photosynthates in non-fruiting tomato plants as affected by leaf positions (n=3). $^{14}CO_2$ was applied to the 4th leaf.

Table 1. Distribution of ^{14}C-photosynthetic products to the young leaves and apex in relation to phyllotaxis in non-fruiting young tomato seedlings (n=3)

Leaf stage[z]	Source leaf[y]	Relative position to the source leaf and ^{14}C-distribution rate to the uppermost and 2nd uppermost leaves										Relative position to the source leaf and ^{14}C-distribution rate to the plant apex						Reference No.
		Same		Side		Opposite		Side		Same		Same		Side		Opposite		
		Position	%	Position	%	Position	%	Position	%	Position		Position	%	Position	%	Position	%	
6	2	5	10.3			6	3.1							7	14.0			1
	3			5	3.2			6	9.6							7	10.8	2
	4			5	1.4			6	6.4			7	11.5					3
7	2					6	4.1	7	20.7					8	10.8			4
	3			6	9.1	7	2.2					8	17.5					5
	4			6	4.6					7	12.2					8	5.1	6
8	2			7	22.3			8	10.0							9	6.7	7
	3					7	1.9			8	19.9			9	4.6			8
	4	7	13.5			8	0.9							9	10.2			9

[z]Uppermost unfolded leaf.
[y]Position of $^{14}CO_2$ fed leaf .

On the other hand, one of them was on the opposite orthostichy to the source leaf, the ^{14}C-distribution rate to the uppermost leaf was apparently lower (Refer Nos. 1, 5 and 9.), or far higher (Refer Nos. 4 and 8.) than that to the next uppermost one, according to whether it was the uppermost or the next uppermost leaf that was situated on the opposite orthostichy to the source leaf.

Although a high proportion of ^{14}C-photosynthates was often distributed into the plant apex because of its high sink activity (Figure 4), the ^{14}C-distribution rate was low when the outermost leaf was on the opposite orthostichy to the source leaf (Refer Nos. 6 and 7 but No. 2 was exceptional.). The outermost leaf also showed a high dry mass distribution rate and this might contribute much to the distribution of ^{14}C-photosynthates to the apex as a whole.

6. Effect of Fruiting and Leaf Position on Export and Distribution of the ^{14}C-Photosynthates in Fruiting Tomato Plants

Leaf position also affects export and distribution of the ^{14}C-photosynthates in fruiting tomato plants [26]. $^{14}CO_2$ was independently applied to each leaf from the 5th to 14th when the plants developed 17 unfolded leaves and 3 inflorescences, and the 3rd inflorescence was in blooming. The ^{14}C-export rate was 47 to 61% and the upper leaves appeared to export higher ^{14}C-photosynthates than the lower leaves (Figure 12) [24].

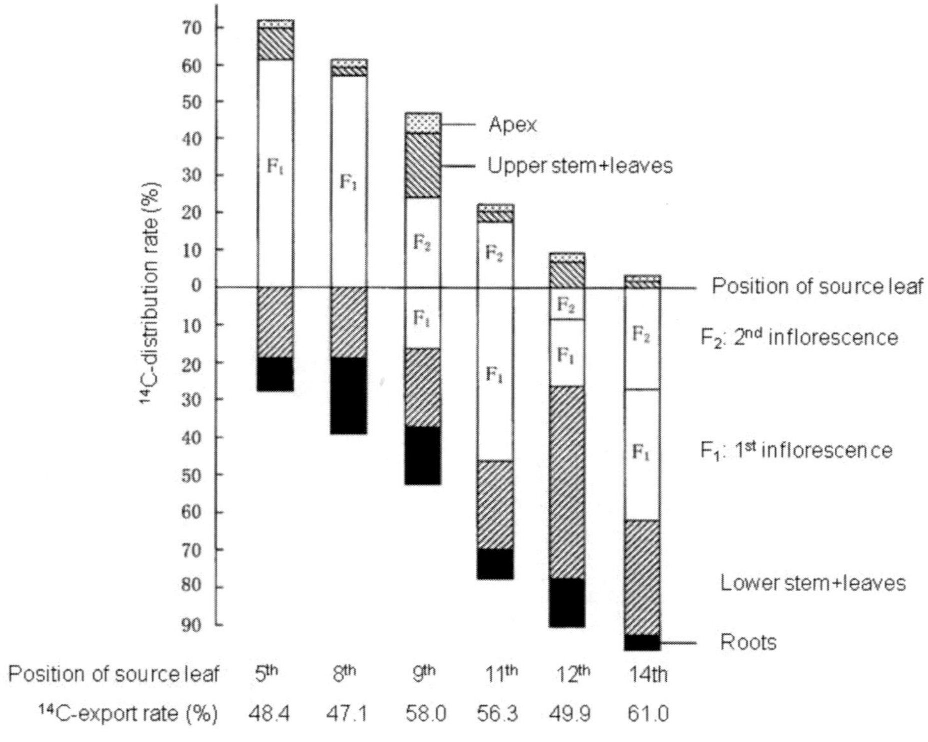

Figure 12. Export and distribution of ^{14}C-photosynthates in fruiting tomato plants as affected by leaf positions (n=3). $^{14}CO_2$ was applied independently to the 5th to 14th leaves.

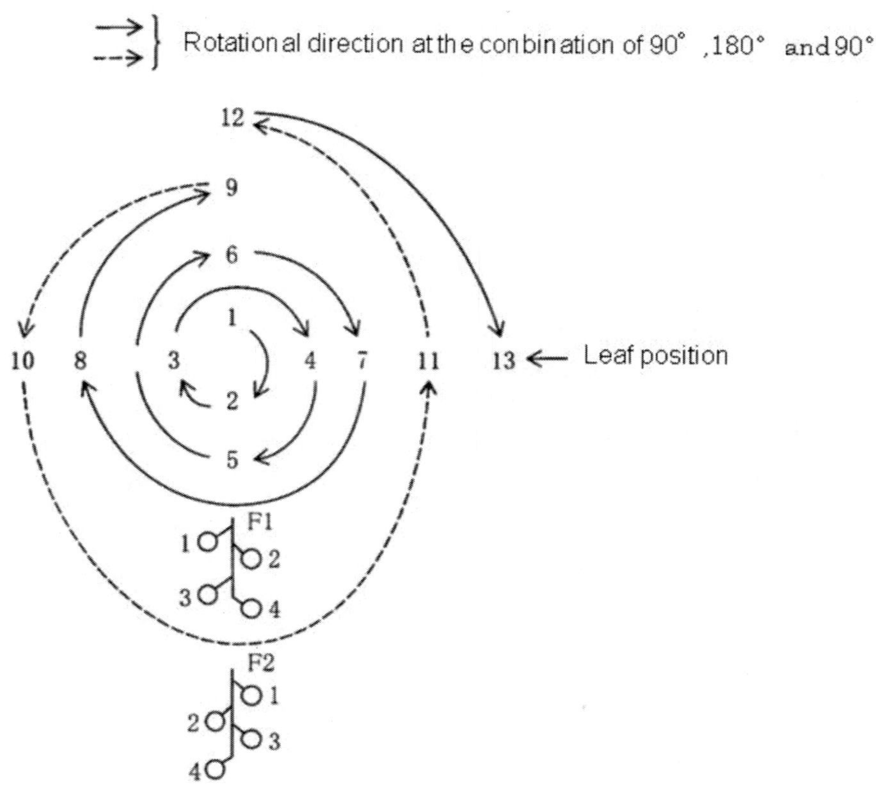

Figure 13. Phyllotaxis of fruiting tomato plants. F_1 and F_2 are the 1st and 2nd inflorescences, respectively. Numbers in each inflorescence show the number of fruits.

The ^{14}C-distribution rate to the 3 inflorescences from the 5th, 8th, 11th and 14th leaves was 58 to 65% while that from the 9th and 12th leaves, which were situated just above and on the opposite orthostichy to the inflorescences, was 44 and 28%, respectively (Figures 12 and 13). To the 1st inflorescence alone, the ^{14}C-distribution rate from the 5th and 8th leaves was approximately 60%.

Moreover, 35 and 48% of the ^{14}C-photosynthates was also obtained from the 11th and 14th leaves, respectively. ^{14}C-radioactivity to the 2nd inflorescence was detected only from the leaves above the 9th leaf (i.e., upper leaves than the 1st inflorescence). Among them, the ^{14}C-distribution rate from the 12th leaf, which was situated just above and opposite to the 2nd inflorescence, was very low, while the ^{14}C-distribution rate from the 9th leaf, which was situated on the same orthostichy with the 12th leaf, was of a similar level or higher than that from the 11th and 14th leaves.

From these results, the following two points are cleared: 1) the Nth inflorescence is supplied with the photosynthates from all leaves situated above the (N-1)th inflorescence, but not from those below it; and 2) the ^{14}C-distribution rate to the inflorescences is particularly low from the leaves situated on the opposite orthostichy to them.

The first point is explained from the relative strength of inflorescences as sinks for the photosynthates. Removal of the 1st inflorescence resulted in the export of ^{14}C-photosynthates to the 2nd inflorescence from the 5th and 8th leaves [24]. In addition, the ^{14}C-distribution rate to the 2nd inflorescence from 11th leaves increased.

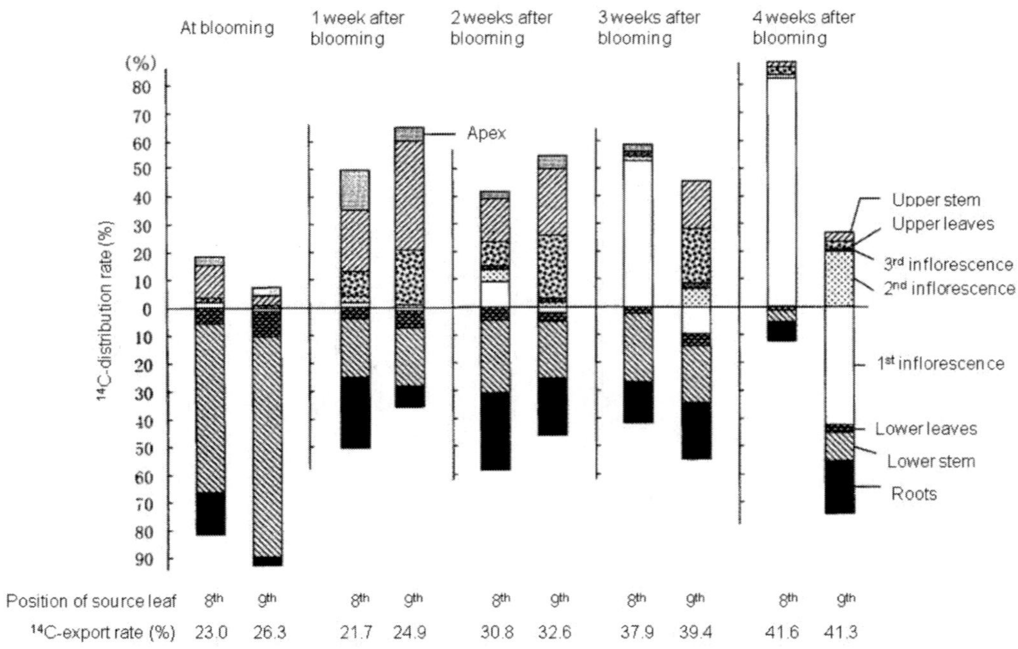

Figure 14. Changes in the ^{14}C-distribution patterns as affected by the developmental stages of the inflorescences (n=3). $^{14}CO_2$ was applied to different developmental stages of the 1st inflorescence.

The second point will be due to the phyllotactic configuration. A similar result was also obtained by Gheorghiev and Gheorghiev who demonstrated that a young developing tomato inflorescence received 48% of the assimilated carbon by a leaf situated on the same side of the stem, while only 3% from a leaf situated on the opposite side [28].

Differing from vegetative and early reproductive stages, export and distribution of the ^{14}C-photosynthates are more greatly influenced as the fruits develop. Figure 14 shows export and distribution of the ^{14}C-photosynthates from the 8th and 9th leaves during four weeks after the onset of anthesis of the 1st inflorescence.

The two leaves mainly exported the ^{14}C-photosynthates downwards at the anthesis. The upward movement surpassed the downward one one week later, but the ^{14}C-distribution rate to the 1st inflorescence was still very low. The ^{14}C-export rate became almost double during four weeks (from 23.0 to 41.6%) and 81% of the ^{14}C-photosynthates were distributed to the 1st inflorescence.

The ^{14}C-distribution rate to the 1st inflorescence from the 9th leaf (downward export) also increased during four weeks as shown in the 8th leaf but the rate was lower than that of the 8th leaf. These results show that both export and distribution rates of the ^{14}C-photosynthates increase as the inflorescences develop as stronger sink organ; nevertheless, the distribution rate is still affected by the phyllotaxis.

To apply these research results to commercial tomato production, tomato seedlings were grown until the 16th leaf stage with three inflorescences (the fruits were green developing stage with 4 to 6 cm in diameter), the shoot apex was pinched, and then $^{14}CO_2$ was independently applied to the leaves from the 4th to 15th [27].

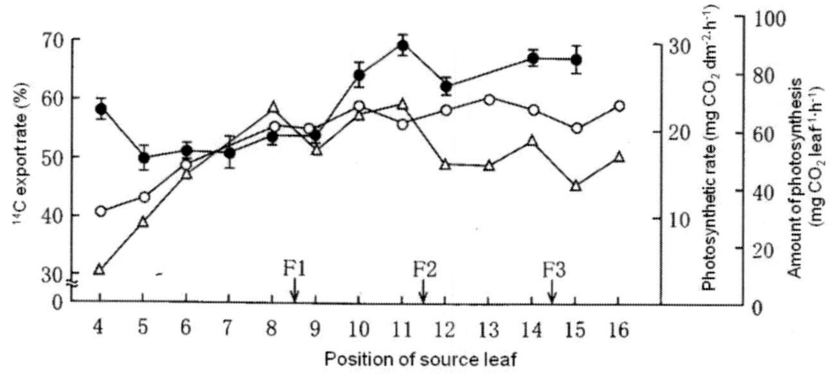

Figure 15. Photosynthetic rate (○), amount of photosynthesis per leaf (△) and ^{14}C-export rate (●) in leaves of various positions (n=3). ↓: Positions of the 1st (F_1), 2nd (F_2) and 3rd (F_3) inflorescences.

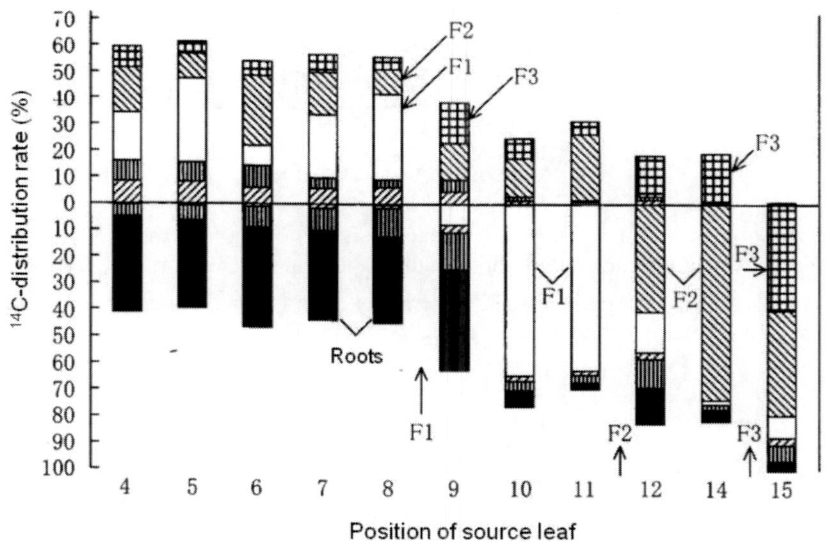

Figure 16. Distribution patterns of the ^{14}C-photosynthates by leaves at various positions (n=3). Zero percent of X axis indicates the position of the source leaf. The ^{14}C-distribution rate above and below 0% (horizontal line) are upward and downward distribution from the source leaf.

The ^{14}C-export rate during the 16-hour dark period attained 50 to 70% and the values were higher in the leaves from the 10th to 15th than in the leaves from the 4th to 8th (Figure 15). Although the photosynthetic rates of each leaf above the 8th showed almost similar levels, the amount of photosynthesis per each leaf was the largest between the 8th and 11th leaves, and then decreased gradually probably due to their smaller areas and younger ages (data not shown).

From the leaves below the 1st inflorescence, the ^{14}C-photosynthates were mainly distributed to both the roots and inflorescences, reaching approximately 80% in total (Figure 16). Among the inflorescences, the highest ^{14}C-distribution rate from the 4th, 5th, 7th and 8th leaves was found in the 1st inflorescence. From the 9th leaf, the ^{14}C-photosynthates was mainly distributed downward, including the roots, while from the 10th and 11th leaves, they were mainly distributed to the 1st inflorescence as shown in Figure 12. From the 12th to 15th leaves, most ^{14}C-photosynthates were distributed to the 2nd and/or 3rd inflorescences.

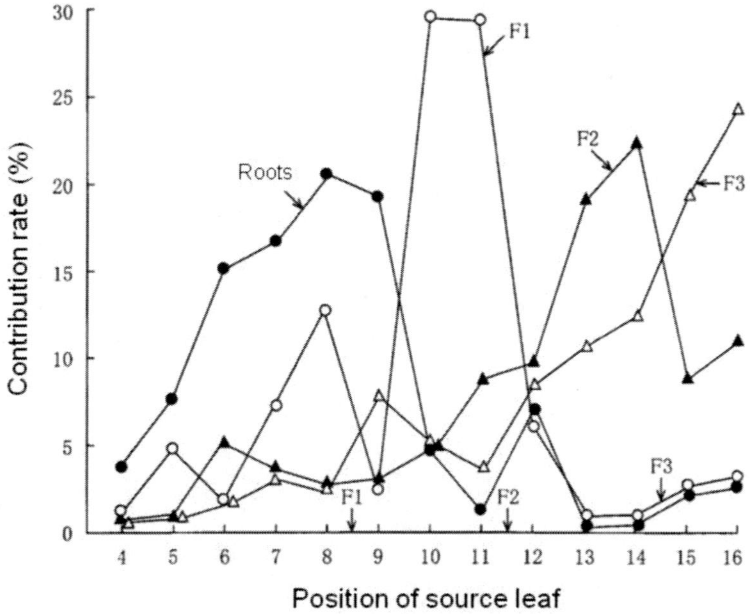

Figure 17. The rate of contribution by leaves at various positions to sink growth (n=3). The rate of contribution to each sink = (The amount of photosynthesis per leaf x ^{14}C-export rate x percentage of ^{14}C-distribution from all leaves to each sink / Σ The amount of photosynthesis per leaf x ^{14}C-export rate x percentage of ^{14}C-distribution from all leaves to each sink) x 100.

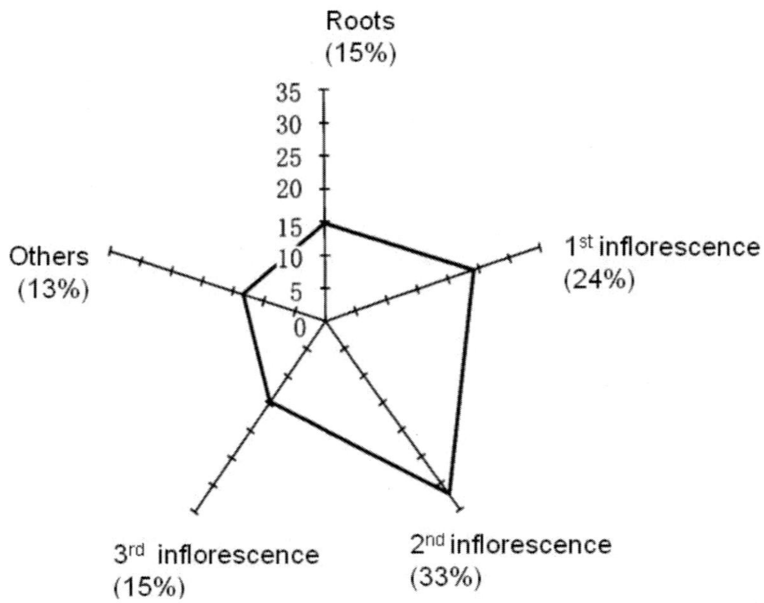

Figure 18. The proportion of photosynthates accumulated in each sink organ.

From Figures 15 and 16, the contribution rate by each source leaf at each position to the growth of the roots and inflorescences was estimated (Figure 17). Although we failed to obtain both export and distribution of the ^{14}C-photosynthates from the 13th and 16th leaves, it was regarded as the same as those from the 15th leaf. As a result, leaves below the 9th mainly contributed to the growth of the roots rather than the inflorescences. Among inflorescences, growth of the 1st inflorescence was mainly supported by both the 10th and 11th leaves, then by the 8th leaves. For the growth of the 2nd inflorescence, such a specifically high contribution rate was observed in the 13th and 14th leaves, but all leaves from the 11th to 16th contributed higher than 9%. For the growth of the 3rd inflorescence, the contribution rate rose as the leaf position became higher, while almost no contribution to the growth of roots was shown especially by the leaves more upper than the 13th.

Furthermore, assuming that respiration rates were the same in all plant organs, relative amounts of photosynthates distributed in each sink organ could also be estimated as follows: 1) 72% of the photosynthates was distributed to the inflorescences, while only 15 and 13% of the photosynthates were distributed to the roots and other sink organs (mainly stem), respectively; and 2) among the three inflorescences, the 2nd inflorescence worked as the most predominant sink organ and 33% of the photosynthates was distributed (Figure 18). Thus, the combined studies of photosynthesis and translocation of photosynthates in each source leaf can reveal which leaf works as the most important carbon source for each sink organ in growing tomato plants.

7. Carbon Transport, Partitioning and Respiratory Loss in Tomato Plants as Affected by Temperatures Using the $^{14}CO_2$-Steady State Feeding (SSF) Technique

The $^{14}CO_2$-pulse feeding method has commonly been utilized for chasing the movement of temporarily assimilated carbon. To investigate the exact carbon budget for a whole day, however, such experiments are often inadequate, mainly because the behavior of the ^{14}C-photosynthates differs by the feeding time and, moreover, the respiratory loss is not taken into consideration [15, 16]. The SSF method based on a semi-closed system of the measurement of photosynthesis/respiration was developed for estimating the carbon budget of the plant for a whole day (Figure 19) [15]. In this system, the source leaf is closed in an assimilation chamber that is connected with an infra-red gas analyzer (IRGA) for monitoring CO_2 gas in the chamber. A constant concentration (400 ppm) of the mixed gas of $^{14}CO_2$ and $^{12}CO_2$ is automatically applied to the source leaf throughout the light period. ^{14}C-radiactivity in the mixed gas of $^{14}CO_2$ and $^{12}CO_2$ that is exhausted from the source leaf is also monitored during the subsequent dark period and used for the calculation of the amount of respired carbon from the source leaf. In addition, the ^{14}C-radioactivity in each plant part is also measured at the end of light and/or dark periods. Thus, the carbon budget of the plant is calculated as mg carbon (abbreviated as C) using the following equations:

$$C_{ass} = C_{acc} + C_{rsp} + C_{tsp} \text{ ---} \qquad (1)$$

$$C_{tsp} = C_{rec} + C_{los} \text{ ---} \quad (2)$$

where abbreviations are the amount of carbon that was:

fixed by the source leaf (C_{ass}),
accumulated in the source leaf (C_{acc}),
respired by the source leaf during the dark period (C_{rsp}),
transported out of the source leaf (C_{tsp}),
recovered from outside the source leaf (C_{rec}), and
respired by the sink organs outside the source leaf (C_{los}).
In equation (1), C_{tsp} is the only unknown, hence:

$$C_{tsp} = C_{ass} - C_{acc} - C_{rsp} \text{ ----} \quad (3)$$

and consequently:

$$C_{los} = C_{ass} - C_{acc} - C_{rsp} - C_{rec} \text{ ----} \quad (4)$$

Tomato seedlings were grown at 25°C until the 5th leaf unfolded, then $^{14}CO_2$ was continuously applied to the 3rd leaf during an 8-hour light period [29]. The photosynthetically active radiation provided by 400 W metal halide lamps at the plant level was 36 W·m^{-2} (≈155 μmol·m^{-2}·s^{-1} and 10,000 lux). The assimilated C in the source leaf during the 8-hour light period was 13.77 mg C leaf^{-1} (leaf area was 65.05 cm^2), which corresponded to a mean rate of 2.65 mg C dm^{-2}·h^{-1} (Table 2).

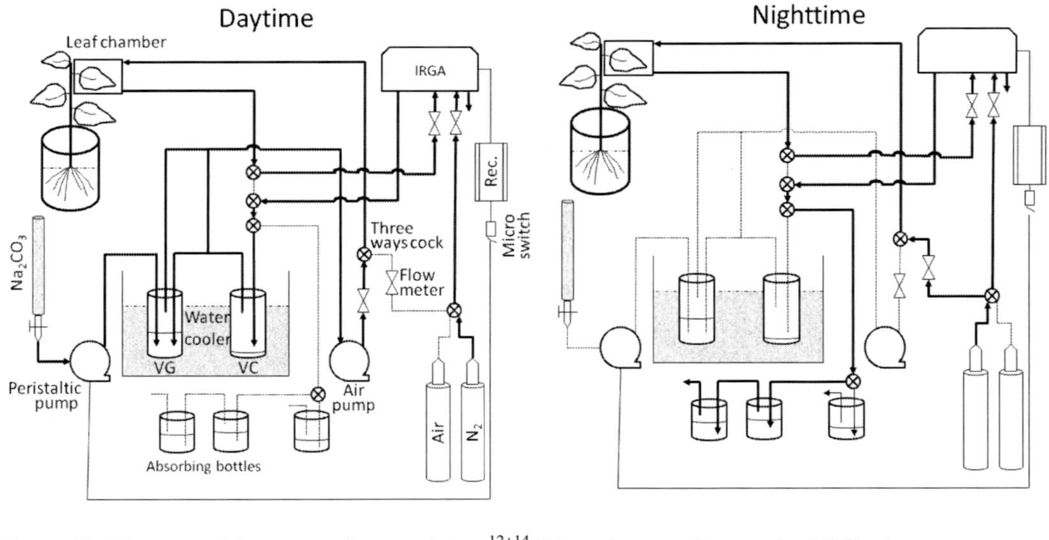

Figure 19. Diagram of the system for supplying $^{12+14}CO_2$ and measuring respired CO_2. A constant concentration (400 ppm) of $^{14}CO_2$ is supplied to the source leaf throughout the light period. Thick lines trace out semi-closed and open systems during daytime and nighttime, respectively. VG and VC are generating vessel containing 10% H_3PO_4 and condensation vessel, respectively.

Table 2. Carbon budget of the source leaf at a day/night temperature of 25°C[z] (n=3)

Sampling time	13:00	17:00	21:00	1:00	9:00	
Day-Night (h-h)	4-0	8-0	8-4	8-8	8-16	
Carbon fixed (mg C·leaf^{-1})	7.64	13.77	—	—	—	(1)
Carbon accumulated (mg C)						
Source leaf	3.65	7.45	5.64	4.49	3.56	(2)*
Sinks	1.15	2.57	3.35	3.75	4.17	(3)*
Total	4.80	10.02	8.99	8.24	7.73	(4)
Carbon exported (mg C)						
Day	3.99	6.32	—	—	—	(5)
Night	—	—	1.41	2.28	2.81	(6)
Total	3.99	6.32	7.73	8.60	9.13	(7)
Carbon respired (mg C)						
Day						
Source leaf	—	—	—	—	—	
Sinks	2.84	3.75	—	—	—	(8)
Night						
Source leaf	—	—	0.40	0.67	1.08	(9)*
Sinks	—	—	0.62	1.10	1.21	(10)
Total	2.84	3.75	4.78	5.52	6.03	(11)

[z]Tomato seedlings were grown at a constant day/night temperature of 25°C until the 5th leaf unfolded, then a constant mount of $^{14}CO_2$ was applied continuously for 8 hours from 9:00 to 17:00 to the 3rd leaf.

*Items were obtained by the direct measurement. Calculation procedures are as follows; (4) = (2) + (3), (7) = ((1) - (9)) - (2), (11) = (1) - (4), (10) = (11) - (8) - (9).

The rate was higher in the morning than in the afternoon (2.92 and 2.36 mg C dm^{-2}·h^{-1}, respectively). In the $^{14}CO_2$-pulse feeding method, we indicated that the time course export rate of temporarily assimilated photosynthates in the morning (09:00) increased more slowly than those in the noon (13:00) and afternoon (17:00) probably because of the shortage of the carbon pool in the source leaf at the onset of the light period (Figure 1). However, the result of the SSF method revealed that the amount of exported C in the morning (09:00-13:00) was larger than that in the afternoon (13:00-17:00), indicating a more predominant ^{14}C-export in the morning rather than in the afternoon from the quantitative viewpoint.

At the end of the light period, 54% (7.45 mg C) of the assimilated C was recovered in the source leaf and 19% (2.57 mg C) was in the sink. Accordingly, the exported C to the sinks and respired C during the 8-hour light period was estimated to be 46% (6.32 mg C) and 27% (3.75 mg C), respectively, of that assimilated C. The recovered C at the end of the subsequent 16-hour dark period decreased to 26% (3.56 mg C) in the source leaf, while it increased to 30% (4.17 mg C) in the sinks. Accordingly, the additionally exported C to the sinks during the dark period was estimated to be 20% (2.81 mg C) and the respired C from the source leaf and the sinks was 8% (1.08 mg C) and 9% (1.21 mg C), respectively. For a whole day, therefore, the ratio of the total exported C between the 8-hour light period and the subsequent 16-hour dark period was approximately 70:30 (6.32 vs. 2.81 mg C), also indicating that the ^{14}C-photosynthates mainly exported during the light period even if the duration of the light period was only 1/2 of the dark period and the plants were grown under low light intensity such as 36 W m^{-2}.

Similarly, 56% (7.73 mg C) of the assimilated C during the light period remained in the plants and 44% (6.03 mg C) was respired. Thus, amounts of accumulated and respired Cs in the sinks during the light period were also larger than those during the dark period (19% and 27% vs. 12% and 9%, respectively), suggesting the importance of the translocation of photosynthates during the light period for promoting the growth of sink organs.

Effects of temperatures on the carbon budget in non-fruiting tomato plants were also investigated using the SSF method. Tomato seedlings were grown at a constant temperature of 25°C until the 5th leaf unfolded, $^{14}CO_2$ was continuously applied to the 3rd leaf for 8 hours, and then the plants were immediately transferred to 15, 20, 25 or 30°C for 16 hours under dark conditions.

Although the exported C to the sink during the dark period increased only a little as the night temperature rose, the total remained C in the whole plants decreased due to the decrease in the remained C in the source leaf (Table 3). Hence, the exported C during the dark period increased 52% (from 2.40 to 3.65 mg C) as the night temperature rose from 15°C to 30°C, but the respired C from the sinks increased 99% (from 0.99 to 1.97 mg C) as compared to 49% (from 0.80 to 1.19 mg C) from the source leaf.

Table 4 shows the distribution patterns of assimilated C in each sink as affected by night temperatures. The recovered C in the sinks increased 7% as night temperature rose from 15°C (3.98 mg C) to 30°C (4.25 mg C). The ^{14}C-distribution rate (based on the accumulated C in the sinks) to the lower parts than the source leaf decreased, while that to the upper parts increased with rising night temperatures.

Table 3. Effect of night temperatures on the carbon budget of the source leaf[z] (n=3-4)

Night temperature (°C)		15	20	25	30	
Carbon fixed at the end of day (mg *C·leaf^{-1})						
	13.77	–	–	–	–	(1)*
Carbon accumulated (mg C)						
Source leaf	7.45	4.24	3.84	3.56	2.61	(2)*
Sinks	2.57	3.97	4.16	4.17	4.25	(3)*
Whole plant	10.02	8.23	8.00	7.73	6.86	(4)
Carbon exported (mg C)						
Day	6.32	–	–	–	–	(5)
Night	–	2.40	2.59	2.81	3.65	(6)
Total	6.32	8.73	8.92	9.13	9.98	(7)
Carbon respired (mg C)						
Day						
Source leaf	–	–	–	–	–	
Sinks	3.75	–	–	–	–	(8)
Night						
Source leaf	–	0.80	1.02	1.08	1.19	(9)*
Sinks	–	0.99	1.00	1.21	1.97	(10)
Total	3.75	5.55	5.77	6.03	6.91	(11)

[*,z]See Table 2 for the details.

Table 4. Distribution pattern of ^{14}C-photosynthates as affected by night temperatures (n=3-4)

Night temperature (°C)	15	20	25	30
Carbon recovered in the sinks (mg C)	3.98	4.16	4.17	4.25
^{14}C-distribution rate in the sink organs[z]				
Apex and 6th leaf	29.7	28.0	30.1	32.3
Leaves above the 6th leaf	2.3	4.9	6.9	4.0
Stem above the 6th leaf	8.7	10.2	11.4	7.8
Subtotal	40.7	43.1	48.4	44.1
Leaves below the 6th leaf	0.3	0.2	0.3	0.2
Stem below the 6th leaf	26.3	27.5	22.6	22.0
Roots	32.8	29.3	28.8	33.7
Sub total	59.4	57.0	51.7	55.9

[z]Respiration rate has been considered equal among plant parts.

Thus, the results of the SSF method were similar to those of the $^{14}CO_2$-pulse feeding method, but only up to 25°C, and not as obviously as in the previous pulse feeding experiments (Figure 3).

The SSF method is a helpful technique for clarifying the carbon budget of the plant for a whole day, but it should be noted that the difference of respiration rate in each sink organ has yet to be taken into consideration as well as the $^{14}CO_2$-pulse feeding method when it is used as a semi-closed system (Figure 19). Consequently, the respiration rate in each sink organ should be measured independently for clarifying the exact carbon budget in each sink organ, especially when the root temperatures were controlled separately from the air temperatures as shown in Figure 8. However, independent measurement of both respired and accumulated C in all sink organs is an impractical idea [30]. Therefore, only the roots were sealed in a glass bottle that was connected with IRGA, and the respired C was monitored for clarifying the role of roots as major sink organs [31]. In this experiment, the plants were also grown at a constant temperature of 25°C until the 5th leaf unfolded, $^{14}CO_2$ was continuously applied to the 3rd leaf during an 8-hour light period at 25°C, and then the root temperatures were kept at 10, 15 and 22°C during the subsequent 16-hour dark period.

During the 8-hour light period, 36.69 to 38.20 mg C was assimilated in the source leaf. Among them, 9 to 18% (3.46 to 7.01 mg C) was distributed to the roots and 2 to 8% (0.79 to 2.87 mg C) was lost as respiration until the end of the dark period (Table 5). Consequently, only 7 to 11% (2.67 to 4.14 mg C) remained in the roots, although the rate became higher as the root temperatures rose. The respired C from the whole sinks was estimated at 27 to 32%. Therefore, the respired C from the stem and leaves was calculated as 23 to 25%. Hence, the total respired carbon measured by IRGA from the roots was 11.21 to 33.26 mg C (note that these values were the sum of the carbon translocated from all leaves). Therefore, assuming that all leaves had the same photosynthetic rates, the total fixed C by the whole plants was estimated to be 253 to 264 mg C. Likewise, assuming that distribution, respiration and accumulation patterns of C were same as those indicated in the source (3rd) leaf, the carbon budget in a whole plant was estimated as follows: 1) the exported C to the roots became almost two times from 10 to 22°C (from 23.8 to 48.4 mg C), but the accumulation does not differ so much (between 12.6 and 15.1 mg C); and 2) more than half of the fixed C would be accumulated in the leaves and stem.

Table 5. Effect of root temperatures on the carbon budget of a whole plant with a simulated calculation (n=3-5)

Amount of carbon (mg)	Root temperature (°C)			
	10	15	22	
*C fixed by the source leaf during the 8 h light period	36.69	36.82	28.20	*(1)
Total C respired from the roots in a whole day[z]	11.21	19.23	33.26	*(2)
*C accumulated in the roots in a whole day	2.67	30.7	4.14	*(3)
	(7.3)	(8.3)	(10.8)	
*C respired from the roots in a whole day	0.79	1.26	2.87	*(4)
	(2.1)	(3.4)	(7.5)	
*C exported from the source leaf to the roots in a whole day	3.46	4.33	7.01	(5)
	(9.4)	(11.8)	(18.4)	
Total *C lost as a whole	(26.6)	(26.1)	(32.0)	*(6)
*C lost by leaves and stem	(24.5)	(22.6)	(24.5)	(7)
The amount of carbon calculated with assumption as a whole plant				
Total C fixed	253.0	253.9	263.5	(8)
C exported to the roots	23.8	29.8	48.4	(9)
C accumulated in the roots	12.6	10.6	15.1	(10)
C lost by leaves and stem	61.9	54.5	64.6	(11)
C accumulated in leaves and stem	167.3	166.6	150.5	(12)

[z]The values were measured by IRGA and were derived from all leaves.

*C was derived from the source leaf.

Calculation procedures are as follows; (5) =(3) + (4), (7) = (6) - (4), (8) = (1) x (total leaf area/leaf area of the source leaf), (9) = (8) x ((5)/100), (10) = (8) x ((3)/100), (11) = (8) x ((7)/100), (12) = (8) - (9) - (11). (1), (2), (3), (4) and (6) were obtained by the direct measurement.

Such a calculation of a carbon budget in a whole plant as affected by root temperatures is not necessarily fit exactly to the real carbon budget of the plants because the photosynthetic rate of each leaf is not same, and the export and distribution of fixed C are strongly influenced by the position of leaves as shown in Figure 11.

However, such a result based on the direct measurement of photosynthesis, translocation and respiration is informative for understanding the carbon budget of a whole plant [32].

8. Carbon Transport, Partitioning and Respiratory Loss in Tomato Plants as Affected by Light Conditions Using the SSF Technique

Among the environmental conditions, the photosynthetic light period is one of the most important factors for the plant growth [33-36]. When $^{14}CO_2$ was applied for 8, 12 and 16-hours to the 3rd leaf of non-fruiting young tomato seedlings as described above, the total recovered C in the source leaf at the end of the light period increased 2.2 times from 13.77 mg C·leaf^{-1} in 8 hours to 30.5 mg C·leaf^{-1} in 16 hours (Table 6) [37].

The mean photosynthetic rate differed little with the light period, being within the range 26 to 29 μg C cm^{-2}·h^{-1}. At the end of 8, 12 and 16-hour light periods, 46, 48 and 55% of the C (6.32, 9.66 and 16.86 mg C, respectively) were exported from the source leaf, respectively, while during the subsequent 16, 12 and 8-hour dark periods, 20, 10 and 10% of the C (2.81, 2.00 and 3.02 mg C, respectively) were exported, respectively.

In a whole day, therefore, the ratio of total exported C between light and dark periods were 69:31, 83:17 and 85:15 for 8, 12 and 16 hours, respectively, indicating that most ^{14}C-photosynthates exported from the source leaf during light periods as shown in Table 2 from the viewpoint of quantitative analysis.

Although the C export rate became higher in longer light periods, the total C export rate for a whole day was similar (66, 58 and 60% in 8, 12 and 16-hour light periods, respectively). The recovered C in the source leaf and the sinks at the end of dark period in 16 hours was 2.7 and 2.0 times larger than those in 8 hours, respectively (9.64 vs. 3.56 and 8.54 vs. 4.17 mg C, respectively). Thus, although the amount of accumulated C in the plants increased as the photosynthetic light period prolonged, the rate of accumulated C differed little in both the source leaf and the sinks (between 26 and 35%). This was also true in respired C (i.e., the respired C from the whole plant in 16 hours was 2.0 times larger than that in 8 hours (12.32 vs. 60.4 mg C)), but the rate of respired C was between 38% and 44% among different light periods.

In addition, the respired C from the sinks during the light period was between 26 and 31% (between 6.32 and 9.66 mg C) of the total fixed C, while that during the subsequent dark period was only less than 8 and 9% from the source leaf and sinks, respectively, indicating that the proportion of the carbon budget as a whole plant was little affected by the prolonged photosynthetic period [38, 39].

Table 6. Carbon budget of the source leaf in different photosynthetic light period (n=3-4)

	Light period					
	8 h		12 h		16 h	
Time	09:00-17:00	17:00-09:00	09:00-21:00	21:00-09:00	09:00-01:00	01:00-09:00
Light-Dark (h-h)	0-8	8-24	0-12	12-24	0-16	16-24
Carbon fixed at the end of day (mg C·leaf^{-1})						
	13.77	—	20.20	—	30.50	—
Carbon accumulated (mg C)						
Source leaf	7.45	3.56	10.54	7.16	13.64	9.64
Sinks	2.57	4.17	4.34	5.43	7.41	8.54
Whole plant	10.02	7.73	14.88	12.59	21.05	18.18
Carbon exported (mg C)						
	6.32	9.13	9.66	11.66	16.86	19.88
Carbon respired (mg C)						
Source leaf	—	1.08	—	1.38	—	0.98
Sinks	3.75	4.96	5.32	6.23	9.45	11.34
Whole plant	—	6.04	—	7.61	—	12.32

In common commercial tomato production, however, photosynthetic light periods do not change suddenly as shown above. On the other hand, light intensity fluctuates frequently as often found in greenhouses and it is a more fundamental factor for leaf photosynthesis than a photosynthetic light period [40].

Therefore, tomato seedlings were grown at 25°C (day/night) under light intensities of 36 or 90 W·m^{-2} until the 5th leaf unfolded for acclimatizing the seedlings to each light intensity. $^{14}CO_2$ was then applied to the 3rd leaf for 8 hours under the different light intensities of 36, 90 and 180 W·m^{-2} (low, medium and high lights, respectively), and the carbon budget under a subsequent 16-hour dark period was investigated using the SSF method for clarifying the effect of temporary changes in light intensity [41].

In the plants that were raised under 90 W·m^{-2}, the photosynthetic rate increased when the light intensity was increased to 180 W·m^{-2}, while no increase occurred in the plants that were raised under 36 W·m^{-2} (Tables 7 and 8).

The total amount of fixed C during the 8-hour light period showed a large difference between plants grown at the two initial light intensities (15.4 to 21.1 mg C·leaf^{-1} and 19.5 to 43.6 mg C·leaf^{-1} for 36 and 90 W·m^{-2}, respectively).

In addition, the amount of exported C during the time course was higher in plants that were raised under 90 W·m^{-2} than those raised under 36 W·m^{-2}, irrespective of the change in light intensity (7.1 to 10.3 mg C for 36 W·m^{-2} vs. 7.5 to 19.6 mg C for 90 W·m^{-2} and 10.2 to 14.2 mg C for 36 W·m^{-2} vs. 13.7 to 27.5 mg C for 90 W·m^{-2} during light and dark periods, respectively).

However, the proportion of exported C during the light period did not differ apparently, irrespective of the change in light intensity: approximately 60 to 70% of the fixed C was exported from the source leaf, 10% was lost by respiration and 20 to 30% remained in the source leaf during the dark period after the 8-hour $^{14}CO_2$ fixation period, irrespective of the changes in light intensity.

These trends suggest that the amount of carbon export from the source leaf and respiratory loss from both source and sink organs depend mainly on the amount of carbon fixed by the source leaf and not on the temporary change in light intensity.

Table 7. Dry mass, leaf area, and specific leaf weight of the plants grown under 36 W·m^{-2} for 30 days or 90 W·m^{-2} for 21 days. $^{14}CO_2$ was fed to the 3rd leaf counting from the basal end (n=4)

Plant growth	Light intensity		Relative value
	36 W·m^{-2}	90 W·m^{-2}	to 36 W·m^{-2}
Whole plant			
Dry mass (mg)	937	2,173	2.32
Leaf area (cm^2)	326	585	1.79
Specific leaf weight (mg·cm^{-2})	2.88	3.72	1.29
Source leaf			
Dry mass (mg)	128	248	1.94
Leaf area (cm^2)	73	92	1.26
Specific leaf weight (mg·cm^{-2})	1.77	2.69	1.52

Table 8. Effects of temporary changes in light intensity on the carbon budget from a source leaf of young tomato seedlings that were grown under 36 or 90 W·m^{-2} (n=4)

Partitioning	Light intensity before $^{14}CO_2$ feeding (W·m^{-2})											
	36						90					
	Light intensity during $^{14}CO_2$ feeding period (W·m^{-2})											
	36		90		180		36		90		180	
	Harvest time from the onset of $^{14}CO_2$ feeding (h)											
	8	24	8	24	8	24	8	24	8	24	8	24
Photosynthetic rate (C μg·cm^{-2}·h^{-1})												
	26.4		36.7		35.7		26.5		48.6		58.4	
C fixed (mg C·leaf^{-1})												
	15.4		21.7		21.1		19.5		35.9		43.6	
	(100.0)		(100.0)		(100.0)		(100.0)		(100.0)		(100.0)	
C accumulated (mg C)												
Source leaf	8.3	4.0	12.2	5.7	10.8	5.4	12.1	4.7	21.8	11.8	24.0	12.2
	(54.1)	(25.9)	(56.0)	(26.5)	(51.2)	(25.6)	(61.8)	(23.8)	(60.8)	(29.0)	(55.0)	(28.1)
Sinks	2.9	4.7	4.2	7.0	4.7	6.7	4.4	7.6	6.2	9.7	9.5	15.9
	(18.7)	(30.3)	(19.5)	(32.4)	(22.4)	(31.5)	(22.5)	(39.1)	(17.4)	(27.1)	(21.8)	(36.4)
Whole plant	11.2	8.6	16.4	12.8	15.6	12.1	16.5	12.3	28.0	20.2	33.5	28.1
	(72.8)	(56.2)	(75.5)	(58.9)	(73.6)	(57.1)	(84.3)	(62.9)	(78.1)	(56.1)	(76.8)	(64.5)
C exported (mg C·leaf^{-1})												
	7.1	10.2	9.6	13.7	10.3	14.2	7.5	13.7	14.1	22.6	19.6	27.5
	(45.9)	(66.3)	(44.0)	(63.0)	(48.8)	(67.2)	(38.2)	(70.1)	(39.2)	(62.9)	(45.0)	(63.0)
C respired (mg C)												
Source leaf	–	1.2	–	2.3	–	1.5	–	1.2	–	2.9	–	3.9
		(7.8)		(10.5)		(7.2)		(6.1)		(8.1)		(8.9)
Sinks	4.2	5.5	5.3	6.6	5.6	7.5	3.1	6.1	7.9	12.8	10.1	11.6
	(27.2)	(36.0)	(24.5)	(30.5)	(26.4)	(35.7)	(15.7)	(31.0)	(21.9)	(35.7)	(23.2)	(26.6)
Whole plant	4.2	6.7	5.3	8.9	5.6	9.1	3.1	7.3	7.9	15.8	10.1	15.5
	(27.2)	(43.8)	(24.5)	(41.1)	(26.4)	(42.9)	(15.7)	(37.1)	(21.9)	(43.9)	(23.2)	(35.5)

Values in parentheses indicate the relative values (%) to the amount of total carbon (C) fixed during the $^{14}CO_2$ feeding period. See the Table 2 for the calculation procedure.

9. Non-Distractive Analysis of the Translocation of Photosynthates Using the Positron-Emitting Tracer Imaging System (PETIS)

Although use of ^{11}C is restricted strictly, the γ-ray radiation from ^{11}C in the plant organs can be detected quantitatively in an intact plant body.

In addition, repeated analyses of one plant are also possible because of the short half-life. Taking advantages of the positron-emitting isotopes, the short-time movement of both ^{11}C-photosynthates and ^{13}N-fixation compounds was studied in the same plant [42], but visualizing the dynamics associated with the long-distance transport of photosynthates was still difficult.

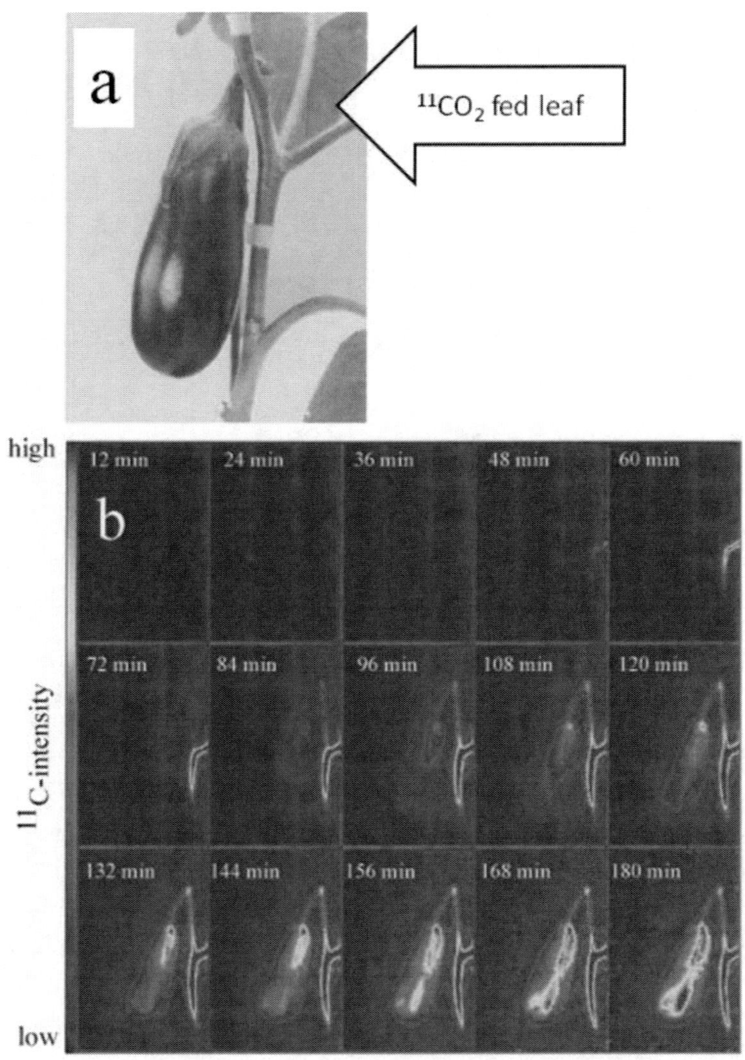

Figure 20. A photograph of eggplant used for the PETIS experiment (a) and serial PETIS images of translocation of the ^{11}C-photosynthates (b). The images were continuously acquired every 10 s for 3 h. Each image represents the integration of 72 serial images.

The positron-emitting tracer imaging system (PETIS) using ^{11}C as a tracer has recently been developed for chasing the real-time movement of the photosynthates in plants [43, 44]. Although PETIS has not been used in tomatoes yet, it should be a powerful technique for revealing the mechanism of translocation and distribution of photosynthates.

Figure 20 shows an example of eggplants used for the experiment and the serial PETIS images of every 12 minutes after a $^{11}CO_2$ feeding; these images were generated from an integration of the corresponding images collected every 10 seconds [45].

Thus, the time course movement of ^{11}C-photosynthates could be visualized. As a result, the ^{11}C-photosynthates began to move downward to the stem via the leaf petiole about 48 minutes after the feeding. About 60 minutes after the feeding, the ^{11}C-radioactivity of the fruit began to increase, and by 120 minutes it reached about 8% of the fixed $^{11}CO_2$, indicating that it took about 60 minutes for the first ^{11}C-photosynthates to reach the fruit. Based on the kinetic analyses of translocation in the ^{11}C-photosynthates, the velocity of translocation through the peduncle of the eggplant was estimated to be approximately 1.17 cm min^{-1}. Thus, PETIS measurement enables us to visualize not only the distribution patterns of ^{11}C-photosynthates inside a fruit but also the translocation velocity through the peduncle and their export rate to the fruit.

CONCLUSION

The supply of photosynthates from a source leaf to sink organs in tomato plants is regulated in part by the leaf itself through the photosynthetic rate and the ability to export photosynthates into sink organs [5]. For instance, the carbon pool in the source leaf is limited at the onset of the light period mainly due to the export and respiration during the night, resulting in the slower export rate of photosynthates in the morning than in the evening. However, the total export rate within a day is similar and the photosynthates start to export immediately after the assimilation, irrespective of feeding time. A comparatively constant export rate of photosynthates is kept even if the environmental conditions such as temperature [1, 12, 29, 31], photoperiod [37], light intensity [41, 47], water potential [46] etc. are suddenly changed after the feeding. On the contrary, distribution patterns of the exported photosynthates from the source leaf are strongly affected by the ability of each sink organ to utilize the available carbon supply. In addition, phyllotaxis also affects the partitioning of photosynthates even if the tomato plants have predominant sink organs such as developing fruits [24, 27, 48]. The $^{14}CO_2$-pulse feeding method is inadequate for understanding the carbon budget within a plant because it can chase only the movement of temporarily assimilated carbon, but not that of a whole day. In addition, respired carbon from both source and sink organs is often unconsidered in this method. The SSF method can overcome this disadvantage to some extent by applying a constant concentration of $^{14}CO_2$ continuously to the source leaf throughout the light period [8, 15, 29, 31, 32, 37, 41]. The analyses by the SSF method revealed the importance of carbon transport during the light period for the carbon accumulation in sink organs because the amount of accumulated carbon in sink organs during the light period was much greater than during the dark period. PETIS is a promising technology for chasing the real-time movement of temporarily assimilated carbon from the source leaf to sink organs such as the velocity of phloem transport, while it is still difficult to conduct the experiment using commercial-size plants because of its strict usage regulations [43-45].

REFERENCES

[1] Y. Shishido and Y. Hori, Tohoku *J. Agr. Res.* 28, 26 (1977).
[2] I.F. Wardlaw, *New Phytol.* 116, 341 (1990).
[3] L.C. Ho and J.D. Hewitt, *Fruit development*, In: J.G. Atherton and J. Rudich eds., *The tomato crop*, Chapman and Hall, New York (1986).
[4] M.M. Blanke, *Acta Hort.* 835, 13 (2009).
[5] I.F. Wardlaw, *Bot. Rev.* 34, 79 (1968).
[6] T. Nishizawa and Y. Hori, *J. Japan. Soc. Hort. Sci.* 54, 467 (1986).
[7] N. Ostel, P. Ineson, D. Benham and D. Sleep, *Rapid Communications in Mass Spectrometry*, 14, 1345 (2000).
[8] Y. Shishido, H. Kumakura and Y. Hori, *J. Japan. Soc. Hort. Sci.* 62, 95 (1993).
[9] D.A. Baker and J. Mooby, *Ann. Bot.* 33, 729 (1969).
[10] A.A. Khan and G.R. Sagar, *Ann. Bot.* 30, 727 (1966).
[11] M. Oda, *Tomato*, In The Japanese Society for Horticultural Sciences (Ed). *Horticulture in Japan 2006*, Shoukadoh, Tokyo (2006).
[12] Y. Shishido and Y. Hori, *Tohoku J. Agr. Res.* 30, 87 (1979).
[13] S. Wolf, A. Marini and J. Rudich, *Ann. Bot.* 66, 513 (1990).
[14] L.C. Ho, A.F. Shaw, J.B.W. Hammond and K.S. Burton, *Ann. Bot.* 52, 365 (1983).
[15] Y. Shishido, H. Challa and J. Krupa, *J. Exp. Bot.* 38, 1044 (1987).
[16] H. Lambers, D.A. Day and J. Azcón-Bieto, *Physiol. Plant.* 58, 148 (1983).
[17] D.R. Geiger, *Bot. Gaz.* 140, 241 (1979).
[18] L.F.M. Marcelis, *J. Exp. Bot.* 47, 1281 (1996).
[19] C.D. Nelson and P.R. Gorham, *Can. J. Bot.* 35, 339 (1957).
[20] F. Bangerth and L.C. Ho, *Ann. Bot.* 53, 315 (1984).
[21] K. Kanahama and T. Saito, *J. Japan. Soc. Hort. Sci.* 57, 215 (1988).
[22] T. Kozai, A. Masuko and K. Kanahama, *J. Japan. Soc. Hort. Sci.* 58, 679 (1989).
[23] P. McSteen and O. Leyser, *Ann. Rev. Plant Biol.* 56, 353 (2005).
[24] Y. Shishido and Y. Hori, *Tohoku J. Agr. Res.* 28, 82 (1977).
[25] O.C. Lehman, *Das Morphologishce System der Kulturtomaten*, Springer-Verlag, Berlin (1955).
[26] C.R. Russel and D.A. Morris, *Ann. Bot.* 52, 357 (1983).
[27] Y. Shishido, C.J. Yun, T. Yuhashi, N. Seyama and S. Imada, *J. Japan. Soc. Hort. Sci.* 59, 771 (1991).
[28] D. Gheorghiev and H. Gheorghiev, *C. R. Acd. Bulg. Sci.* 16, 313 (1963).
[29] Y. Shishido, N. Seyama, S. Imada and Y. Hori, *Ann. Bot.* 63, 357 (1989).
[30] S. Tachibana, *J. Japan. Soc. Hort. Sci.* 51, 299 (1982).
[31] Y. Shishido and H. Kumakura, *J. Japan. Soc. Hort. Sci.* 63, 81 (1994).
[32] Y. Shishido, H. Kumakura and T. Nishizawa, *Physiol. Plant.* 106, 402 (1999).
[33] N.J. Chatterton and J.E. Silvius, *Plant Physiol.* 64, 749 (1979).
[34] N.J. Chatterton and J.E. Silvius, *Physiol. Plant.* 49, 141 (1980).

[35] N.J. Chatterton and J.E. Silvius, *Ann. Bot.* 46, 739 (1980).
[36] L.C. Ho, *J. Exp. Bot.* 27, 87 (1976).
[37] Y. Shishido, N. Seyama, S. Imada and Y. Hori, *Ann. Bot.* 66, 729 (1990).
[38] A.J. Gordon, G.J.A. Ryle, D.F. Mitchell and C.E. Powell, *J. Exp. Bot.* 33, 241 (1982).
[39] A.J. Gordon, G.J.A. Ryle and G. Webb, *J. Exp. Bot.* 31, 845 (1980).
[40] T.R. Sinclair, *Improved carbon and nitrogen assimilation for increased yield,* In H.R. Boerma and J.E. Specht (Eds.). *Soybeans: improvement, production and uses,* Amer. Soc. Agr, Madison, 537 (2004).
[41] T. Nishizawa, Y. Shishido and H. Kumakura, *Physiol. Plant.* 136, 351 (2009).
[42] C.D. Caldwell, D.S. Fensom, L. Bordeleau, R.G. Thompson, R. Drouin and R. Didsbury, *J. Exp. Bot.* 35, 431 (1984).
[43] H. Uchida, T. Okamoto, T. Ohmura, K. Shimizu, N. Satoh, T. Koike and T. Yamashita, *Nucl. Instrum. Meth. A.* 516, 564 (2004).
[44] S. Matsuhashi, S. Fujimaki, N. Kawachi, K. Sakamoto, N.S. Ishioka and T. Kume, *Soil Sci. Plant Nutr.* 51, 417 (2005).
[45] K. Kikuchi, S. Ishii, S. Fujimaki, N. Suzuki, H. Matsuhashi, I. Honda, Y. Shishido and N. Kawachi, *J. Japan. Soc. Hort. Sci.* 77, 199 (2008).
[46] Y. Shishido, *Agr. Hort.* 83, 306 (2008).
[47] H. Hamamoto, Y. Shishido, T. Uchiumi and H. Kumakura, *Envion. Cont. Biol.* 38, 63 (2000).
[48] Y. Shishido, N. Seyama and Y. Hori, *J. Japan. Soc. Hort. Sci*, 57, 418 (1988).

In: Tomatoes: Cultivation, Varieties and Nutrition
Editor: Tadahisa Higashide

ISBN: 978-1-62417-915-0

Chapter 3

SENSING, CONTROL AND INSTRUMENTATION DURING TOMATO GROWTH

Federico Hahn*

Universidad Autonoma Chapingo, Estado de Mexico, Mexico

ABSTRACT

During tomato growth several variables should be controlled to make it profitable. In greenhouses the climate should be controlled to provide the adequate environment for the plants. Controlled irrigation and fertigation optimizes yield, saves water and fertilizers increasing producer's revenue. Radiation control for screen automation is essential to avoid tomato cracking and fruit water loss on hot summers. Integrated plant management detection was reviewed with trendy technology which can help growers to obtain better harvests. New humidity, radiation and electrical conductivity sensors are compared in base of their accuracy, range and output signals so producers can select the best option. Their application in climate and ferti-irrigation control with some up to date fertigation mechanisms show last technological advances. Implementation of closed loop fertigation can be economically implemented avoiding soil and water pollution.

1. INTRODUCTION

It is the intent of this chapter to study and analyze different monitoring and control techniques carried out during tomato cultivation. Control techniques are better suited for protected environment greenhouses. Tomato cultivation in greenhouses can produce fruit throughout the year which is not possible in open field farming due to heavy rainfall and wind, especially in tropical regions [1]. Greenhouse technology can contribute to solve environmental pollution, energy and water shortage in various ways. Agricultural crops can be produced with increased land productivity [2], with low water input and with higher

* E-mail address: Irrigation Department, km 38.5 Carr. México - Texcoco, 56230, Chapingo, México; fhahn@correo.chapingo.mx.

fertilizer use efficiency. However field produced fruits present better quality that protected environment fruits [3]; field-grown fruits are more tasteful and acid being its ascorbic acid contents higher than fruits grown in a protected environment.

1.1. Tomato Automated Production

Automation is required to produce high yields with better fruit quality. Automation provides the ability to produce tomato in the right moment with energy saving systems. Tomato prices depends on worldwide production markets and can increase or decrease daily; if the greenhouse produce tomato when they are expensive, higher profits will be obtained. Clean and alternative energies should be used to reduce fruit pollution and decrease production costs.

Parral or multispan greenhouses control is mainly focused on fertigation and climate systems. Integrated pest management requires climate control and some method for killing insects, molds or pathogens. Fertigation and climate systems can be represented as event-based control systems where the proper dynamic evolution of the system variables decides when the next control action will be executed [4]. Actuator life increases with event-based control systems which generate a new control signal when a change is detected [5].

Changes in light intensity, temperature and relative humidity within protected environments affect production and fruit composition [6, 7]. The main concern in humidity and temperature control is to provide the best condition for water and nutrient movement through the plant. Air and leaf temperature, humidity, photosynthetic active radiation (PAR) and CO_2 sensors obtain data from the greenhouse environment; fans, heaters, fogg systems, curtains and lights are common actuators for controlling the environment (Figure 1).

Fertigation provides an excellent opportunity to maximize yield and minimize environmental pollution [8] by increasing fertilizer use efficiency and minimizing fertilizer application. Drip fertigation applies mineral nutrition together with irrigation water [9]. Drip fertigation is effective in tomato cultivation saving 30% fertilizer and 23% of the applied water [10]; plant growth and yield decreases with excess fertilizer due to plant stress [11]. Fertigation equipments use EC and pH sensors to detect and control the proper feeding of plants, Figure 1. Soilless culture horticulture lives a transition from open to closed fertigation systems but salinity and toxic ion concentration increments are encountered in the recycled solution. Crops absorb nearly 50% of fertilizer and the remaining leaches; the crop uses 70% of the water for growth and transpiration [12]. To protect public health the European Member States agreed an EC Regulation, setting limits for nitrate in lettuce and spinach [13]. Limits to maximum levels of nitrate for trade in other vegetables are set in some European countries; fertigation can control the amount of nitrates applied.

In older greenhouses it was enough to have one cabled measurement point in the middle to provide the system dynamic information for greenhouse automation. The system was simple but was unable to control local heating, lights, ventilation or some other activity, which affected the greenhouse interior climate. Modern greenhouses are much bigger and more measurements are needed to make the system work properly; installations requiring large amount of wire to distribute sensors and actuators become complex and expensive. Greenhouses should provide several options to make local adjustments to lights, heating and other greenhouse support systems.

A solution to easily change the location of the measurement points according to the greenhouse structure and to plant placement inside the greenhouse requires of wireless sensors. Wireless Sensor Networks (WSN) is a collection of sensor and actuators nodes linked by a wireless medium performing distributed sensing and acting tasks [14]. The sensor nodes collect data and communicate over a network environment to a computer base station. Based on the information collected, the base station takes decisions and then the actuator nodes perform appropriate actions.

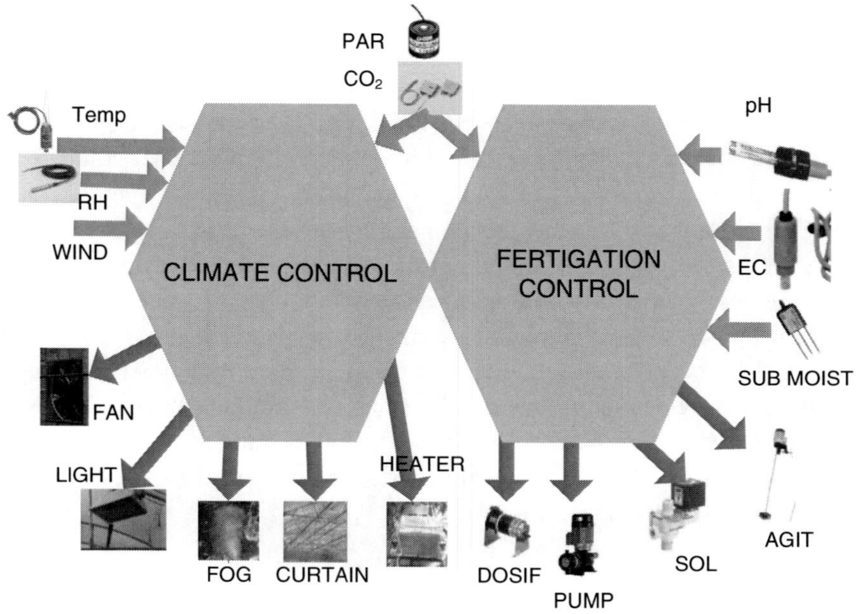

Figure 1. Climate and fertigation control with sensors and actuators.

2. Climate Control

A good day-climate is the most important factor for a optimal growth being a perfect combination of temperature, light, humidity and CO_2. A poor greenhouse climate reduce quality and yield; small tomato size, fruitskin problems (cracking), botrytis and other fungi-problems will result. At sunrise when minimum outside temperatures are recorded, the crop must be heated to 17°C. Afterwards it most be followed by a gradual raise during the morning towards 18°C until its peak value of 26°C in the afternoon. During peak temperature pollen is released and evaporation is stimulated to uptake water and nutrients, especially when lightlevels are low. Rail pipe should be 15-20°C greater than greenhouse air temperature to create air movement.

2.1. Temperature Control

Temperature is the most important single parameter in greenhouse environmental control as temperature has a significant role in plant growth and development. Optimal greenhouse temperatures for tomato production should vary between 21-28°C during the day and 16°C at night. Too high temperature reduces plant growth, resulting in plant wilting and death; over 30°C lycopene no longer develops. Leaves respond more rapidly than air to sudden increases in solar radiation [15]; thereby leaves quickly heat up becoming hotter than the surrounding air. Greenhouse control operations based on air temperature gives an imprecise control of plant canopy temperature, especially in summer, being leaf temperature influenced by transpiration. Greenhouse climate should be regulated based on plant temperature as top canopy temperature drops much faster than air temperature when ventilated. Temperature data from weather stations used to predict phenological events in the crop canopy may lead to large errors.

Time taken for fruits to mature decreased when air temperatures exceeded 26°C. This shorter crop production time or days to harvest (DTH) produced fruits 46% smaller and lower yields. Fruits took 66 days to ripen in a 15°C night temperature regime, compared with 74 days when the night temperature was 11°C [16]. Time of ripening was dependent on the fruit temperature [17], presenting little effect the temperature of other organs (leaves, truss). Plants grown at 26°C had poor appearance and abnormal trusses presented reduced bud numbers; tomatoes were seedless or had low seed numbers. Plants grown between 18-22°C produced normal fruits and had a normal canopy structure. Tomato growth decreased at 14°C and trusses had many flowers and were prone to splitting [17]; nutrient deficiencies may occur at this temperature showing leaf purpling due to poor phosphorus uptake. Air temperature should be integrated over a period of 45 days between anthesis and harvest (750-1100° C); R^2 correlations of 0.77 and 0.91 were encountered between fruit firmness and phenolic compounds with cumulative temperature [18].

2.1.1. Temperature Sensors

Sensors for measuring temperature are of the resistive type, using platinum (Figure 2 a) and standard ceramic NTC (thermistor) devices [19]. It is a good practice to use a four wire connection between the resistance sensors and the measuring system. One pair of wires conducts the measuring current to the resistance sensor; the other pair connects the measured voltage to a differential instrumentation amplifier with high input impedance. Traditional greenhouses use thermostats that should be located near the center of the greenhouse for good temperature control. Thermostats do not have cables that could affect the reliability of the resistance measurement. Erroneous temperatures will be measured by the thermostat when sun shines directly over it or when wind speed exceeds 3 ms^{-1}; so it should be enclosed in an aspirated box. Temperature sensors casing should be thin-walled to minimize heat conductance from the temperature sensor to the wall. Oil or petrol fired furnaces are activated by a thermostat when the air temperature within the greenhouse drops below 15°C.

Leaf temperature is generally measured with non-contact infrared sensors (Figure 2 b and Figure 2 d) which are less accurate than T type copper-constantan thermocouples (wire diameter 0.1 mm- Figure 2 c) using a reduced temperature span; thermocouples should be adhered to the underside of the leaves. Typical infrared thermometers (IR) are used to measure surface temperature and consist of a lens that collects infrared radiation emitted by the target object and directs it towards a detector. The IR sensor should be calibrated once a year and checked daily against a portable secondary standard to evaluate stability. The IR sensor is generally mounted 0.80m above the crop canopy to view the crop diagonally across the rows [20]. Individual measurements can be expected to vary as much as ± 0.334° C owing to sensor repeatability error. J or K type thermocouples provide such small signals throughout the measuring range that if poorly instrumented will lose accuracy caused by noise. Thermocouple signals can be converted to 4-20 mA signals by transmitters and should use the slighter temperature span to increase processing accuracy.

Figure 2. Leaf: (a) Pt100 sensor, (b) infrared sensor, (c) T-type thermocouple sensor, (d) manual infrared probe.

2.2. Cooling Control

In greenhouses without cooling systems, air temperature often exceeds the optimal range for plant production during hot seasons; such high air temperatures can be reduced with evaporative cooling (pad & fan, fog and mist) systems. Fog systems are commonly used in nurseries where roots are small and moisture should be applied through the leaves. The lowest temperature will determine the cooling method which should employ roof and side vents control. Several published books and articles [21, 22, 23, 24, 25] explain evaporative cooling systems in detail discussing its performance.

Mist systems spray with low-pressure (0.3 MPa) systems in most commercial greenhouses producing large drips leading to wet leaves. Mist increased tomato yield by 22%, average fruit weight by 11% and tomato leaf area by 38% [26]. The high pressure (12 MPa) fogging system sprays 20 μm water droplets to increase the water surface in contact with the air; as water evaporative cooling efficiency increases, foliage remains dry. Fog systems include a storage tank, a high-pressure pump with an adjustable valve, high-pressure tubes, and electrically operated solenoid valves to switch the fog system on and off [27]. Fog application resulted in a 17.8% higher tomato yield, an increase in fruit quality, and a lower tendency for fruit to crack [28]. Variable pressure fog systems are flexible, presenting water and energy savings [29].

Fog systems produces a better cooling, a lower air temperature and less nutrient leeching from leaves than mist systems [30]; fog systems optimize medium aeration and minimizes medium cooling. Greenhouse shading is important to prevent solar heat build-up meanwhile a good fan ventilation will remove heat from the leaf as the leaf surface isn't wetted directly. The advantage of mist and fog systems over wet pad systems is the uniformity of conditions throughout the greenhouse, thereby eliminating the need for forced ventilation and airtight enclosure. These systems combined with wireless sensor monitoring could control one hectare greenhouses using precision agriculture techniques; temperature control of localized zones using sliced pads will be inaccurate.

The control can be implemented using two electric timers to provide the on and off periods preset by the user; this control can cause excessive or fog shortage under rapid weather changes. Fogging operated by an on-off control presented fixed spraying cycles of 20 s followed by fogging-off periods being the duration adjusted in proportion to the temperature difference between the measured value and the upper air temperature limit. A shorter operation cycle period led to smaller changes in air temperature and relative humidity (14-17%) inside the greenhouses; air temperature would decrease 2.1- 2.6°C. The timers can be energized by a photoelectric cell, which charges a condenser to lengthen or shorten the interval between misting or fogging. In greenhouses "normally closed" solenoid valves are used when power failures are infrequent replacing manual bypass valves. Leaf temperature is the main signal and humidity works within ranges to save water and produce a similar yield than a fog system using a 50% of duty cycle; water was applied during 38% of the time, optimizing water use and energy consumption, Figure 3. Climatic changes require that humidity and leaf temperature variables are controlled dynamically. Event controlled systems can fix temperature set points based on previous week measurements optimizing fog system performance.

Researchers at the University of Arizona studied microclimate changes caused in tomato plants by evaporative cooling systems [31]. The greenhouses presented shade curtains 3.4 m above the floor; the pad and fan system (15-cm thick cellulose) and a high-pressure-fog system of 8960 kPa were analyzed, comparing canopy temperature and water used. Factors affecting the temperature along the length of the greenhouse are: ventilation speed, transpiration of plants, soil evaporation, shading and water evaporation from the pad. Leaf temperature increased with increasing distance from the cooling pad for two ventilation rates. The best spatial uniformity throughout the greenhouse was obtained when the fog overhead line was operated for ventilation rates of 4.5 $m^3 s^{-1}$. Air exchange rate was too high for the high pressure-fog system for retaining the cooled air inside the greenhouse, at a ventilation rate of 9.4 $m^3 s^{-1}$. Some of the fog droplets do not evaporate, causing temperature differences along the greenhouse. Increasing the ventilation rate for fog systems did not necessarily improve greenhouse climate (Table 1).

At 4.5 $m^3 s^{-1}$, the irrigation system used 30% more water than the pad-fan system, but for a ventilation rate of 9.4 $m^3 s^{-1}$, the pad-fan system used 40% more water than the irrigation system. The high-pressure-fog system in this greenhouse used more water than the open irrigation system. The amount of water required to cool (9.2 L $m^{-2} d^{-1}$) at a ventilation rate of 9.4 $m^3 s^{-1}$ represents two-thirds of the total greenhouse water consumption per day or twice the amount of water used for open irrigation. Water used by the 4.5 $m^3 s^{-1}$ ventilation rate represented close to 60% of the total greenhouse water use.

For the pad-fan system, the lower ventilation rate produced a mean greenhouse air temperature (Table 1) and relative humidity close to the desired conditions. For the high pressure fog system, both ventilation rates maintained the setpoint temperature, but at the expense of higher water consumption (Table 1).

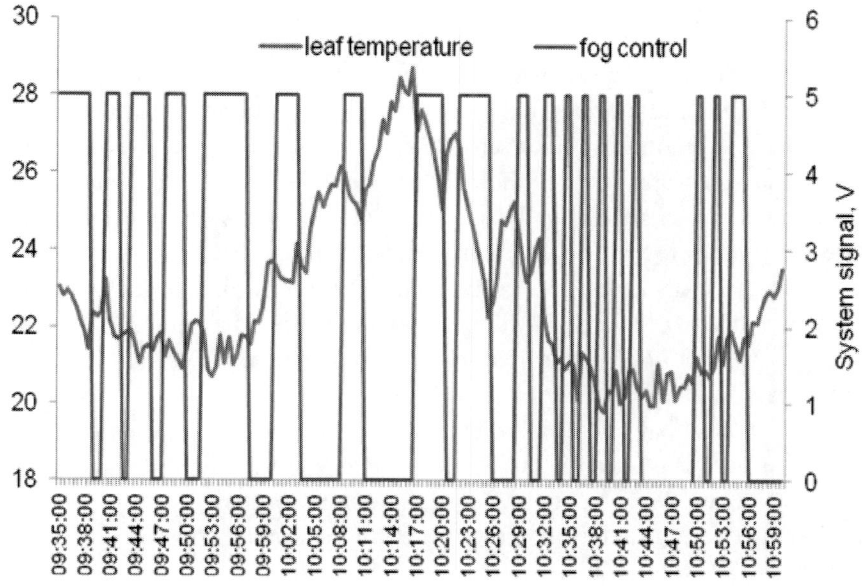

Figure 3. Leaf temperature operated under an automatic fog control.

Table 1. Fad-pan and fog system comparison at the leaf temperature at three sections of the greenhouse, open irrigation water use and total water use by the ventilation system for two different ventilation rates [31]

	Q, m3/s	Leaf temperature, °C			Open irrigation System, L m^{-2} d^{-1}	Water use by system, L m^{-2} d^{-1}
		5.5 m	11 m	17.5 m		
Pad-fan	4.5	21.1	21.9	22.1	4.5	3.2
Pad-fan	9.4	18.7	19.1	19.5	4.5	6.3
Fog	4.5	19.3	19.2	19.1	4.5	7.3
Fog	9.4	21.6	19.4	18	4.5	9.2

2.3. Heating Control

Greenhouse heating is essential to maximize crop production yield and quality. Heating costs have a critical influence on the profitability, but in the long term may also determine the survival of the greenhouse industry. Apart from the costs problems associated with high energy consumption heating are associated with environmental problems through the emission of noxious gases. Generally, greenhouses are heated by hot water circulating in pipes or by hot air that is distributed by ducts or by a combination of both systems. Less electricity is consumed in steam-heated greenhouses as no pumps for distributing heat are needed. Heating energy released was 6% higher with steam than with hot-water.

Pipes heat greenhouse air meanwhile leaves are radiated directly. As warm air rises, it creates a climatic environment over the plants maintaining the crop above its dew point; when leaf surface temperature exceeds the dew point temperature no condensation occur preventing plant diseases. Buried heating pipes stimulate air movement through the crop and enhance crop transpiration. Heating requirement increases 5-10% with overhead pipes and heating pipe position has little influence on the total crop evaporation; but does influence the distribution of the evaporation within the crop [32].

Air heaters are generally used as the heating source in greenhouses due to their quick response to temperature changes; air heaters present low energy efficiency. When the air heater inside the greenhouse is turned on, air temperature close to the greenhouse roof was higher than near the crop zone. The crop is cooler than the inside air with air heating, whereas, with pipe heating the crop is warmer than the surrounding air [33]. In an arc tunnel the tomato plants were 0.77°C cooler than air with pipe heating and 1.57°C cooler than air with the additional use of the air heater [34].

Floor heating control has to consider the large thermal mass of the floor causing the floor to respond over a period of several hours. The water supplied for floor heating should be controlled according to the temperature of the floor or of the growing substrate; water temperature can also be adjusted using air temperature as variable. Pipes laying on the surface of the soil required 3°C higher input water temperature than buried pipes to obtain the same substrate temperature [35]; soil thermal conductivity has little effect on surface temperature but has a direct impact on the heat flux to the soil below. Economic benefits of installing insulation under the floor arise. As temperature uniformity is important for crop quality, a lower pipe position should be used [36]. After evaluating tubes on top of the floor, 13 mm diameter pipes provided a higher average substrate temperature than the 19 mm diameter pipes; pipes placed closer together provided a more uniform temperature distribution throughout the substrate.

A new innovating system for heating is composed of a water-filled cylindrical polyethylene (PE) sleeve [37] with a diameter of 0.45 m combined with two perforated tubes (diameter of 0.2 m) glued on the top of the water filled tube [38]. The sleeve-tubes combination serves as gutter-channel for drained nutrient solution collection; each 7 m long sleeve contains 1287 l of water. The perforated tubes had holes of about 0.01m every 0.50 m and air coming from the greenhouse environment was pumped to inflate the tubes and mix the greenhouse air (Figure 4 a); air heats the rockwool slabs. The water temperature in the sleeves was also measured by a PT-100 type temperature sensor with deviations of ± 2%, at the center of the sleeve. The greenhouse air temperature and humidity were measured by four temperature-humidity sensors. Stored heat in the water contained in the PE sleeves is transmitted by natural convection and radiation to the greenhouse environment when the air temperature inside the greenhouse is lower than that of the water [38]. At night minimum substrate temperature was 18.6°C and during the day it was 25.07°C; night substrate temperature at plants without water sleeves was 16.9ºC, meanwhile during the day it reached 29.1ºC. Maximum air temperature in sleeve heated plants reached 29.50°C, while over the non-sleeve grown plants it reached 32.76°C[38]. Water temperature was 1.5-2 °C higher than air temperature (Figure 4 b); vertical air temperature inside the greenhouse presented null variations during the first 2 meters [39].

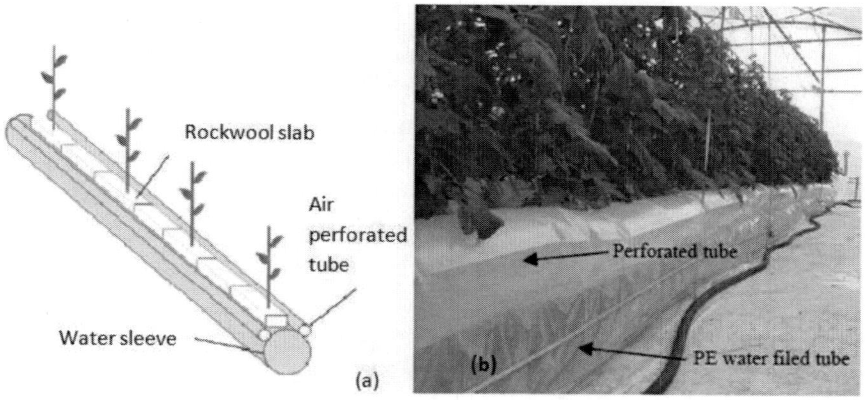

Figure 4. Heating gutter (a) main diagram [36] and (b) implemented system [37].

2.4. Moisture Control

Humidity is the most difficult environmental factor to control being for healthy plants between 50-70% RH; humidity levels fluctuate when air temperature changes, and during plant transpiration. Humidity control is very important as high humidity decreases transpiration and induces disease development; low humidity causes hydric stress, closes the stomata reducing photosynthesis due to a poor CO_2 assimilation [40]. Under saturated humidity conditions (dew-point) plants cannot evaporate water from their leaves so the uptake of nutrients such as calcium and boron is limited. With high air moisture, plant irrigation requirements decreases.

Relative humidity (RH) and vapor pressure deficit (VPD) are used to measure humidity. Relative humidity describes the ratio of water vapor compared to the total amount of water that could be held in the air at saturation. VPD is often used to measure plant/air moisture relationships being a better indicator of what the plants ‘feel’. At the same relative humidity levels, but at different temperatures, the transpiration demand for water from the leaves may be double. Stomata open and close in response to vapor pressure deficit (VPD), opening wider as humidity increases. Stomata close to 50% when humidity levels drop to 12 mb VPD avoiding wilting; the exchange of C02 is reduced affecting photosynthesis.

Controlling excessive humidity in plastic-covered greenhouses is still unresolved; condensation and dripping over the crops cause disease problems. High air humidity (low VPD) causes reduction of growth and leaf area of tomato plants due to a low Ca content in leaves [41]. High humidity reduces fruit yield and tomatoes quality [42]. Tomato plants under low humidity regimes grow stronger with more trusses than those under high humidity (high VPD) regimes. Marketable fruit yield was 26% higher for plants having low humidity regimes than under high humidity regimes. High fruit yield was attributed to both fruit size and fruit number [43]. Better growth and higher fruit yield and quality under low humidity regimes were attributed to the promotion of nutrients absorption such as N, P, K, Ca and Mg [43].

2.4.1. Humidity Sensors

Humidity sensor requirements are: 1) sensitivity in a wide relative humidity (RH) range, 2) rapid response to RH variation and good reproducibility, 3) resistance to contaminants, 4) predictable dependence on temperature, 5) easy connection to circuits and 6) minimum calibration and low cost. RH sensor response time can be determined using medium (magnesium chloride: $MgCl_2$ - 50% RH) and high-humidity (sodium chloride: NaCl - 75% RH) saturated salt solutions. The most common humidity instrument is the wet-dry bulb psychrometer. The sensing element of the dry bulb is exposed to the air and the wet bulb is covered with a sheath obtaining accuracies of 5% RH. The manual RH measurement takes one minute and its accuracy is slow due to the 1°C temperature increments (Table 2). The psychrometer cannot be used in closed loops and requires a great amount of labor to refill the water reservoir and to replace the wick.

Capacitive sensors are characterized by low temperature coefficients, ability to operate at 200°C, full recovery from condensation, and reasonable resistance to chemical vapors. The dielectric constant of capacitive relative humidity (RH) sensors changes with the relative humidity of the environment; as RH raises by 1%, capacitance increases from 0.2 to 0.5 pF. Response times between 30-60 s result from a 63% RH step change. Commercial capacitive humidity instruments have both sensors (temperature and RH) in the same chip; both measurements are obtained at the same position of the airstream.

The HMP45C is a rugged and accurate probe that uses a capacitive polymer H chip to measure RH and a platinum resistance thermometer to measure temperature [44]; the probe should be recalibrated annually. The relative humidity sensor accuracy is within 2% RH in the 0<RH<90% RH range, and temperature accuracy is within 0.5 °C (Table 2). The humidity sensor has a long term stability of 1% RH per year, a temperature dependence of 0.05% RH/°C and a response time of 15 seconds with a membrane filter. A radiation shield is required when the HMP45C is exposed to sunlight. Vaisala makes the highest quality relative humidity sensors available on the market based on its accuracy, reliability, and durability being its market share of 90%. The CS500 exhibited more variable water vapor pressure measurements mainly due to its increased variation in temperature measurements compared to the HMP45C [45]; long-term stability and response time are similar than the HMP45C probe. A dew point hygrometer can be made with a capacitive humidity instrument having maximum relative humidity and temperature error magnitudes of 2% RH and 0.3 °C, respectively.

In the Testo 6631 humidity sensor, a humidity-sensitive polymer serves as the dielectric between the capacitor electrodes (Table 2). The temperature is recorded by a separate, accurate temperature probe Pt1000 class A. The upper electrode is completely permeable for water vapor which penetrates through to the polymer dielectric; to protect the sensor it becomes impermeable and resistant to oil, dirt and condensation [46]. A long-term measurement stability of +/-1 %RH is guaranteed for 5 years with a response time of 6 seconds. In the Testo 6681 instrument the humidity probe is heated at the rear, creating a 5 degrees warmer micro-climate around the probe. The moisture probe sinks from saturation (100% RH) to 73% RH; the response time becomes better. The transmitter computes the actual humidity conditions providing 4-20 mA and 0-10 V analog outputs; the equipment displays the values on a LCD screen, but is the most expensive instrument of the market.

Table 2. Relative humidity sensors are compared according to accuracy, range, current consumption and cost (year 2011)

	Model	RH range	Temp range °C	Accuracy	Time res-ponse, s	Avg. Current /control, mA	Cost, US $
	SLING PSI CHROMET	0-90	-5-50	±5 % 1°C	60		50
	CS500 CSI	0-90 90-100	-40-60	±3 % ±6 % 0.2%	15	<2 consum 0-1 V output	335
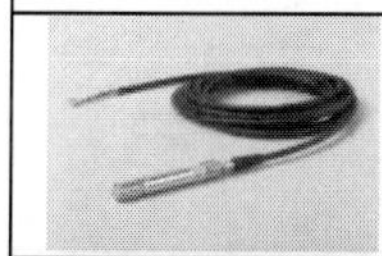	INTERCAP HMP60 VAISALA	0-90 90-100	0-40	±3 % ±5 % 0.6°C	2	1 4-20 mA	300
	HMP45C VAISALA	0-90 90-100	-39.2-60	±2 % ±3 % 0.5°C	15	<4 4-20 mA	850
	6631 TESTO AG	0-90 90-100	-10-60	±2.5 % ±4 % 0.5°C	6	RELAY 4-20 mA 0-10 V	1072
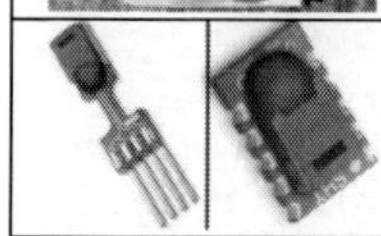	SHT75/ SHT11 SENSIRION	0-100	-40-120	±3 % 0. 4°C	8	0.027mA DIGITAL	30/10

Wireless grids using cheap Sensirion humidity sensors require calibration. Each sensor is calibrated in a precision humidity chamber and the calibration coefficients are programmed into the memory. These coefficients are used internally during measurement to calibrate the signals from the sensors. Sensirion SHT11 chip present temperature compensation for the capacitive humidity sensor and a heater prevents condensation (Table 2); its output is a digital serial signal with a predefined protocol [47]. Sensor accuracy remained within ±1.5% RH tolerance at different levels of relative humidity and of temperature.

2.4.2. Humidity Controllers

Humidity control has to consider that air temperature within the greenhouse and its relative humidity is inversely related and that the same actuators are used for controlling both temperature and humidity [40]. Greenhouse metal and cover internal surfaces can be at different temperatures than air temperature; those surfaces cooler than air at high RH will condense water vapor. Temperature control is prioritary over humidity control because it affects crop growth directly; humidity can be kept within a determined range where RH acts over the temperature set-point. Relative humidity levels can be reduced with a combined strategy of venting (exchange moist air by outside dry air), and heating raising the temperature of plant surfaces. Raising humidity levels without creating excessive free water requires of fog units that add water vapor to the air, or screens that help to retain the evaporated water from the plant canopy. Fog systems cool the air, increase the humidity and reduce the water being evaporated from the leaves. Screens reduce leaf temperature and help to trap the large amount of water that the plants are evaporating; hot air should be displaced by venting.

Four different strategies for controlling humidity in greenhouses [48, 49] are shown in Table 3 comparing air temperature and humidity. The first model is a practical one which employs five rules easily implemented in embedded systems or PLC (programmable logic controllers):

- RULE 1: Open vents up to 25% for air relative humidity within 80<RH<90; Increase the opening linearly by 5% per 2% RH increments.
- RULE 2: Close vents below 80% RH.
- RULE 3: During the first month of growth tomatoes present less biomass and transpire less, so the maximum opening should be 10%.
- RULE 4: If daily temperatures during winter decrease drastically (countries at latitudes over 30°) maximum opening should be reduced to 10%.
- RULE 5: When high humidity cannot be controlled by venting alone heating should be applied. Heat from 25°C to 35°C linearly starting at 85% RH.

The other three models employ different air venting conditions together with curtain shading to control humidity in greenhouses. The second model employed a rolled curtain with closed vents so air became hotter by 3.1°C and RH humidity decreased by 4.9% [49]. As the curtain was unrolled and vents opened (model Curtain-vent 2) air temperature dropped off by 2.6 °C, meanwhile humidity decreased by 6.5% RH (Table 3). When fog was applied under the previous model conditions humidity increased by 10% RH and air was 2°C cooler than curtain-vent 2 treatment.

These rules work for tomato, but for huitlacoche a fungus inoculated on maize requires of a closed loop control [50]. Convection cooled Parral greenhouses should analyze the wind behavior in the zone. Lateral vents should be opened at higher RH values when wind remains constant all day long.

Table 3. Models used to control humidity in greenhouses

Model	Rules	Air temperature, °C	RH humidity, %
Simple 5 rules	RH setpoint	Linear change	-3.1
Curtain-vent 1	Roll curtain Close vent	3.1	-4.9
Curtain-vent 2	Unroll curtain Open vent	-2.6	-6.5
Curt-vent-fog	Unroll curtain, fog, open vent	-4.6	+3.8

2.5. Solar Radiation Control

Light quantity is affected by a combination of day length, solar angle, cover materials, plant density, canopy structure, meanwhile the spectral distribution of light received at a given point depends on solar angle, atmospheric conditions, transmission through leaves, and reflection from nearby plants and other objects, including the soil surface [51]. Changes in plant photoperiod and light quality are sensed by photoreceptors such as the phytochrome involved in stomatal opening and leaf color. A rule of thumb estimating the effects of light on production is the one-percent rule, stating that 1 % reduction in light will reduce production by 1 %. This rule often gives a close estimate of light loss consequences on tomato yield. For a single-truss production 1.94 MJ m^{-2} is required for a positive yield and 50 MJ m^{-2} of global radiation is required for producing 1.0 kg of fresh tomato fruit.

Light influences air temperature around the plant canopy and leaf temperature; leaves absorb solar radiation and it responds quicker than air temperature to solar radiation increments. Correlation coefficients between air and leaf temperature with total irradiance are 0.7167 and 0.7801, respectively. After sunset, plants continue transpiring for almost 4 hours indicating the influence of temperature and wind. Net photosynthesis saturates when the radiation exceeds 300 Wm^{-2} and summer horticultural crops cannot survive when solar radiation drops from the trophic limit 8.4 MJ m^{-2} day^{-1} [52]. Light levels below 1.5 MJ m^{-2} of solar radiation result in an increased incidence of reduced fruit set and poor flower quality.

Cumulative values of daily PAR were integrated over a period of 45 days before harvest [18]; correlations between cumulative PAR with firmness, dry weight, soluble solids content and phenolic compounds were lower than those encountered with cumulative air temperature. Producers obtain tomatoes of similar quality under lower PAR than those provided by natural sunlight. As incoming PAR drops by 30 and 50%, tomato yield decreases by 14.3 and 38.1%, respectively [18]. Fluctuations in the weekly fruit number and yield for tomatoes grown in greenhouses during summer and fall were strongly affected by solar radiation changes 10 days before anthesis [53].

Two sensor types are being used to measure solar radiation: pyranometers and quantum or photosynthetic active radiation (PAR) sensors. These sensors are used for shade and illumination control, placing one sensor over and one sensor beneath the shade curtain. The CM 11 and TSP-700 expensive pyranometers (Table 4) use a thermal detector sensitive to long wave infrared radiation which quantifies the total power absorbed; the thermal detector is non-selective to the spectral distribution of the radiation. Radiant energy is absorbed by a black disk, flows through a thermal resistance to the heatsink; the temperature difference across the detector thermal resistance is converted to voltage. The CM11 is the most precise pyranometer in the market with an accuracy of ±2%; analog output ranges between 0 and 5 mV so a high quality transmitter is required. The LI200X pyranometer uses a silicon photovoltaic cell mounted in a cosine-corrected head to measure sun and sky radiation on a horizontal surface. Sensors should be cosine corrected to maintain its accuracy when radiation comes from different angles. The output current of the sensor is transferred to voltage, which facilitates direct connection to the datalogger. The LI200X should not be used under vegetation or artificial lights because it is calibrated for the daylight spectrum (400 to 1100 nm). The LI200X response time is extremely quick compared to thermal pyranometers as CM11, but is less accurate (Table 4). One of the two common calibration methods: the pyrheliometer and the artificial sun should be employed every two years. In the calibration from a narrow field of view instrument (pyrheliometer) to a wide field of view instrument (pyranometer), the direct component of the solar radiation should be eliminated by shading the instrument outer dome with a disk. There is no thermal equilibrium with this method and pyranometers can present zero-offset drift. The artificial sun embodies a 1000 W tungsten-halogen lamp with compact filament fed by an AC voltage stabilizer. Behind the lamp is a diffuse reflector with a diameter of 7.5 cm and a built-in ventilator allows continuous operation [54]. Pyranometer periodic calibration is recommended at least every 2 years. PAR sensors as the SQ-100 are commonly used due to its reduced cost, but measurements between sunny and cloudy days vary less than 0.5 percent. A quantum sensor would give equal emphasis to all photons between 400 and 700 nm and would exclude photons above and below these wavelengths (Table 4); the SQ-100 sensor measures green light (500-550 nm) accurately, so it can be used to detect the radiation at the bottom of plant canopies. Quantum meters are accurate to within ±3 percent for common light sources and its output tends to decrease over time as the detector ages; recalibration is recommended every 3 years.

2.6. Shade Control

A thermal screen adds an additional barrier between the greenhouse and its surrounding and reduces air convection and air vent movement [55]. When movable, it has less impact on light transmission compared to fixed screens or double covering materials. About 80% of the greenhouse area in The Netherlands and 35% in France are equipped with thermal screens [56]. Theoretically, screens may reduce energy consumption by more than 35-40%, depending on the material. In practice, movable screens are closed only a part of the entire period depending on humidity and light restrictions. Shading is the ultimate solution for cooling greenhouses as it affects tomato yield and decreases plant transpiration [57]; a better quality can be obtained from shading. Leaf temperature should control shading in greenhouses.

Table 4. Radiation and PAR sensors compared according to accuracy, range, current consumption and cost

	Model	Spectral range	Temp range °C	Range, accuracy W/m^2	Time response, s	Signal provided	Cost, US $
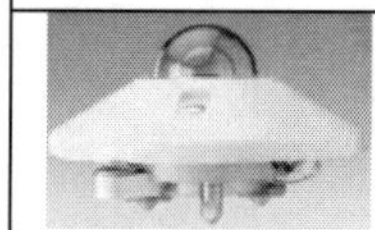	CM11 KIPP&ZONEN	350-2200	-10-40	0-1400 ±2%	5	0-5 mV	3666
	LI-200x CAMPBELL SCIEN	400-1100	-40-65	0-3000 ±5 %	10μs	0.2kWm^{-2} mV^{-1} 0-15V	250
	461105 STARPYRANO FISCHER	300-3000	-40-60	0-1500 ±3 %		4-20 mA	2000
	TSP-700 YANKEE ENV SYST	300-3000	-50-50	0-1400 ±5 %	1	0-4 V	
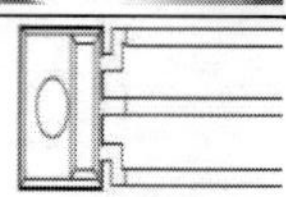	TAOS TLS260R	800-1100	-40-120	±3 % 0. 4°C	260 μS	111 mV/(μW/cm^2) 2.5-5.5 V	10
	SQ-100 APOGEE	400-700	-25-55	±5 %	< 1 ms	mV 5.00 μmol m^{-2} s^{-1} per mV	139

An automated Raschel net system was developed to provide shade to a tomato crop [58]. Raschel nets are made of flat, black polypropylene filaments, 1.0 mm wide and 0.1 mm thick; it is fabricated in a triangular weave and can be produced in varying mesh densities. The system after measuring solar radiation, canopy and substrate temperature irrigated the tomato plants and opened/closed the net automatically (Figure 5); an embedded fuzzy system optimized its operation (Figure 6 a). A gear motor rolls the net at low speed being its movement controlled by a photocell, which is less sensible than a phototransistor (Figure 6 b); when air temperature exceeds 25° C the curtain closes reducing fruit cracking (Figure 6 c). On this particular cloudy day radiation felt beneath 8,000 luxes (69 Wm^{-2}) after 14:00 hr when clouds appeared; the phototransistor presented a better response. An approximation for converting luxes to Wm^{-2} requires of dividing the quantity of luxes by 116 [59]. A Raschel net providing 35% coverage, allowed that 30% of the water from the fog pass through it.

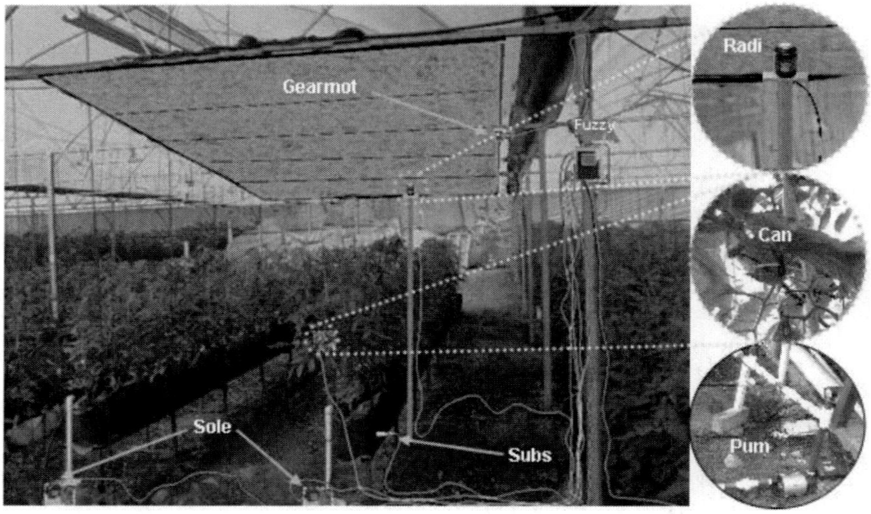

Figure 5. Tomato plants shading with a Raschel net.

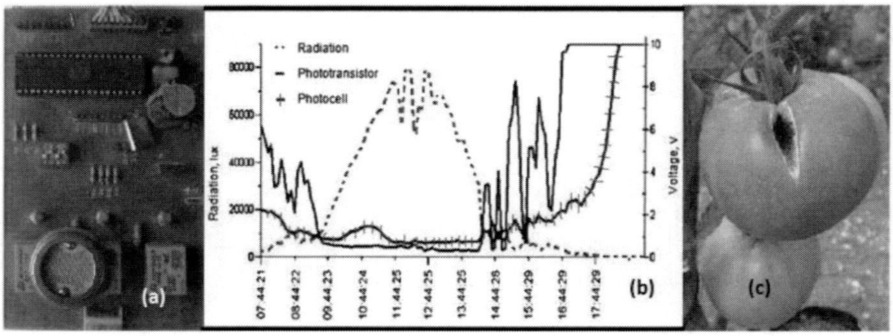

Figure 6. Solar radiation (a) measurement, (b) graph and (c) causes tomato cracking.

2.7. Light Control

In greenhouses, the amount of PAR available to the plants and the extent to which this radiation is scattered or diffused is of considerable importance for plant production. Diffusing light takes the scattered light in all directions, and distributes it equally over an area; plants grow more efficiently under diffused light conditions. Dry glass transmitted radiation very efficiently, but 81% of the transmitted radiation was scattered when condensation was present. Polyethylene dry film scattered 70% of the radiation, increasing to 78% with condensation [60]. Crop net photosynthesis increased with scattering, being better in polyethylene dry film; condensation reduced the net photosynthesis by 3 to 6%.

Supplementary lighting increases productivity and quality of greenhouse grown crops. Acidity and sugar content of tomato fruit is highest in summer; sugar content decreases with shading treatments. The source of supplemental lighting most used for crop production is the energy efficient high-pressure sodium (HPS) lamps; 26-30% of the electric energy is converted into useful light for photosynthesis. Light quality from HPS lamps differ from that of the sun, and as the percentage of total light received increase, variations on plant response are expected. Supplemental light for 8 hours increased fruit yield and fruit weight; tomato fruits were harvested earlier [61]. Adverse effects with HPS lighting were purpling of young leaves and flower development arrested [62]. Plants in production exhibited leaf "burn out" and fruits failed to set; these symptoms were overcome after increasing phosphorus levels in the nutrient solution. Lighting systems release heat energy increasing greenhouse air temperature; if air temperature exceeds a desired setpoint (for example 24 °C), hot air is vented out.

3. Nutrient and Water Control

A significant proportion of horticultural crops are grown in soilless culture. In 1995, 75% of the vegetables crops under greenhouses in the Netherlands were grown in substrate other than soil, predominantly rockwool. As plants cannot benefit from rain when they are grown inside a greenhouse, water should be supplied. Irrigation systems can be done either by overhead irrigation such as drip irrigation (mostly vegetables) or by sub-irrigation. Plants cool themselves during transpiration and water transport the minerals that roots absorb; plants use 5 to 10 % of the water for their growth [63]. The amount of transpiration depends of solar radiation, humidity, air temperature and wind speed.

Salinity is one of the major environmental problems confronting agriculture especially in arid and semi-arid regions. As water evaporates from the soil, salts accumulate in the topsoil, so to prevent soil deterioration water should be properly managed. Salts should be removed by leaching (200-300 liters of water per square meter), as plant water and nutrient uptake becomes more difficult when the soil contains high quantities of salts. Salt accumulation in the root environment is noted at the volcanic rock (tezontle) substrate were tomatoes are cultivated (Figure 7 a); once water-leached fertilizer salts disappear (Figure 7 b). Salts adhere in the root hairs as shown by the red elongated circles over the microscopic image (Figure 7 c). When the soilless culture is washed [64] the image shows lack of minerals (Figure 7 d). The accumulation of high soluble salts in the soil can significantly decrease the value and productivity of agricultural lands. Salt concentration is measured with an EC (electrical conductivity) meter, being a high EC value indicative of a high salt content. Soil salinization can be prevented using rapid and reliable techniques for monitoring in situ volumetric soil water content and electrical conductivity of soils.

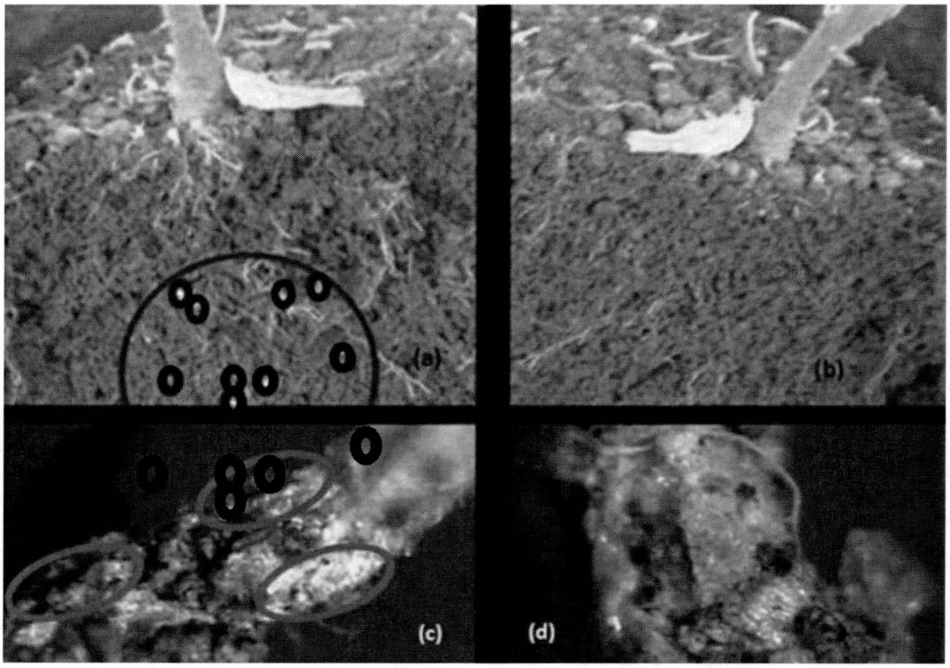

Figure 7. Root environment shows (a) salt accumulation, (b) salt elimination after leaching, (c) microscopic image of root hair with salt and (d) image without salt.

Clay minerals in soil and corrosive substances are negatively charged; components of fertilizers having positive charge are electrically retained in the soil and will not be easily washed away by rain. As phosphoric ions are negatively charged, they are not attracted by negatively charged soil particles; phosphoric ions combine with aluminum, iron, and calcium in the soil to form compounds that are barely soluble [65]. Thus, phosphorus is less likely to leach into groundwater. Positively charged potassium ions do not leach into groundwater as these ions are retained by the soil particles. However, nitrogen fertilizers tend to leach into groundwater. Nitrogen fertilizer is taken into plants as ammonium irons or nitrate ions. With excessive quantities of nitrogen applied, vegetables will absorb too much nitrate ions and accumulate them in the leaves.

Nutrients are absorbed by plant roots from the soil during crop growth. Each farmer should diagnose the remaining nitrogen prior to the start of cultivation, and determine the necessary amount of fertilizer required by the crop; this amount varies significantly according to the type of soil and water drainage [64]. Simple soil diagnosis is carried out by measuring the electrical conductivity (EC) of soil that is related to the nitrate – nitrogen concentration, or by directly measuring the nitrate ion concentration in the soil.

Fertigation techniques use localized irrigation to apply water and minerals properly. Growers are familiar with the concepts of pH and electrical conductivity (EC) of the nutrient solution. The automated irrigation equipment takes watering decisions according to the amount of solar radiation and volume of water transpired (Figure 8). Automated fertigation systems should provide real-time data on substrate moisture and electrical conductivity (EC), for better management decisions on a daily basis. Irrigation should be automatically scheduled based on plant-available moisture measurements. In situ substrate EC data should be acquired, to ensure that nutritional requirements are being met by hydroponic systems. It is possible to control the EC of the applied nutrient solution without the need for taking and analyzing samples from the medium.

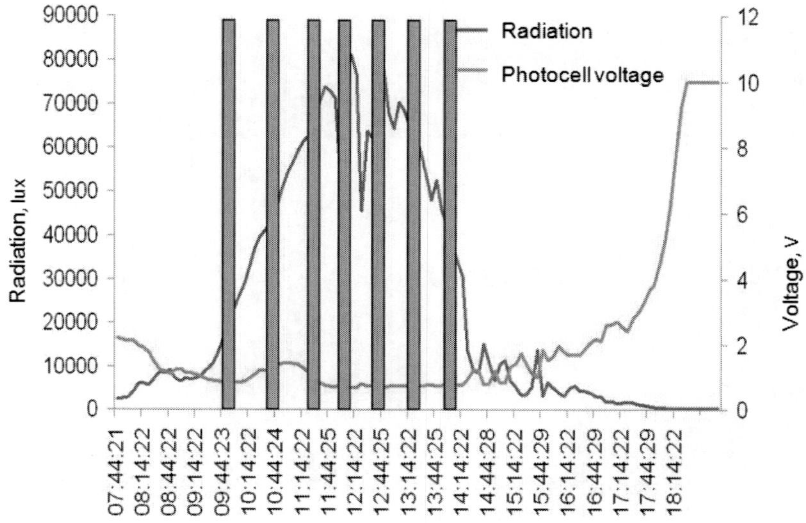

Figure 8. Irrigation based control according to solar radiation.

3.1. Water Management Control

As water supplies become scarce and polluted, there is a need to irrigate more efficiently in order to minimize water use. Agricultural fields are effectively over-or under-irrigated due to spatial variability of crop water use, water infiltration and runoff of rainfall and irrigation. Under-irrigated areas are subject to water stress, resulting in production loss; over-irrigated areas suffer from plant disease and nutrient leaching [66]. Automatic irrigation systems are based on the feedback of the soil water status to maintain soil water content within a specific range; water consumption decreases without reducing crop yield.

Air pulled by air-vents to the hydraulic system does not allow that the proper quantity of liquid is applied by the emitters during pulsed or high frequency irrigation. As the main irrigation tube becomes thicker, a greater air volume is stored in the hydraulic system. The time required to withdraw the air by the emitters depends on the number of emitters; during two-minute irrigation pulses emitters only 40-60% of the required water is applied. A simple equipment was designed to measure the water applied to the tomato plants and it was placed randomly in one of the laterals beneath an emitter [67]. The equipment measures the water level within the tube being the solenoid activated by a micro-switch when it arrives to the desired water level (Figure 9 b); a wireless signal stops the irrigation pump. Once the energy supplied to the solenoid is removed, water within the tube is drained-off, getting ready for the next measurement (Figure 9 c). The micro-switch will not be activated when only air is withdrawn from the emitter.

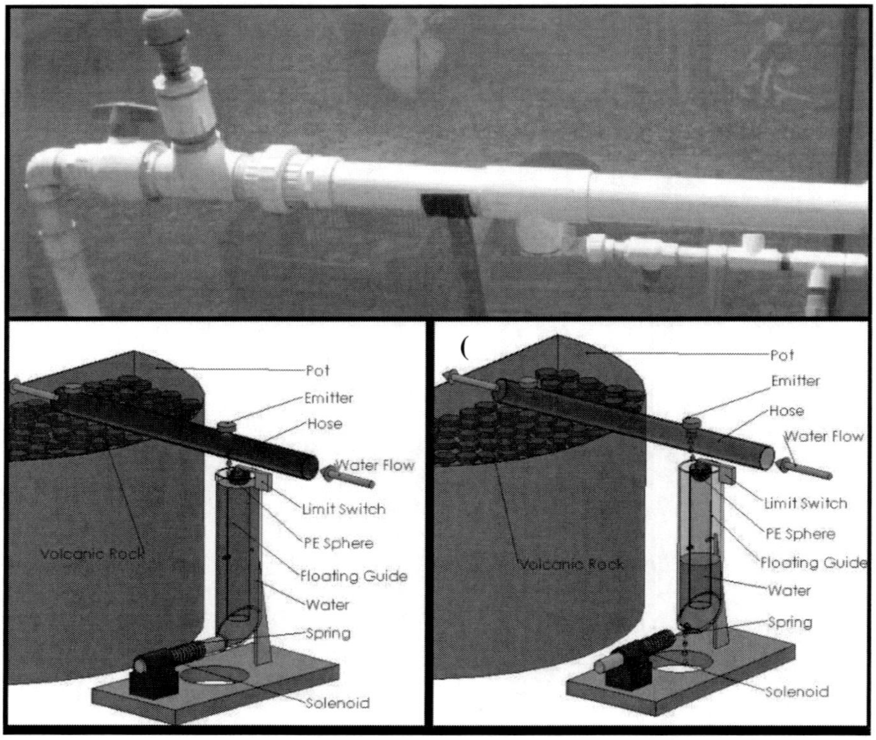

Figure 9. Proper water applied to pots after (a) passing an air vent; (b) during filling; and (c) after discharging.

The most common and simple system uses timed irrigation, allowing the grower to match water supply with evapotranspiration rate; the number of events per day are previously preset. Climate data and a computer may be used in some irrigation control based on light integral. This reliable system is simple to manage using an inexpensive photocell sensor to measure solar radiation, Figure 8; several sensors located around the greenhouse can estimate light levels [68]. The light-based approach does not function well under low light conditions [69] as it does not consider factors such as humidity and temperature. The light-based system presents a great frequency of irrigation events including in the early morning and late evening [69].

Tensiometers are sealed water-filled tubes placed in contact with the soil through a permeable ceramic porous cup. Water inside the tube comes into equilibrium with the soil solution creating a vacuum within the tube. Some of the common problems of tensiometers during irrigation are entrapped air, organic growth on ceramic cups and need for recalibration [70]. Optimal irrigation at 10kPa increased yield, improved fruit quality and reduced nutrient leaching [71]. A switching tensiometer-controlled drip irrigation system set at 15 kPa on tomatoes, reduced irrigation by 70% compared to typical farmer practices in sandy soil while maintaining similar yields. In spite of these results, tensiometers have not been widely adopted for vegetable production in coarse soils, due to the frequent maintenance required [72].

3.2. Soil Water Content Sensors

It has long been recognized that reliable, robust and automated techniques for the measurement of soil moisture content can be extremely useful in agricultural applications. Among the widely used automated soil moisture measurement techniques are neutron scattering, gamma ray attenuation, soil electrical conductivity, tensiometry, hygrometry and soil dielectric constant.

Granular matrix sensors (GMS) consist of a porous medium with two embedded wires that read the electrical resistance in the medium between them. Often, some gypsum (calcium sulfate) is included to buffer the soil water solution and decrease the effect of salinity on the resistance value. With greater soil water tension, less water is in the porous medium increasing the electrical resistance. Installation and contact problems in sandy and clay soils are similar to those encountered by tensiometers or gypsum blocks. At tensions beneath 30 kPa, tensiometers responded quicker than GMS sensors [73] in silt loam, loam, and coarse sand. Gypsum blocks reading require of an alternating current to minimize effects of capacitive charge build up and ionization.

The humidity sensor controls an automated capacitive irrigation system which varies its capacitance when soil water content (SWC) changes. The sensor provides a DC signal proportional to the SWC which is compared with a preset value given by the SWC required by the plant; the comparator activates a relay which turns-on a pump. Automated irrigation treatments using tensiometers and capacitive probes presented the highest marketable yields. Treatments with soil water feedback mechanisms saved on average 72% irrigation water compared to grower scheduling and 65% compared to historical (long-term) evapotranspiration based methods [74].

Moisture sensors based on dielectric properties are a powerful tool for real-time, simultaneous measurement of soil water content and bulk electrical conductivity on the same in soil sample; measurement problems arise with soils having significant salts. Instruments that use the capacitance technique or time domain reflectometry (TDR) to determine volumetric water content θ rely on the high dielectric constant of water [80] relative to mineral soil (3–5) and air (1), because water in the soil will influence the propagation of an electric signal through the soil medium [75]. The calibration is strongly dependent on the properties of the selected probe and on the soil sampled properties. If the site-specific probe calibration given by the manufacturer differs from laboratory tests, it should be recalibrated [76]. The probe values obtained from the laboratory calibration for the specific soil should be cross correlated against tensiometer readings during the first week of the season.

Reliable EC-20 low-cost Decagon dielectric probes (100 US) are now available being a low-maintenance alternative to tensiometers (Figure 10 a); probes require of a power supply and data loggers to store continuous information from 4 or 8 capacitive sensors (Figure 10 b). Capacitive sensor values depend on the electric field frequency used by the system [77] and the electrical conductivity of the material. The EC-20 Decagon probe provides an output signal ranging from 250 to 1000 mV at 2500-mV excitation. The manufacturer provides an equation having an accuracy of 3% Θ if the soil does not have "high" sand content, clay content, or electrical conductivity; the probe does not require temperature. Water content measurement of gravel or volcanic rock is very imprecise as too much air surrounds the plate. When the substrate pore water EC levels are <1.0 dS m^{-1}, ECH2O-10 probes can accurately measure the water content of the substrate. However, as EC level is usually 2.0-4.0 dS m-1 in nutrient solution this cheap sensor has to be recalibrated in field being recommended for greenhouse use [78].

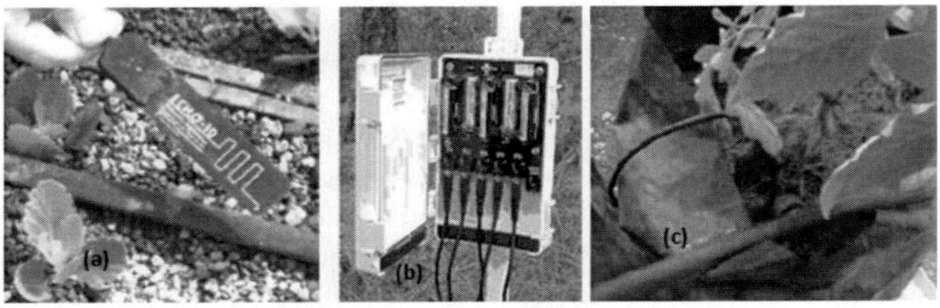

Figure 10. Measuring moisture (a) with ECHO-10; (b) data logger; and (c) in coco coir substrate.

The Campbell Scientific CS615 water content reflectometer consists of two 30 cm long stainless steel waveguides connected to a printed circuit board, and measures the soil moisture content using the TDR technique. The CS615 reflectometer is specified to have an accuracy of ±2.5% v/v when applied to typical mineral soils using the manufacturer's standard calibration relationship; both accuracy and stability of the sensor are affected by the soil's electrical conductivity and temperature. Output sensor changes are noted when soil electrical conductivity increases over 2 dS m^{-1}; for electrical conductivity over 20 dS m^{-1} it becomes unstable. The CS615 reflectometer has a waveguide diameter of 3.2 mm but due to the moist state of the soil during setting up it can be easily installed without pre-forming holes. Dry soil conditions during TDR sensor installation may produce air or fluid filled gaps which introduce measurement errors, especially under wet soil conditions. The CS615 reflectometer has diurnal variations in the soil moisture measurements, and do not exhibit high frequency noise; the sensor predicted soil moisture contents that exceeded the soil porosity during periods of saturation [79]. The CS615 cost 150 US and is accurate but less precise than EC-20, presenting high variation in the data [75].

Table 5. Average fruit weight, substrate EC after four days of application and biomass production after ten days

Tanks mixing	**Frequency cycles**	**Fertilizer/water cycles**	**Fruit weight, g**	**EC, dS m^{-1}**	**Biomass, g**
Both	6	6/0	128	2.1	15
Both	10	10/0	137	2.1	19
Both	12	12/0	122	4.1	22
Both	12	8/4	130	2.1	23
Both	12	10/2	139	2.3	22
Separate	12	10/2	138	2.2	23
Both	16	16/0	114	6.1	26
Both	16	8/8	130	2	26
Both	16	10/6	141	2.2	29
Separate	16	10/6	141	2.3	29

The CS616-L is a self-contained water content sensor that uses high-speed electronics in the body of the sensor to generate a square wave output with a period that is proportional to water content. Measurement of the square wave by a datalogger's single-ended channel gives precise, high-resolution water content measurements in less than 0.1 seconds—even with long cables.

3.3. Fertigation Control

The application of fertilizer through irrigation systems is referred to as fertigation; fertilizers are directed towards the plant root zone by the irrigation water. Fertigation not only reduces production costs but also lessens groundwater pollution caused by fertilizer leaching. Fertigation allows adapting the amount and concentration of the applied nutrients in order to meet the actual nutritional requirement of the crop throughout the growing season. Other advantages of the fertigation are saving in energy and labor together with a higher flexibility of the moment of nutrient application; the supply of nutrients could be carefully controlled and monitored.

The same nutrient solution concentration was applied to plants in soilless systems having different frequency of irrigation; it affected yield and fruit quality [80]. After 30 days of fertigation, peat substrate with an EC of 4.5 dS m^{-1} increased peat salinity by four times decreasing photosynthetic rate and leaf water potential [81]. Six, ten, twelve and sixteen irrigation cycles were applied to tomatoes cultivated in volcanic rock (tezontle) substrate. The solution applied every 45 minutes (12 high-frequency cycles per day) increased EC to 4.1 dS m^{-1} after 4 days, reducing fruit size [64]. Ten irrigation cycles per day spaced by one hour did not increase the substrate EC and presented an average fruit weight of 137 g (Table 5). As the number of cycles increased to 16, biomass increased and produced the minimum average fruit weight of 114 g. Combined cycles of fertigation and irrigation (only water) provided better yields. Twelve irrigation cycles (10 of fertigation and two of only water) produced tomatoes weighting 139 g while maintaining an EC of 2.3 dS m^{-1} after 4 days. Maximum fruit weight of 141 g was obtained with 16 irrigation cycles (10 of fertigation and six of water) producing also the highest biomass with an average value of 29 g.

Only water-soluble dry fertilizer should be used; poor soluble fertilizers as urea, ammonium nitrate and mono-ammonium phosphate take 20 minutes to dissolve into water meanwhile potassium salts dissolve in 5 minutes. Any undissolved material may cause plugging problems if allowed to reach the emitters. The fertilizer solution may require dilution with water to provide the final nutritional liquid distributed to the crop. The fertilizer used should be tested with the irrigation water before being injected into the micro-irrigation system. It is recommended to dissolve the fertilizer in question in irrigation water at the concentration that would be provided to the plants in a clear glass jar; let this mixture set for 12 hours in the dark, and if the mixture becomes cloudy or forms precipitants, it may clog your micro-irrigation system. When irrigation water has an EC > 2 dS/m and the crop is sensitive to salinity the accompanying ions added with N or K should be reduced.

Adding or extracting fertilizer salts will impact directly on the electrical conductivity (EC) of the solution. An increase in EC decreased total and marketable (> 150 g) fruits fresh weight regardless of plant transpiration. The lower fresh weight observed for EC= 8 dS m^{-1} is due to a reduced water flow into the tomatoes during development. Fruit firmness decreased with increasing salt concentrations [82] from 44.5 kPa (EC=3) to 43.3 kPa (EC=6.5). High EC can reduce Ca uptake, even if high calcium concentration is encountered in the nutrient solution; it induces a local Ca deficiency in tomato fruit called blossom-end rot [83]. As EC increased from 2.6 to 4.6 $mS \cdot cm^{-1}$ with NaCl (up to 12 $mmol \cdot L^{-1}$), tomato fruit quality improved without affecting total yield and plant development [84]. The use of NaCl reduced the cost of fertilizer and environmental pollution from nitrate and other plant nutrients.

3.4. Fertigation Equipment

Passive injectors (venturi, pitot tube) use the energy supplied by the irrigation system or the atmosphere to inject the chemicals. Venturi injectors are very economical ($50 - $500 US) and require no electric power to operate. As a rule of thumb about a 20% pressure drop [85] is required to effectively operate a venturi injector. Active injectors use an external energy source or a mechanical moving part to inject the fertilizer solution at a higher pressure than the irrigation mainline pressure. Expensive positive displacement pumps present null head loss, high discharge rate and good concentration control; electrical and water powered positive displacement pumps can be accurately metered. Piston and diaphragm pumps require of system pressure to operate but do not employ electricity. The application rate obtained from a suction line injector cannot be accurately metered; the fertilizer solution enters the pump, contacts the seals and shortens its life. Periodic calibration is needed to ensure that an injector is operating properly. Final fertilizer concentration is checked by the volume ratio method and/or EC sampling detecting injector malfunction [86].

The different components of the fertigation system are shown in Figure 11. The use of two fertilization tanks allows separating those fertilizers that interact and cause precipitation, placing in one tank the calcium, magnesium and microelements and in the other tank the phosphorus and the sulfate. Some producers prefer to use four tank deposits, three containing fertilizer solutions and the fourth acid, which is used to control the pH of the nutrient solution. One palette pump is fed from each tank (Figure 11 a), injecting the nutrient solution to the irrigation mainline tube. A check valve prevents chemicals returning to the water source and should be located between the water source and the injection point. When only two tanks are used, the acid is added in the nutrient solution. The fertilizer solution should be injected up-stream towards a filter disk system (Figure 11 d), to eliminate any precipitants before arriving to the emitters. The 3 hectare hydroponic system was divided in six 0.5 ha zones; a solenoid controlled which zone was irrigated (Figure 11 e). The system uses a pH and an EC controller for feeding the correct nutrient solution to the plants (Figure 11 b).

A new fertigation hydraulic head used only one pump for injecting the solutions from both tanks (Figure 11 a). The new system uses a servomotor (Figure 11 c) which turns an aluminum structure having four flat pieces (sills) fixed to the control valves, opening or closing them mechanically; when valve-1 is opened, valve-2 is closed and vice versa (Figure 12). These bronze valves are connected through a "T" connection to the pump entrance. The valves in both extremes return (ret tank) some of the nutrient solution pumped to their respective tank controlling the flow being injected to the main water tube. The mechanism resulted reliable and injected only one tank during an irrigation application, or the mixed fertilizer solution of both tanks. When the nutrient solution of each tank was applied in subsequent irrigation cycles tomato fruit weight decreased by one gram (Table 5).

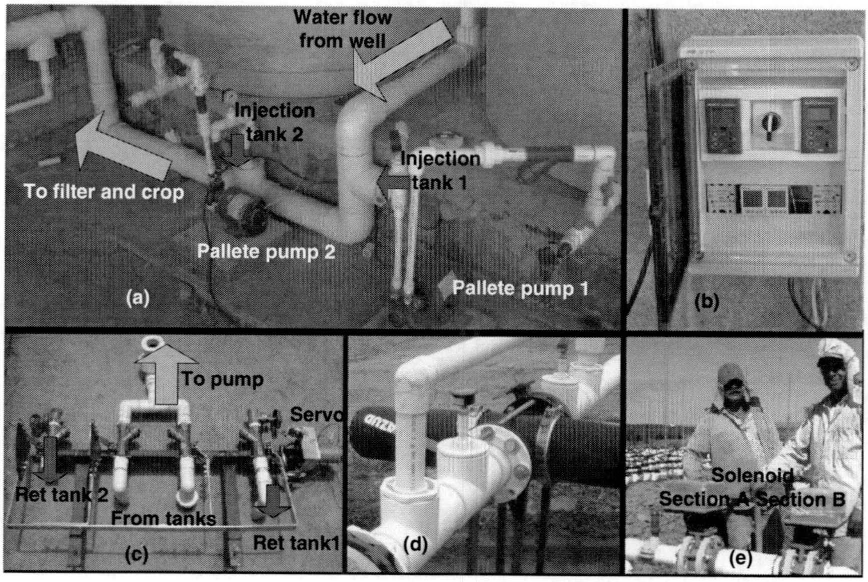

Figure 11. Fertigation (a) injection system, (b) control, (c) special injector mechanism, (d) disk filter and (e) section solenoids.

When the nutrient solution from each tank is prepared, personnel should avoid the application of dry fertilizers in excess of 20%; otherwise an accurate control will be impossible. After monitoring weekly EC measurements from each tank it was found that 23% of the times, EC values were higher or lower by 20%. An automated fertilizer dosing system was developed using a programmable logic controller (PLC), which weights the exact dose from each dry fertilizer that will be applied to 100 lt of water for preparing the nutrient solution (Figure 13).

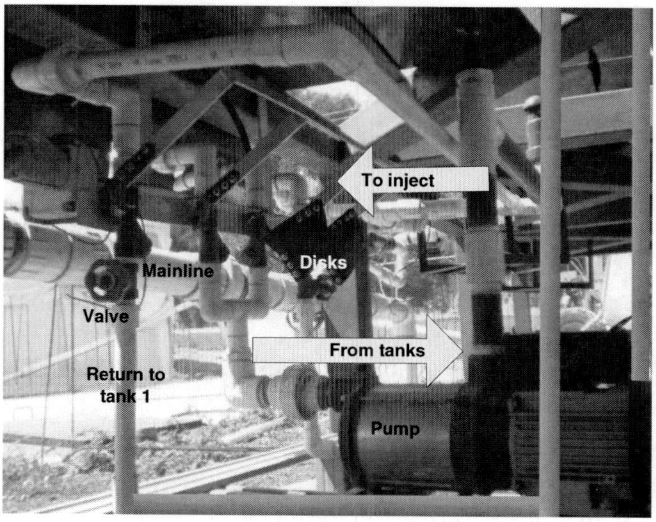

Figure 12. Fertigation head injects the nutrient solution from two tanks using one common pump.

Figure 13. Equipment for (a) applying fertilizers; (b) balance for weighting fertilizers.

A vertical worm drived by a 24VDC gear-motor, delivered powder from a conical closed hopper to the weighting gauge (Figure 13 b). When the weighted powder reaches its setpoint it is blown into the motor-stirred tank being dissolved into water. After applying the first fertilizer to the tank, the top base turns 45° until the weighting gauge gets synchronized with the next funnel. This weighting operation is repeated 4 times before stopping, preparing the solution for one irrigation day.

3.5. EC Sensors

The HI 943500 is a panel-mounted analog EC controller which can be used for dosing fertilizers to the crop. The instrument presents membrane keypads and a large LCD on the front, having a series of auto-diagnostic functions. The output current varies from 4 to 20 mA and is proportional to the measured conductivity value (Table 6). Fertigation on-line applications use the HI 7635 potentiometric probe, presenting automatic temperature compensation from 0-50°C with a NTC sensor; the glass operating temperature can vary from 0 to 120°C at a maximum pressure of 5 bars. Deposits on the conductivity probe can be removed by immersing the probe in hydrochloric acid for 30 minutes. The probe is connected to the EC controller being calibration carried out using air as the lower value and a known electrical conductivity solution as the higher (Table 6). The 750 Series II Myron controller presents a digital display with an analog output signal ranging between 0-10V. It presents automatic temperature compensation and a solid state output for control purposes.

Sensors based on dielectric properties can provide simultaneous measurements of soil water content and bulk electrical conductivity of the same *in situ* soil sample. EC substrate bulk measurements indicate how well the substrate as a whole [87] conducts electricity; from a plant's perspective it is important to determine the EC of the pore water (PoreEC). Inoue *et al.* compared salinity sensitivity of twelve commercial dielectric sensors in a sandy soil [88] and reported that the output of ECHO10 probe from Decagon Devices, exhibited the highest sensitivity from soil solution electrical conductivity. Many sensors can monitor EC in soilless substrates, but the measurement of real time EC is not trivial [89]; soil WC-EC (water content EC) relationships depend on temperature, texture, salinity, layering, and their interactions.

Table 6. Electrical conductivity sensors compared according to sensor type, accuracy, range, time response, signal provided and cost

	Model	Type/f, Mhz	EC range,	Accuracy,%	Time response, s	Signal provided	Cost, US $
	HI 943500 HANA	Analog	0-19.9	±2% full scale		4-20 mA Relay 2A	740
	750 Ser II MYRON L	Analog	0-1 μS/ppm	±2% full scale		0-10 V Solid relay	850
	WET DELTA T	FD/ 20	0-300	±10%	5 s	RS232	1200
	GS3,CAMPBELL SCIENTIFIC	FD/70	0-23 0-10	±10%	150 ms	Serial TTl, 0-3.6 V	
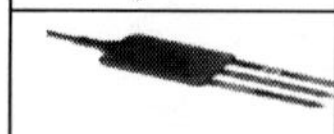	5TE DECAGON	FD/70	0 – 23 0 - 7	±10%	150 ms	RS232	600

Most sensors measure bulk EC, which is the total electrical conductivity of the surface soil/substrate ionic charge plus the ions in solution. Temperature also affects ionic concentration measurements, so sensors should have temperature compensation for precise measurement. WET is a frequency domain dielectric sensor that measures water content, EC and temperature in an all-in-one 3-pin sensor [90]. The sensor presents a high repeatability (1%) once inserted into a growing medium and left untouched (Table 6). The sensor needs a specific calibration for the type of growing medium to obtain good absolute readings. Another detector available in the market is the WCM sensor which is a derivative of the original WET. WET sensors measures bulk EC and now supplies extended calibration curves up to 5 dS m^{-1}.

The 5TE sensor (Decagon Devices, Pullman, WA) is a low-power sensor that measures soil water content and EC using a RS232 output to send data to an external micro-computer; it cannot be connected to standard data loggers. The 5TE is supplied by a 3.6 – 15 VDC voltage, consuming 10 mA during the 150 ms measuring period [90]. The 5TE has minimal salinity and textural effects, having a bulk EC operating range up to 23 dS m^{-1}; its accuracy is ±10% in the 0 – 7 dS m^{-1} range and user calibration is required above 7 dS m^{-1}. In the 5TE capacitance sensors, EC measurements are taken by two small stainless steel screws embedded in the sensor serving as the electrodes. In a fertigation system using 5TE as sensor, PoreEC control results difficult [87]. Factors that contributed to large fluctuations in PoreEC and lack of accuracy were: (1) no control as plants were not fertigated until reaching the setpoint. (2) Deficient contact between electrodes and substrate creates air pockets surrounding the sensor, measuring low bulk EC and a low calculated PoreEC. (3) As substrate dries out following irrigation PoreEC increases because the solution becomes more concentrated. (4) Fertilizer appears to move much more slowly through the substrate than water. After analyzing several bulk EC sensors, accuracy was lost beneath a substrate water content of 35% [91]. After corrections for temperature and permittivity the frequency domain (FD) sensor gave an accurate EC estimation up to 6 dS m^{-1} in rockwool, useful for horticultural practices [92]. When coco peat was used, EC measurements lost sensibility over 3 dS m^{-1} [93].

3.6. Open Fertigation Systems

Nutrient application systems can be classified as open or closed. Open systems commonly used due to it easy management do not re-circulate the drain solution (25-30%). Open soilless systems present 25 and 30% waste of water and nitrogen, respectively; this is responsible for higher production costs and deep water pollution. In a rockwool tomato culture during 4 months 30% of nitrogen was drained together with 25% of water [25]. Water supply is normally adjusted using on-line measurements of volume and EC of the drainage solution [25]; also it should be estimated according to climatic conditions in the greenhouse.

3.7. Recirculation Closed Systems

Growing plants in containers allows the collection of leaching water and its comparison with irrigation water. Measurements of pH, EC and nutrients concentration in the leached solution indicates if fertilizers are being applied in excess or with deficiency and allows correcting the fertigation routine. A lower EC value in the leached solution than in the applied solution indicates that the plant absorbs more nutrients than water and a greater amount of water should be applied. If the difference between both EC values is greater than 0.4 dS m^{-1}, leaching irrigation should be applied to wash the excess of salts. Alkaline pH over 8.5 in the leaching solution indicates phosphorus precipitation and decrease in micronutrients availability. Closed systems re-circulate the drained water after proper adjustment of pH and electrical conductivity (EC); drained water has to be disinfected to prevent plant diseases [94]. Closed systems avoid environmental problems in deep water bodies due to nutrient runoff. The uptake of water and important macro-ions is regulated by the plant itself. Above a certain threshold the plant nutrient uptake is independent of water potential and nutrient concentration in the root zone; the threshold is determined by kinetic and transport considerations [95]. A system that keeps the drain flow constant by manipulating the irrigation automatically compensates for the water uptake demand of the crop including evapotranspiration losses [96]. As substrate water content remains high, no water stress will occur. Such a feedback system has advantages over radiation based dosage, as it is independent of leaf area changes and moisture conditions within the greenhouse. The development of smart low cost chemo-sensors for continuous monitoring of the ion concentration provides an alternative to conventional EC feedback controls [25]. Continuous measurements taken by sensors are higher than daily 24-hour based data. Generally closed recirculation systems are considered expensive, but can be implemented in a cheap way (Figure 14). The greenhouse soil is laser-leveled so that rows present a 5-10° slope (Figure 14 b). A polyethylene film (PF) is inserted in the soil letting a triangular shape pattern, so that the drained water run-off; this film is placed beneath the polyethylene bags filled with substrate (Figure 14 b). At the end of each row the PF is fixed to a tube glued to the main 4" tube which takes the water out from the greenhouse (Figure 14 a). Drainage from a randomly chosen bag is studied taking volume and EC measurements four times a day (Figure 14 d); values are averaged and used for producing the next day nutrient solution. Actual concentration of nutrient ions can be monitored weekly using quantitative quick tests which are easy-to-use and cheap [97].

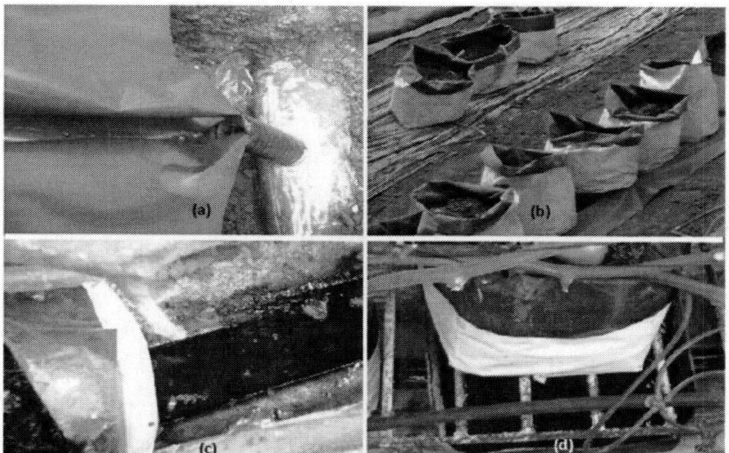

Figure 14. Simple recirculation system showing (a) connection to main tubes; (b) pot distribution; (c) drained nutrient movement; (d) sampling section.

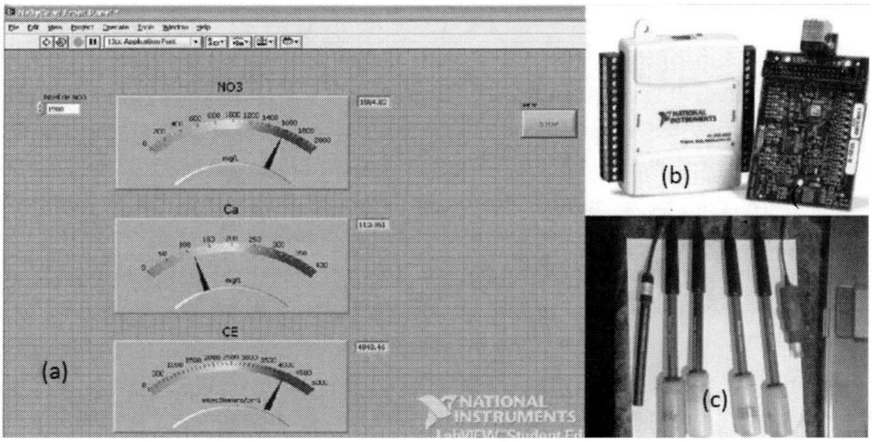

Figure 15. System (a) user interface, (b) NI-USB 6008 acquisition board, and (C) ISE sensors.

LabView is a widely used graphical programming environment which allows designing systems in an intuitive block-based manner [98]. The drain fertilizer monitoring interface of NO_3, Ca++ and EC are acquired for dosing control, Figure 15 a. A multifunction DAQ NI-USB 6008 acquisition board (Figure 15 b) acquired the values and displayed them in graphs. Graphs analyzed by the producer allow growing high quality tomato, as calcium increases shelf-live. Ion selective sensors (Figure 15 c) were employed but require of continuous calibration mainly if liquids passing through the membrane are not filtered; algae can produce measurement errors.

4. Integrated Pest Management

Greenhouses have rain-free, warm, humid and windless climates, ideal for growing crops but also ideal for disease and pest development [99]. Increased gutter roof heights have been beneficial for integrated pest management (IPM) because they increase the height at which insects and pathogens must be transported by the wind to find their way into greenhouse vents. Covering materials present also an effective influence on IPM, as admixtures within their polyethylene films block UV wavelengths necessary for fungi sporulation; blockers efficacy decreases as films age. Thermal and shade curtains are beneficial for IPM because they reduce extreme climates that stress the crop and biological controls. Thermal curtains reduce leaf net radiation, increasing leaf temperature and decreasing condensation on leaves.

Environmental conditions, such as temperature, moisture and sunlight affect the physiology of the host plant and subsequent disease development [100], as well as alter the interactions among plant, pathogen, and bio-control agent in various ways. High risk of infection with *Botrytis cinerea* causing grey mould on a tomato crop occurred when RH >90% for more than 9 h per day [101]. Deficiencies and excesses of macro- and micronutrients, and imbalances in relative amounts of fertilizers can predispose plants to most diseases [99]. Excess fertilizers that increase foliage density at the expense of flowers and fruit not only reduce yield but increase the risks of infection. Soil-borne diseases are no less prevalent in soilless substrates than in soil [99]. All substrates must be substantially free of insects and pathogens during planting and must be kept so throughout the life of the crop. The whole greenhouse can be closed in sunny conditions for solarization of both substrate and superstructure; this operation sterilizes the entire greenhouse.

Symptoms of *Fusarium oxysporum* diseases are rots, leaf spots, blights or wilts [102] (Mace et al., 1981); once the soil is infested, the pathogen is difficult to eliminate. Optimum temperature for *F. oxysporum* growth is 28-29°C, but no infection occurs if soil temperature drops below 20°C or exceeds 30°C [103]. Rhizopus rot is a fungal soft rot requiring injuries caused by insects, hail, or cracking for infection to occur. Rot progression is temperature related with rapid fungal growth at 27°C, and no spore germination at 4°C. Rhizopus rot is more likely to be a problem when fruit is allowed to fully ripen on the plant and when poor sanitary conditions are found on field bins and at the packinghouse. Other diseases that attack tomato plants are anthracnose, early and late blight, Septoria leaf spot.

4.1. Pest and Pathogen Detection

Spectral analysis and machine vision techniques have been used for the detection of pathogen, diseases and pests in the tomato crop. *F. oxysporum* was detected on tomatoes using spectral Fourier signatures with an accuracy of 91% [104]; a Fourier program provided different harmonics, which varied between healthy and diseased plants. A sensor detected *Rhizopus stolonifer* by the peduncle scar, which acts as a membrane to decrease water loss in harvested fruits [105]. Air is sucked and the relative humidity measured; this monitoring method is non-dependent on tomato maturity stage.

Multi-spectrum imaging detected the diseases and insect pests of tomato [106]. The experimental results showed that the near infrared band (0.7-1.3 μm) was more significant than the visible light band in monitoring crop diseases and insect pests. The 0.75-0.93 μm band is useful for inspecting tomato late blight, while the 750-930 nm spectral bands are the best for remote sensing of tomato diseases [107]. Spectral reflectance decreases significantly with the increasing severity level of leaf miner infestation at near infrared wavelengths of 800–1,100 nm. Changes at the individual water-dependant bands of 1,450 and 1,900 nm are noted as the infestation severity increases [108].

Digital infrared thermography has been shown to be a useful tool for the prediction of diseases, but digital infrared thermography alone is not suitable for the quantification of diseases in the field. Two or three days after inoculation, infected leaves transpiration rate was significantly higher than healthy leaves transpiration [109].

4.2. Disease Warning Systems

Disease-warning systems are the most effective IPM option for tomato disease control. These systems use research-based information about weather conditions that increase the risk of fungal disease outbreaks. Disease-warning systems allow growers to spray fungicides only when the risk of an outbreak (and accompanying potential economic loss) is sufficiently high. Daily weather data including maximum and minimum air temperature, hours of leaf wetness, maximum and minimum air temperature during the wet period, hours of relative humidity > 90% and daily rainfall were the environmental inputs to the disease warning software.

In Ontario, Canada, growers use a model called TOM-CAST, which provides disease-risk ratings as Daily Severity Values (DSV's) to apply fungicide spray only when the sum reaches a predetermined threshold. At each site, temperature, relative humidity, rainfall amount, and duration of wetness periods were recorded by electronic sensors at 5-minute intervals and summarized hourly by an automated datalogger. Data were downloaded to a PC via modems and telephone so that DSV's for TOM-CAST could be calculated each day. It saves from 2 to 3 sprays per season which corresponds to fungicide sprays up to 50%, saving 125 US/ha [110] (Gleason et al., 1995). The TOM-CAST system can schedule fungicide sprays for controlling early blight when weather data is available. The first fungicide application after transplanting should occur when 25 DSV's have accumulated [111] (Gleason et al., 1994); if 25 DSV's have not accumulated before mid-July, the first fungicide should be applied at that time. Subsequent applications should occur when 18-22 DSV's have accumulated since the previous application.

4.3. Greenhouse Fly Control

Rapid decisions are essential to control diseases and pests avoiding permanent infestation, due to temperature and hygrometric conditions in greenhouses [112]. For instance, a white fly *Trialeurodes Vaporariorum* female may lay about 130 eggs when proper temperature conditions are set [113]. Length of the life cycle depends on temperature, ranging from four weeks at 27 °C to over eight weeks at 14° C degrees. Damage is caused by flies and larvae sucking the tomato leaf sap, which can cause stunting, leaf drop, and reduced yield. A cognitive vision system combining image processing, learning and knowledge-based techniques was used to detect and count white flies at mature stage [114].

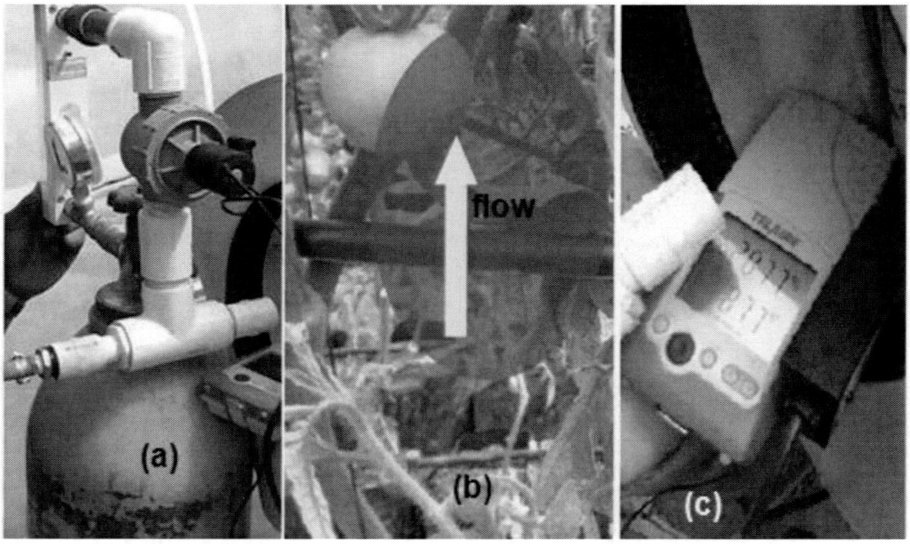

Figure 16. Fly Control using (a) carbon dioxide tank, (b) sprinkler and (c) gas sensor.

Changing the composition of the atmosphere in the greenhouse by either reducing oxygen or increasing carbon dioxide appears to provide some control of greenhouse flies, especially adults [115]. In a small experiment carried out inside small anti-aphid chambers, carbon dioxide was applied at different hours noting that at 5,000 ppm, 95% of flies fainted; at noon, 14:00 and 16:00 hr flies got a second opportunity of life. When the gas was applied early in the morning, 70% mortality was encountered on the sticky yellow plates over the floor. Carbon dioxide (Figure 16 a.) was applied through hoses with sprinklers placed between two consecutive tomato plants (Figure 16 b); the 360° micro-jet provided 30 gallons per hour and covered an area of 3 m. A CO_2 sensor (Figure 16 c) was used to monitor gas permanence in the environment, encountering at sunrise lower gas assimilation by the leaves. Air convection brings the carbon dioxide being sprinkled upwards; taking the leaves some of the gas and the rest attacks the flies.

Carbon dioxide application will be more effective if flies are detached from the leaves, so a vibration system was implemented. Vibration moved the tomato plants ±20 cm causing insects flight. Vibration was generated by a gear motor (Figure 17 a) with an eccentric pulley rotating at a speed of 10 RPM. Tomato plants are tied to wire guides, which are connected to the eccentric pulley (Figure 17 b). Whiteflies standing on the leaves move away as the oscillation become stronger. A glued-plastic was introduced between two tomato rows to evaluate how many flies remained sticked (Figure 17 c). It was noted that only 40% of the flies that hit the plastic remained attached, 30% fainted and the rest escaped. The system becomes more efficient after applying carbon dioxide to avoid flies escape, obtaining a mortality of 79%.

Insect antennas of the flies being sprayed with carbon dioxide were damaged as shown with the red rings in the microscopic images (Figure 17 d); those which revived lost their way back flying to the space. Carbon dioxide was applied for ten minutes with a concentration of 5,000 ppm. CO_2 concentration measured at a height of 30 cm over the hose presented peaks of 4,000 ppm; concentration felt to 2000 ppm at a height of 1 m over the hose (Figure 18).

Figure 17. Vibration system showing (a) gearmotor, (b) guides, (c) glued plastic and (d) dead flies.

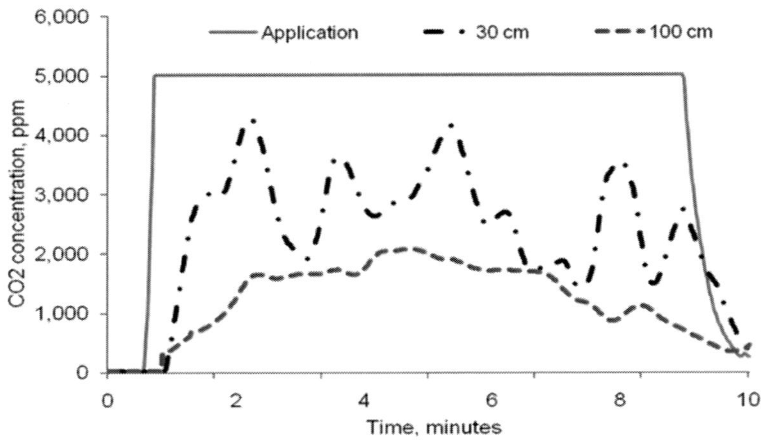

Figure 18. Carbon dioxide concentration 30 and 100 cm away from the hose.

REFERENCES

[1] Von Zabeltitz. *Greenhouse Structures, Ecosystems of the World's Greenhouses*. Elsevier, Amsterdam (1999).

[2] Y. Hashimoto, ICAME 2000 Proc Third International Conf on Agric Mach Eng vol. I, Seoul, South Korea, pp. 1–13 (2000).

[3] F.R. Brunele, D.J. Henriques, P.C. Stringheta, P.C. Rezende, G.R. Moreira, E. Chartuni, *IDESIA* 28(2), 75 (2010).

[4] J.H. Sandee, W.P. Heemels, P.P.J. van den Bosch, *Proc. American Control Conference*, Portland, Oregon, USA (2005).

[5] L. Gonda and C.E. Cugnasca, Proc of 4thWorld Congress Conf Comp Agric Natural Resources, Orlando, Florida, USA (2006).

[6] P.F. Martinez, *Acta Horticulturae* 357, 29 (1994).

[7] J.C. Bakker, *Acta Horticulturae* 399, 25 (1995).

[8] J. Hagin, A. Lowengart, *Fert Res.* 43, 5 (1996).

[9] Hartz, T.K. *Acta Horticulturae* 542, 103 (2001).

[10] Araki, K.; Yamaguchi, H. *Acta Horticulturae* 761, 417 (2007).

[11] P.R. Johnstone, M.A. Nichols, K.J. Fisher and J. Reid, *Acta Horticulturae* 542, 143 (2001).

[12] S.Y. Choi, Y.B. Lee, J.Y. Kim, *Acta Horticulture* 548, 543 (2001).

[13] P. Santamaria, *J Sci Food Agric.* 86, 10 (2006).

[14] Y.W. Zhu, X.X. Zhong and J.F. Shi, *J. Phys.* 48, 1195 (2006).

[15] F.A. Langton and P.J. Hamer, *Grower* Feb 2001, 18-19 (2001).

[16] R.G. Hurd and C. J. Graves, *Journal of Horticultural Science* 60, 359 (1985).

[17] S.R. Adams, K.E. Cockshull, C.R.J. Cave, *Annals of Botany* 88, 869 (2001).

[18] P. Riga, M. Anza, C. Garbisu, *J Sci Food Agric.* 88, 158 (2008).

[19] T.H. Gieling, *Acta Horticulturae* 421, 19, (1998).

[20] M.S. Borhan, X. Hao, J. Zheng, *ASABE Paper No. 074154*, St. Joseph, Mich. ASABE, (2007).

[21] A.M. Abdel-Ghany, T. Kosai, *Environ. Control Biology* 45(1), 9 (2007).

[22] A. Arbel, O. Yekutieli, M. Barak, *J Agric Engng Res.* 72, 129 (1999).
[23] K. Handarto, M. Hayashi and T. Kozai, *Acta Horticulturae* 710, 165 (2006).
[24] A. Reilly, Proc. Int. Conf. on Greenhouse Systems: Automation, Culture, and Environment, New Brunswick, New Jersey, 20-22 July 1994, 176-188 (1994).
[25] G. Vox, M. Teitel, A. Pardossi, A. Minuto, F. Tinivella and E. Schettini, *Sustainable greenhouse systems.* Nova Science Publishers. Inc. (2010).
[26] R. Romero, T. Soria, J. Cuartero, *J Amer Soc Hort Sci.* 127(4), 644 (2002).
[27] U. Schmidt, C. Huber, T. Rocksch, *Acta Horticulturae* 801, 1301 (2007).
[28] S. Sase, M. Ishii, H. Moriyama, C. Kubota, K. Kurata, M. Hayashi, N. Sabeh, P. Romero and G.A. Giacomelli, *Acta Horticulturae* 719, 385 (2006).
[29] F. Villareal, M.E. Kacira, R. Rodriguez, A. Linker, A. Arbel, C. Kubota and G. Giacomelli, ASABE Paper No. 1009388, ASABE (2010).
[30] H.A. Jackson, D.E. Darby, Report M-6704, Canada Plan Service (1990).
[31] N.C. Sabeh, G.A. Giacomelli and C. Kubota, *ASABE Paper No. 074013*, ASABE (2007).
[32] F.L.K. Kempkes and N.J. van de Braak, *Agr Forest Meteorol.* 104, 133 (2000).
[33] M. Teitel, I. Segal, A. Shklyar and M. Barak, *J Agr Engin Res.* 72, 259 (1999).
[34] T. Bartzanas, T. Tchamitchian and C. Kittas, *Biosys Eng.* 91, 487 (2005).
[35] S. Kurpaska and Z. Slipek, *J Agric Eng Res.* 76(2), 129 (2000).
[36] E. Reiss, D.R. Mears, T.O. Manning, G.J. Wulster and J. Both, *Trans ASABE* 50(1), 275 (2007).
[37] M. Kenisarin, K. Mahkamov, *Ren. Sust. En. Rev.* 11(9), 1913 (2006).
[38] G.K. Ntinas, P.G. Kougias, Ch. Nikita-Martzopoulou, *Int. Agrophys.* 25, 257 (2011).
[39] N. Katsoulas, C. Kittas, D. Feidaros, T. Batrzanas and K. Baxevanou, Symp for Greenhouse Systems GreenSys 2009, Université Laval - Québec City – Can. 14 -19 June (2009).
[40] A. Pawlowski, J.L. Guzman, F. Rodríguez, M. Berenguel, J. Sánchez and S. Dormido, *Sensors* 9, 232 (2009).
[41] R. Holder and K.E.Cockshull, *Journal of Horticultural Science* 65, 31 (1990).
[42] Joliet, O.; Baily, B. *Agric Forest Meteorol.* 58, 43 (1992).
[43] H.L. Xu, D. Iraqi and A. Gosselin, *Acta Horticulturae* 761, 85 (2007 a).
[44]Campbell Scientific. HMP45C Temperature and Relative Humidity Probe. www.campbellsci.com (2011).
[45] M.J. Savage, *Sensors* 10, 7748 (2010).
[46] TESTO, 2011 Product Catalog, pp 45-47. www.testo.com (2011).
[47] B.A. Van Tuyl, E.A. van Os, E.J. van Henten, *Acta Horticulturae* 801, 547 (2008).
[48] G.D. Pasgianos, K.G. Arvanitis, P. Polycarpou, and N. Sigrimis, *Computers and Electronics in Agriculture* 40, 153 (2003).
[49] A. Perdigones, V. Valiño, J. L. García, F. Baptista, J. I. Montero, R. M. Benavente and S. de la Plaza. *Spanish Journal of Agricultural Research* 6(2), 189 (2008).
[50] F. Hahn, T. Espinosa, I. Lopez, C. Villanueva and J.Madrigal, ASABE Paper No. 064208, ASABE (2006).
[51] E. Heuvelink and M. Dorais, Tomato. CABI Publishing. CAB International. UK, Pp 85-144 (2005).
[52] FAO. *Protected cultivation in the Mediterranean climate*. Rome, FAO (1990).
[53] T. Higashide, *HortScience* 44(7), 1874 (2009).

[54] Kipp & Zonen. *CM 11 pyranometer instruction manual*. Delft, Holland (2000).
[55] M.P.N. Gent, *HortScience* 42(3), 514 (2007).
[56] J.C. Bakker. Energy Saving Greenhouses. *Chronica Horticulturae* 49(2), 19 (2009).
[57] M.A. Sandri, J.L. Andriolo, M. Witter and T. Dal Ross, *Hort. Brasileira* 21, 642 (2003).
[58] F. Hahn, *Computers and Electronics in Agriculture* 77, 21 (2011 a).
[59] A.J. Both , D. R. Mears, E. Reiss, and W. J. Roberts. *Hort. Engineering* 17(3), 1 (May 2002)
[60] I.V. Pollet, P.G.M. Eykens, J.G. Pieters, *Acta Horticulturae* 534, 101 (2000).
[61] T.J. Blom, F.J. Ingratta, *Acta Horticulturae* 148, 905 (1984).
[62] R.J. McAvoy, H.W. James, *Acta Horticulturae* 148, 877 (1984).
[63] S. Naika, J van Lidt, M de Goffau, M. Hilmi, B. van Dam M. Peet, *Cultivation in tomato*. ISBN Agromisa: 90-8573-039-2, Wageningen, Netherlands. Fourth edition (2005).
[64] F. Hahn, M.A. Peña, P. Coras and M.Vazquez, *Canadian Biosystems Eng.* 51, 1.1 (2009).
[65] H. Fukushima, *Science and Technology Trends* 23, 23 (2007).
[66] L.S. Pereira, T. Oweis and A. Zairi, *Agricultural Water Management* 57, 175 (2002).
[67] F. Hahn, A. Martínez and R. Arteaga, *Agrociencia* IN PRESS (2012).
[68] F. Hahn, *Scientia Agropecuaria* 2, 31 (2011 b).
[69] T.J. Shelford, A.K. Lau, D.L. Ehret and S.T. Chieng, *Can Biosystems Eng* 46, 1.1 (2004).
[70] A.G. Smajstrla and R.C. Koo, *Proc Fla State Hort Soc*. 99, 51 (1986).
[71] Y. Li, R. Rao, H. Bryan and T. Olczyk, *Proc Fla State Hort Soc.* 111, 58 (1998).
[72] R. Muñoz-Carpena, M.D. Dukes, Y. Li, W. Klassen, *HortTechnology* 15(3), 584 (2005).
[73] H.G. Taber, V. Lawson, B. Smith, D. Shogren, *Int. Water Irrig.* 22, 22 (2002).
[74] R. Muñoz-Carpena, M.D. Dukes, Y. Li and W. Klassen, *Applied Eng. Agric.* 24(2), 183 (2008).
[75] N. Czarnomski, G.W. Moore, T.G. Pypker, J. Licata, B.J. Bond, *Can J. For Res*. 35, 1867 (2005).
[76] T. Saito, H. Fujimaki and M. Inoue, *Am J Environ Sci.* 4(6), 690 (2008).
[77] T.J. Kelleners, D. Robinson, P. Shouse, J.E. Ayars, T.H. Skaggs, *J Soil Sci Soc Am.* 69, 67 (2005).
[78] K.S. Nemali, F. Montesano, S.K. Dove and M. van Iersel, *Scientia Horticulturae* 112, 227 (2007).
[79] J.P. Walker, G.R. Willgoose and J.D. Kalma, *Journal of Hydrology* 293, 85 (2004).
[80] M.M. Peet and D.H. Willits, *HortScience* 30, 65 (1995).
[81] H.L. Xu, L. Gauthier and A. Gosselin, *J Japanese Soc for Hort Sci* 65, 777 (1997).
[82] S. Krauss, W.H. Schnitzler, J. Grassmann and M. Woitke, *J Agric Food Chem*, 54, 441 (2006).
[83] L.C. Ho and P. Adams, *Acta Horticulturae* 396, 33 (1995).
[84] S. Chrétien, A. Gosselin and M. Dorais, *HortScience* 35(4), 627 (2000).
[85] R.T. Burns, Report IR-02-98. University of Tennessee Agricultural Extension Service (2008).

[86] R. Kessler, B. Pennisi, Report ANR-1243. Alabama Cooperative Extension System (2004).

[87] J. Miralles, M.W. van Iersel, S. Bañón, Third International Symposium on Soil Water Measurement Using Capacitance, Impedance and TDT, Paper 2.10, Murcia, Spain (2010).

[88] M. Inoue, B.A. Ould Ahmed and M. Irshad, *Am J Environ Sci.* 4, 367 (2008).

[89] F.R. Arguedas-Rodriguez, J.D. Lea-Cox and A.G. Ristvey, Proc. Southern Nursery Assoc. Res. Conf. 54, 216 (2009).

[90] A. Pardossi, L. Incrocci, G. Incrocci, F. Malorgio, P. Battista, L. Bacci, B. Rapi, P. Marzialetti, J. Hemming, J. Balendonck, *Sensors* 9, 2809 (2009).

[91] H.L. Scoggins, M.W. van Iersel, *HortScience* 41, 210 (2006).

[92] U.K. Saha, A.P. Papadopoulos and X. Hao, *HortScience* 43(2), 484 (2008).

[93] R. Baas and N.A. Straver, *Acta Horticulturae* 562, 295 (2001).

[94] W.T. Runia and J.J. Amsing, *Acta Horticulturae* 548, 215 (2001).

[95] G. Van Straten and Th.H. Gieling, *CIGR Journal* 2004, Paper BC 03 016 (2004).

[96] Th.H. Gieling, H.J. Janssen, G. van Straten, M. Suurmond, *Comp Elect in Agric* 26, 361 (2000).

[97] R.B. Thompson, M. Gallardo, M. Joya, C. Segovia, C. Martinez-Gaitan and M.R. Granados, *Spanish Journal of Agricultural Research* 7(1), 200 (2009).

[98] D.G. Pereira, A. Tonso, B.V. Kilikian, *Brazilian Journal of Chemical Eng.* 25(2), 247 (2008).

[99] W.R. Jarvis, *Managing Diseases in Greenhouse Crops*. APS Press, St Paul, Minn (1992).

[100] R.P. Larkin and D.R. Fravel, *Phytopathology* 92(11), 1160 (2002).

[101] F.J. Baptista, B.J. Bailey, J.F. Meneses, *Acta Horticulturae* 893, 1263 (2011).

[102] M.E. Mace, A.A. Bell and C.H. Beckman, *Fungal Wilt Diseases of Plants*. Academic Press, New York, NY, U.S.A. (1981).

[103] P. Holliday, *Fungus Disease of Tropical Crops.* Cambridge University Press. Cambridge, U.K. (1980).

[104] F. Hahn, *J. Biosyst. Eng.* 81, 249 (2002).

[105] F. Hahn, *J. Biosyst. Eng.* 95, 171 (2006).

[106] M. Zhang, Z. Qin, X. Liu and S. Lu, *Int J Appl Earth Obs Geoinf.* 4, 295 (2003).

[107] M. Zhang and Z. Qin, Proc Int. Geosciences and Remote Sensing Symp. IEEE Alaska VI, 4091 (2004).

[108] H.R. Xu, Y.B.Ying, X.P. Fu and S.P. Zhu, *Biosystems Engineering* 96, 447 (2007 b).

[109] E.C. Oerke, U. Steiner, H.W. Dehne, M. Lindenthal, *J Exp Bot.* 57, 2121 (2006).

[110] M.L. Gleason, A.A. MacNab, R.E. Pitblado, M.D. Ricker, D.A. East and R.X. Latin, *Plant Disease* 79, 113 (1995).

[111] M.L. Gleason, S.E. Taylor, T.M. Loughin, K.J. Koehler, *Plant Disease* 78, 1011 (1994).

[112] L. Lapchin and D. Shtienberg, *Integrated Pest and Disease Management in Greenhouse Crops*. Kluwer Academic Publishers vol. 14, 82 (2002).

[113] M. Malais and W. Ravensberg, *Knowing and Recognizing. The biology of glasshouse pests and their natural enemies.* Koppert B.V. The Netherlands (2003).

[114] P. Boissard, V. Martin, and S. Moisan, *Computer Electronics in Agriculture* 62(2), 81 (2008).

[115] K. Tripp and M. Peet, American Vegetable Grower, pp. 43 (Nov 1993).

In: Tomatoes: Cultivation, Varieties and Nutrition
Editor: Tadahisa Higashide

ISBN: 978-1-62417-915-0

Chapter 4

LIBRARIES OF TOMATO MUTANTS INDUCED BY HEAVY-ION BEAM IRRADIATION: NEW EFFECTIVE METHOD FOR ANALYZING FUNCTIONAL GENOMICS

Shunsuke Imanishi*, Arisa Noguchi and Ichiro Honda

National Institute of Vegetable and Tea Science, National Agriculture and Food Research Organization, Japan

ABSTRACT

Mutational approaches are highly instrumental for the study of the genetics and molecular bases of traits. Mutant plants are useful tools not only as parents of new cultivars but also as materials for assessing physiological mechanisms. To obtain information on these mechanisms, we induced mutations by irradiation with accelerated heavy ions in tomatoes. Irradiation with accelerated heavy ions, which was recently established as an effective method for inducing mutations in plants, could be a new approach for studying the functional genomics of tomato and those of the Solanaceae family.

INTRODUCTION

Tomato (*Solanum lycopersicum* L.) is an excellent model system for the analysis of those plant mechanisms that cannot easily be studied in *Arabidopsis* or rice, such as fruit physiology and accumulation of secondary metabolites. The availability of mutants collected over a period of several decades is also a big advantage to the analysis of these plant mechanisms [1]. About 1000 monogenic stocks from spontaneous and induced mutants as well as natural variants and wild relatives of tomato are available at the CM Rick Tomato Genetic Resource Center, University of California, Davis. Zamir et al. developed a tomato mutation library derived from ethyl methanesulfonate (EMS) and fast neutron mutagenesis in the inbred tomato variety M82 [2]. Further, we used the determinate miniature *S.*

* E-mail address: phppublish@ml.affrc.go.jp.

lycopersicum cultivar Micro-Tom obtained using EMS treatment [4] and γ-ray irradiation [5] for arranging of new mutation libraries for large mutagenesis. We used Micro-Tom because of its small size and rapid life cycle [3]. Emmanuel and Levy [6] indicated that the size of a saturating tomato mutant population, where every gene is mutated, is in the range of about 10^5. However, for studying tomato gene functions, analysis of several alleles is required to identify the entire gene function, because a single mutation often describes only a part of the gene function [7]. A novel mutagen may create a mutation spectrum different from those of well-known mutagens, and thus, become valuable for studying mutagenesis in tomato. In this chapter, we review a new application for mutagenesis, namely, accelerated heavy-ion irradiation (HII) technology.

IRRADIATION WITH ACCELERATED HEAVY-ION (HII)

Accelerated ions are a form of ionizing radiation. Characteristics of HII are as follows: (1) HII is 10-fold more effective per unit dose in creating somatic and germline mutations in plants than that by other sources of radiation such as X-rays, γ-rays, or electrons [8]. (2) HII causes more localized and denser ionization within cells than that by the other ionizing radiations [9]. (3) HII produces more closely positioned and clustered DNA damage sites than that by the other ionizing radiations [10,11,12]. (4) HII appears to not only induce mutations more frequently per unit dose but also to generate mutations of a different nature than those produced by the other types of ionizing radiations [13].

MUTATIONS INDUCED BY HII IN *ARABIDOPSIS*

In *Arabidopsis*, Tanaka and colleagues isolated HII-induced monogenic recessive mutants and examined the mutational events at the genome level [14,15,16,17,18,19]. They reported that half of the mutants had small mutations, such as base changes and deletions of a few bases, whereas the other half had large DNA alterations, such as inversions, translocations, and deletions [20,13].

MUTATIONS INDUCED BY HII IN SOLANACEAE

In Solanaceae, several mutants have been isolated using HII technique. In petunia [21], HII-induced cultivars are available commercially, but studies of the mutational events from a genetic and biological perspective have not been conducted.

Abe and colleagues irradiated ovaries of tobacco containing embryos 24–108 h after self-pollination by using various accelerated ion beams [22]. By using ^{12}C and ^{14}N ion beams of low irradiation doses of 10–50 Gy, morphological mutations were observed in M_1 plants (directly grown from irradiated seeds); however, high lethality was observed after irradiation by ^{20}Ne and ^{36}Ar ion beams. Treatment with ion beam gave rise to needle leaves and ageotropic plants, which were characteristics that were not observed when the ovaries were subjected to chemical treatment. The changes in the M_1 seeds obtained from these irradiation experiments indicate that ion beams are more effective than chemical treatment for mutagenesis. High effects of mutation rates were obtained in the absorbed dose levels with a germination rate of 60–80%. Under these established irradiating conditions, a novel albino mutant [23] and salt/cobalt-tolerant mutants [22] were isolated from M_2 populations that were derived from a self-pollinated M_1 plant that was irradiated with ^{14}N ion beam.

Effects of heavy-ion bombardment on mutagenesis in sweet pepper (*Capsicum annuum* L.) were examined [24]. Because the growth was inhibited in M_1 seeds irradiated with more than 30 Gy of either ^{12}C or ^{20}Ne ion beams, mutants were obtained from 10 Gy irradiated seeds. Two putative dwarf and one *xantha* mutant were found in M_1 plants directly grown from 10 Gy ^{20}Ne ion irradiated seeds, and the phenotypes did not segregate in the M_2 progeny. Genetic studies indicated that homozygous monogenic recessive mutations in nuclear genes had occurred in these M_1 plants.

The isolation of monogenic homozygous recessive mutants in the M_1 generation was neither described in other mutational studies that used many plant species nor in the HII mutational studies. Jain et al. [25] isolated dwarf mutants in M_1 tomato plants treated with several chemicals. Although Jain et al. did not report the segregation of the traits in M_2 plants, they showed that the homozygous mutation occurred in M_1 plants. Therefore, the results are the first definitive data of a homozygous monogenic recessive mutation in a nuclear gene in M_1 pepper plants. These findings may suggest that HII has unique properties that induce homozygous recessive mutations, even in the M_1 generation.

MUTATIONS INDUCED BY HII IN TOMATO

Masuda et al. [26] irradiated dry tomato seeds (cv. First) with ^{12}C and ^{4}He ion beams (50–300 Gy) and monitored the effect of the irradiation on the survival ratio and plant characteristics. They analyzed the mutation spectrum of M_2 progenies and found that the rate of occurrence of chlorophyll mutants due to ^{12}C ion beam irradiation was relatively lower than that from ^{4}He ion beam irradiation, which resulted in several chlorophyll mutants and mutants with short internodes. In this case, by performing radiation-dosage analysis, they found that less than 50 Gy of ^{12}C and 150–200 Gy of ^{4}He were considered to be optimum levels for irradiation of tomato seeds.

Micro-Tom, determinate miniature *S. lycopersicum,* is well suited for large-scale mutagenesis and functional genomics studies because of its small size and rapid life cycle. We have irradiated dry or imbibed seeds of Micro-Tom by ^{12}C or ^{20}Ne ion beams accelerated to 135 MeV/nucleon with a RIKEN Ring Cyclotron within the dose range of 1–300 Gy. The linear energy transfer values of the ^{12}C and ^{20}Ne ions were 22.7 and 64.2 keV/μm, respectively [27,28]. The irradiated seeds were sown in a greenhouse and transplanted to the field after 3 weeks. M_2 seeds were collected from individual plants to generate independent lines.

About 20% of ^{20}Ne 50 Gy and ^{12}C 200 Gy irradiated M_1 plants showed the *xantha* phenotype, which is observed often with HII [22]. The Micro-Tom demonstrated low sensitivity to higher HII dosages [27,28] than that by tomato cv. First [26] and sweet pepper [24] (Figure 1A). The survival rate could be determined for some physiological and/or irradiation conditions (such as a difference of water content in the seeds or a difference of dose rates), revealing differences in the type of DNA damage generated or DNA repair.

Figure 1. NIVTS assorted Micro-Tom libraries Irradiated by Heavy-ion beams (NaMIHei). (A) Seedlings were grown for 3 weeks after irradiation by 50, 100, 200, and 300 Gy (left to right) of Ne. (B) One M_1 plant with mutant inflorescence. All inflorescences were similar to those of broccoli. (C–E) M_2 mutants showing *pale flower* (C), *big calyx* (D), and *orange fruit* (E).

Screening 6,794 M_1 plants, we found various mutations, including mutations that affected the size of plants, leaves, and fruit, and a mutation that conferred a broccoli-like inflorescence [27,28].

This M_1 mutant (Figure 1B) resembles the tomato *anantha* mutant, in which the development of floral primordia is blocked before organogenesis [29,30,31]. Because we could not obtain seeds from this mutant, the inheritance and segregation of the trait are unknown; however, it was reported that *anantha* was caused by a recessive mutation [30]. The isolation of monogenic homozygous recessive mutants in the M_1 generation has not commonly been described in other mutational studies using many plant species, even though our group isolated such mutants in sweet pepper by M_1 plant selection in HII mutational studies [24].

Furthermore, we found several morphological mutants in the M_1 generation. Most of these mutations are not inherited in the M_2 generation, as previously reported [24], whereas about 3% of the M_2 progenies derived from seeds irradiated with an ion beam showed abnormalities that were similar to those observed in the M_1 generation. We continued to analyze the mode of inheritance of these mutations.

In the M_2 generation, 5–8 seeds from each line were sown and transplanted in a field. A phenotype was determined on the basis of morphological characteristics. Overall, about 35,000 plants were grown in the field and subjected to phenotypic analysis. M_3 seeds from the lines that were segregated on the basis of visible mutation were collected from each plant. Seeds of the other lines were collected in bulk from siblings to create large seed resources for further experimentation.

A total of 4,642 lines have been screened to date (Table 1; Figures 1C, D, E), and 789 candidate mutant lines have shown variations in one or more phenotypic characteristics (Figure 2) [32]. To test the genetic nature of the mutants in the libraries, we sowed a selected subgroup of 50 mutants that were identified in the M_2 generation and scored the M_3 generation plants for mutations. Of these 50 mutants, 33 mutants showed the same phenotype as that in the M_2 generation [32].

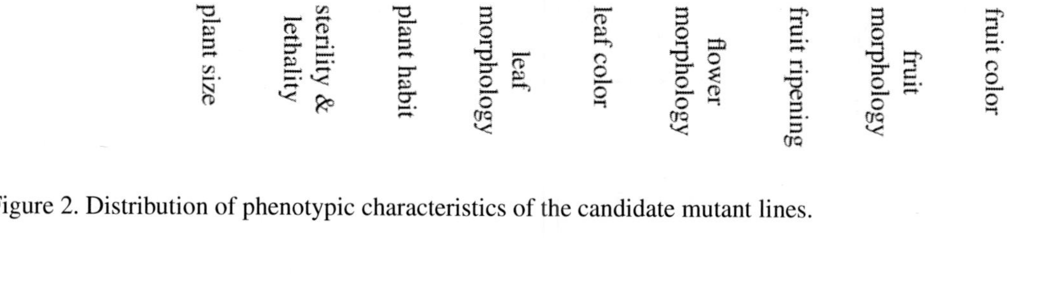

Figure 2. Distribution of phenotypic characteristics of the candidate mutant lines.

Table 1-1. Composition of libraries of dry seeds irradiated with heavy−ion beam

		Radiation Dosage (Gy)								total		
		1	3	10	30	50	100	200	300			
Ne	M_2 lines stock[a]	131	414	495	1,488	1,015	732	157	0	4,432		
	M_2 lines tested		24	268	526	150	216	50			1,234	
	M_3 families stock[b]		24	263	518	149	213	49				1,126
C	M_2 lines stock[a]	119	366	1,037	1,392	542	819	149	63	4,487		
	M_2 lines tested		24	314	568	175	243	75	25		1,424	
	M_3 families stock[b]		24	308	563	175	240	73	24			1,407
										8,919	2,658	2,533

[a] No. of M_1 plants set the selfed seeds

[b] No. of M_2 lines set the selfed seeds

Table 1-2. Composition of libraries of imbibed seeds irradiated with heavy− ion beam

		Imbibed Time - Radiation Dosage (Gy)									total		
		6 h - 5	12 h - 5	24 h - 5	6 h - 10	12 h - 10	24 h - 10	6 h - 20	12 h - 20	24 h - 20			
Ne	M_2 lines stock[a]	83	91	81	84	75	74	25	27	19	559		
	M_2 lines tested	60	87	74	83	68	66	18	22	15		493	
	M_3 families stock[b]	58	85	73	80	68	66	17	16	14			477
		6 h - 10	12 h - 10	24 h - 10	6 h - 20	12 h - 20	24 h - 20	6 h - 40	12 h - 40	24 h - 40			
C	M_2 lines stock[a]	343	330	171	367	360	183	74	40	81	1,949		
	M_2 lines tested	280	284	160	226	253	150	51	28	59		1,491	
	M_3 families stock[b]	280	276	101	218	150	87	50	24	29			1,21:
											2,508	1,984	1,69:

[a] No. of M_1 plants set the selfed seeds

[b] No. of M_2 lines set the selfed seeds

CONCLUSION

In this chapter, we introduced the concept of developing comprehensive tomato mutant libraries, which can serve as resources for functional genomic studies. Compared to mutant libraries obtained by using EMS and fast-neutron mutagenesis, the mutant libraries obtained by using our approach showed lower rates of sterility and lethality and a different spectrum of mutations [2]. Because our libraries of mutations induced by HII could include various types of novel alleles, successive examination of the mutants for effective mutagenesis will contribute to an understanding of the physiological mechanisms of plants.

REFERENCES

[1] S. Moore, J. Vrebalov, P. Payton and J. Giovannoni, *J. Exp. Bot. 53*, 2023 (2002).

[2] N. Menda, Y. Semel, D. Peled, Y. Eshed and D. Zamir, *Plant J. 38*, 861 (2004).

[3] R. Meissner, Y. Jacobson, S. Melamed, S. Levyatuv, G. Shalev, A. Ashri, Y. Elkind and A. A. Levy, *Plant J. 12*, 1465 (1997).

[4] S. Watanabe, T. Mizoguchi, K. Aoki, Y. Kubo, H. Mori, S. Imanishi, Y. Yamazaki, D. Shibata and H. Ezura, *Plant Biotech. 24*, 33 (2007).

[5] C. Matsukura, I. Yamaguchi, M. Inamura, Y. Ban, Y. Kobayashi, Y. Yin, T. Saito, C. Kuwata, S. Imanishi and S. Nishimura, *Plant Biotech. 24*, 39 (2007).

[6] E. Emmanuel and A. A. Levy, *Curr. Biol. 5*, 112 (2002).

[7] K. D. Hirschi, *Trends Plant Sci. 8*, 205 (2003).

[8] H. H. Smith, *Induced Mutations and Plant Improvement*, International Atomic Energy Agency, Vienna 75 (1972).

[9] E. A. Blakely and A. Kronenberg, *J. Radiat. Res. 150*, S12 (1998).

[10] D. T. Goodhead, *Int. J. Radiat. Biol. 65*, 7 (1994).

[11] D. T. Goodhead, *J. Radiat. Res. 40*, S1 (1999).

[12] J. F. Ward, *Int. J. Radiat. Biol. 66*, 427 (1994).

[13] N. Shikazono, C. Suzuki, S. Kitamura, H. Watanabe, S. Tano and A. Tanaka, *J. Exp. Bot. 56*, 587 (2005).

[14] A. Tanaka, S. Tano, T. Chantes, Y. Yokota, N. Shikazono and H. Watanabe, *Genes Genet. Syst. 72*, 141 (1997).

[15] A. Tanaka, *Gamma Field Symposia 38*, 19 (1999).

[16] Y. Hase, A. Tanaka, T. Baba and H. Watanabe, *Plant J. 24*, 21 (2000).

[17] A. Sakamoto, L. T. T. Vo, Y. Hase, N. Shikazono, T. Matsunaga and A. Tanaka, *Plant Cell 15*, 2042 (2003).

[18] N. Shikazono, Y. Yokota, S. Kitamura, C. Suzuki, H. Watanabe, S. Tano and A. Tanaka, *Genetics 163*, 1449 (2003).

[19] S. Kitamura, N. Shikazono and A. Tanaka, *Plant J. 37*, 104 (2004).

[20] N. Shikazono, A. Tanaka, H. Watanabe and S. Tano, *Genetics 157*, 379 (2001).

[21] K. Miyazaki, K. Suzuki, T. Abe, Y. Katsumoto, S. Yoshida and T. Kusumi, *RIKEN Accel. Prog. Rep. 35*, 130 (2002).

[22] T. Abe, C. H. Bae, T. Ozaki, JM. Wang and S. Yoshida, *Gamma Field Symposia*, *39*, 45 (2000).

[23] C. H. Bae, T. Abe, N. Nagata, N. Fukunishi, T. Matsuyama, T. Nakano and S. Yoshida, *Plant Sci. 151*, 93 (2000).

[24] I. Honda, K. Kikuchi, S. Matsuo, M. Fukuda, H. Saito, H. Ryuto, N. Fukunishi and T. Abe, *Euphytica 152*, 61 (2006).

[25] H. K. Jain, R. N. Raut and Y. G. Khamnkar, *Heredity 23*, 247 (1968).

[26] M. Masuda, S. G. Agong, A. Tanaka, N. Shikazonoa and Y. Hase, *Acta Hort. 637*, 257 (2004).

[27] S. Imanishi, A. Noguchi, M. Nagata, I. Honda and A. Abe, *Acta Hort. 745*, 485 (2007).

[28] S. Imanishi, A. Noguchi, T. Hiraga, M. Nagata, Y. Hayashi, N. Fukunishi, H. Ryuto, T. Abe and I. Honda, *RIKEN Accel. Prog. Rep. 41*, 1231 (2008).

[29] J. Helm, *Züchter 21*, 89 (1951).

[30] E. F. Paddock and L. Alexander, *Ohio J. Sci. 52*, 327 (1952).

[31] K. D. Allen and M. Sussex, *Planta 200*, 254 (1996).

[32] S. Imanishi, A. Noguchi, N. Yokotani, Y. Kazama, T. Hirano, Y. Hayashi, K. Tsuji, I. Honda and T. Abe, *RIKEN Accel. Prog. Rep. 44*, 272 (2011).

In: Tomatoes: Cultivation, Varieties and Nutrition
Editor: Tadahisa Higashide

ISBN: 978-1-62417-915-0

Chapter 5

APPLICATION OF MOLECULAR METHODS FOR TOMATO BREEDING: THE SOLANUM/CLAVIBACTER PATHOSYSTEM

***José Pablo Lara-Ávila*[1], *June Simpson*[2] and *Ángel Gabriel Alpuche-Solís*[3]**

[1] Laboratorio Nacional de Genómica para la Biodiversidad, Cinvestav Campus Irapuato, Guanajuato, México

[2] Departamento de Ingeniería Genética, Cinvestav Campus Irapuato, Guanajuato, México

[3] IPICYT, División de Biología Molecular, Instituto Potosino de Investigación Científica y Tecnológica A.C., SLP, México

ABSTRACT

Clavibacter michiganensis subsp. *michiganensis* (*Cmm*) is a Gram-positive phytopathogen causing bacterial wilt and canker of tomato (*Solanum lycopersicum*), an economically important disease producing economic losses worldwide. This chapter will review the current state of research on the interaction between *C. michiganensis* and *S. lycopersicum* with emphasis on the exploitation of molecular methods to study the genetic basis of resistance through gene expression profiles. A detailed example of gene expression profiling conducted at the early stages of bacterial wilt disease in tomato, using several resistant tomato-related wild species, including *S. peruvianum* LA2157, *S. peruvianum* LA2172, *S. habrochaites* LA2128, and a susceptible tomato species, to identify genes involved in the disease response will be presented. The potential of these findings and those from other recent studies for the identification of new candidate bacterial wilt disease resistance genes which could be subjected to further functional analyses to determine the molecular basis of incompatibility between wild species of tomato and *Cmm* will be discussed with the aim of providing a contribution to the development of more effective and sustainable *Cmm*-control methods.

INTRODUCTION

It is the intention of this chapter to describe some of the molecular mechanisms associated with plant-pathogen interactions that lead to plant disease and plant resistance. The study of wild species as novel sources of plant disease resistance by molecular methods applied to gene expression profiling will be emphasized. The example described in this paper is the *Solanum-Clavibacter* interaction. The genus *Solanum* includes the susceptible species *S. lycopersicum* (tomato) and the resistant species *S. habrochaites* LA2128 and *S. peruvianum* LA2172 and LA2157, therefore we have the opportunity to compare in related species, the plant responses during compatible and incompatible interactions.

WORLDWIDE IMPORTANCE OF TOMATO

Tomato is second most important vegetable after potato. In 2009 the FAO estimated tomato production at 154 million tons cultivated on 4.5 million hectares, worldwide. The top five leading tomato-producers are China, United States of America, India, Egypt and Turkey [FAO STATS]. The United States, despite being the second major tomato producer, channels most of its production into the processed tomato industry; consequently the market segment of fresh tomato suffers a deficit, and depends on imports to meet the demands of consumers.

The American consumer considers the tomato as their fourth favorite vegetable for fresh consumption, behind potatoes, lettuce and onions. Part of the tomato´s appeal is the perception by the American public that they are easily and quickly prepared, common to many dishes and have an attractive presentation and taste. Increased consumption of fresh tomatoes by the population is achieved by their use in salads and sandwiches.

In general terms, we can affirm that although there is a market for fresh tomatoes, both from greenhouse and open field culture, this is a mature, competitive market, which is not growing at the pace that governments and agencies promoting consumption of vegetables would like. The growing demand for fresh tomato and the inability to meet this demand within the U.S. implies a continued need for imported tomatoes in the coming years. This is an advantage for countries such as Mexico who largely fill the gap in demand.

Among specialists in water resources, there is a growing conviction that the shortage of irrigation water is a growing reality. This situation together with the increasing demand for food in combination with environmental deterioration, forces us to use production techniques that exploit available resources more efficiently and sustainably. A good choice of crop production method, which helps to reduce the effects mentioned, is in a greenhouse system that can be operated with different levels of technological input.

The function of the greenhouse is to control the environmental conditions, according to the state of plant development, in addition to irrigation and application of fertilizer, leading to increased production, both in quality and quantity. It should also be noted that, although the environmental conditions inside the greenhouse are favorable for the crops to express their maximum yield potential, they are also favorable for propagation of phytopathogens and the resulting damage caused by them, therefore strict hygiene and curative and preventive measures must be taken, since pests and diseases can wipe out an entire crop in a short time.

BACTERIAL CANKER AND WILTING OF TOMATO

Despite the technical advances in greenhouse design, construction and crop management, the presence of phytopathogens and pests is a constant challenge. It has been reported that tomato is affected by at least 24 fungi, 7 bacteria, 10 viruses, 3 viroids and multiple nematodes [1]. Among bacterial diseases, tomato bacterial canker and wilting, caused by the actinomycete *Clavibacter michiganensis* subsp. *michiganensis* (*Cmm*) is considered one of the most destructive diseases of tomato. *C. michiganensis* is a quarantined organism in the European Union (Commission Directive 95/4/EC), the United States (NAPPO) and Mexico (NOM-007). Tomato bacterial canker is a devastating disease due to systemic movement of the pathogen, shown by the generation of a bioluminescent *Cmm* strain [2, 3, 4]. Infection initiates through wounds, stomata or contaminated seeds, the latter being the most common cause of outbreaks [5, 6, 7]. *Cmm* is considered a biotrophic organism, but can survive for long periods of time associated with plant debris, providing a source of infection for the next agricultural season. Chemical control by antibiotics, ammonium salts or copper derivatives, is ineffective with adverse effects on the environment. Alternatively, the use of endolysins produced by bacteriophages CMP1 and CN77, has been proposed as a control strategy. These proteins show a high specificity for *Cmm* with no adverse effects, however this biological control method has not yet been widely deployed under field conditions [8]. A decrease in disease incidence has also been achieved through the use of certified seed and cultural practices. Little is known about the virulence strategies employed by *Cmm* during tomato infection. An exopolysaccharide (EPS) is thought to be responsible for symptoms and wilting of leaves, but is not an important factor of pathogenicity [3, 4]. The *Cmm* strain NCPPB382 harbors two plasmids pCM1 (27 kb) and pCM2 (70 kb), that carry the *celA* and *patI* genes, both important factors of pathogenicity [9, 10]. Recently, the chromosomal genes *chpC* and *tomA* were also identified as important virulence factors that can affect the tomato-*Cmm* interaction [11, 12]. The function of these genes is still not clear, but they could play roles during plant colonization and signaling. The *Cmm* genome sequencing revealed a large pathogenicity island (PAI) located near the chromosomal replication origin and with a low G+C content, which is required for full pathogenicity. This island contains genes involved in sugar metabolism and some serine proteases, in addition to cell wall degrading enzymes that could interfere with plant signaling and be involved in plant colonization [7, 13]. Recently, a correlation between the expression levels of several chromosomal genes encoding serine proteases and cell wall degrading enzymes with the plasmid-borne genes *celA* and *patI*, suggests the probable evolution of *Cmm* from an endophytic to a virulent state, through horizontal acquisition of virulence factors [14]. However, other genes required for virulence, plant defense suppression, signaling and colonization are located on the chromosome, as shown by *Cmm* strain NCPB382 CM00, a plasmid-cured strain that can colonize tomato without producing symptoms [15].

TOMATO GERMPLASM RESOURCES

Tomato belongs to the *Solanaceae* family that includes pepper, potato, eggplant, *Petunia spp*, *Nicotiana spp* and *Datura spp*. The section *Lycopersicon* includes *S. lycopersicum* (the domesticated species) and 12 wild relatives [16]. The wild tomatoes have a high genetic diversity that has been investigated for specific traits and is being exploited in breeding programs, by contrast the poor level of genetic diversity in cultivated tomato has been demonstrated by using sensitive molecular markers, and is thought to be a consequence of domestication. Tomato domestication initiated with the intensive selection of genotypes from existing germplasm, which were taken from America to Europe, causing a genetic bottleneck leading to a decrease in genetic diversity [16]. In 1940, Charles Rick (University of California) observed that crosses between wild and domesticated species produced offspring with novel traits and effectively initiated tomato breeding programs based on wild germplasm [16, 17] [Figure 1].

Taken from S.D. Tanksley, *The Plant Cell,* **16,** 181 (2004).

Figure 1. Collage depicting the wide variation in size and shape of tomato fruit.

(A) The large-fruited tomato cv Giant Heirloom common to modern agriculture (right), and the typical fruit of a related wild species (*L. pimpinellifolium*).

(B) The range of variation in fruit size and shape found in tomato.

(C) Cross-section of the fruit from a plant homozygous for a mutation at the *fasciated* locus causing multiple locules.

(D) Alternate allele of *fasciated* associated with unfused carpels.

(E) Fruit from cv Long John, which carries mutations at both the *sun* and *ovate* loci causing extremely long, narrow fruit.

(F) Bell pepper–type fruit produced by cv Yellow Stuffer.

(G) Fruit from two different cultivars homozygous for a mutation at the *ovate* locus. In the variety at left, the *ovate* mutation results in the production of fruit that are both elongated and constricted at the stem end of the fruit (hence, the pear shape). However, in the processing variety at right, the *ovate* mutation causes elongated fruit but has a much reduced effect on neck constriction. The angular or square shape of the variety at left can be attributed to the *fs8.1* locus.

Conventional tomato breeding involves crosses of *S. lycopersicum* with wild relatives, carrying the trait of interest, to produce F1 interspecific hybrids. By using molecular marker analysis some progenies are used to obtain permanent populations of near-isogenic (NIL) and introgression (IL) lines. NIL populations are homozygous lines each containing a marker from a wild genome in a uniform genetic background, allowing the identification of quantitative trait loci (QTL) and the mapping of genes involved in trait determination [18]. However, genetic variation in wild germplasm may produce adverse phenotypes in cultivated tomato, under commercial production conditions. Here the challenge is to identify, select and introgress only the advantageous alleles.

In spite of the drawbacks, the first breeding programs focused on traits such as flavor and processing qualities have been highly successful. Given the global economical importance, there is a constant interest to generate new tomato varieties by employing modern technologies, such as molecular marker assisted selection (MAS) and genome sequencing. Among the most valuable traits than can decrease production costs are: fruit quality, yield, heterosis and resistance to biotic and abiotic stress. In addition selection for synthesis of healthy metabolites such as antioxidants is a new topic which is currently being exploited [16, 17, 18].

Despite extensive breeding programs and selection, tomato diseases are a constant challenge in several parts of the word. Although several commercial tomato cultivars resistant to diseases caused by bacteria, fungi, nematodes and virus have been developed [19, 20, 21], to date there is no a commercial cultivar resistant to *Cmm*.

Potential sources of resistance to *Cmm* in wild relatives such as *S. peruvianum* LA2157 [22], and *S. habrochaites* LA2128 [23], with 2 and 3 QTL´s associated with resistance have been reported, however, the molecular mechanisms underlying this resistance have not been described. Obtaining *Cmm*-resistant tomato cultivars has not been successful due to: i) ineffective crossing between tomato and wild species [22] and ii) *Cmm* shows a high degree of genetic variability, implying that different strains could show different degrees of virulence on different tomato cultivars [24].

Plant Disease Resistance Mechanisms

Much of the current knowledge of plant disease resistance is based on proteobacteria (*Agrobacterium, Erwinia, Pseudomonas, Ralstonia, Xanthomonas*) [25, 26, 27, 28], whereas limited information on diseases produced by Actinomycetes is available [13, 29, 30]. With the completion of the *Cmm* genome sequence and that of other important Actinomycete phytopathogens, it will be possible to study in detail the molecular mechanisms of pathogenesis of this group and make comparisons with the proteobacteria [7, 31].

During plant-pathogen interactions, the pathogen focuses on colonization of the plant tissues whereas the plant concentrates resources on pathogen detection and defense mechanisms. When the plant responses are successful in halting infection, symptoms are restricted and the interaction is called incompatible. If the plant responses are unsuccessful, the pathogen invades plant tissue, generalized symptoms are produced and the interaction is called compatible [32].

Plants and pathogens are co-evolving species, this means that plant barriers can be broken by a particular pathogen, but can successfully block infection by others [32, 33, 34]. In general terms, a few microorganisms can overcome the defense systems of specific plant species or cultivars, leading to significant economic losses, as is the case for *Cmm* in tomato. However, it is still unclear why a highly virulent microorganism is non-pathogenic in other species, even within same genus.

In order to prevent infection, plants have developed a defense system both at the individual cell level and based on systemic signals emanating from the infection site [27]. Plants have three levels of defense: (a) the passive defense provided by physical or chemical barriers (waxy cuticle, cell wall, cytoskeleton, trichomes, saponins, phytoalexins), (b) the induced resistance mediated by phytohormones (jasmonate and salicylic acid) and (c) R-gene based defense.

Passive defense or non-host resistance is the most common in plants. This is a complex response system, consisting of chemical and mechanical barriers, providing a durable and effective resistance against general pathogens in most plant species, however it is the least understood mechanism. Detailed knowledge of non-host resistance is key to the development of effective strategies for crop disease control [35]. It remains unclear whether the durability of non-host resistance is due to recognition of several microbial elicitors, and/or by the simultaneous action of specific genes [35]. Non-host resistance is based on the fact that the phytopathogen is unlikely to counteract the action of multiple defense mechanisms acting simultaneously.

Infection by a phytopathogen can trigger responses in both proximal and distal tissues. Systemic acquired resistance (SAR) and induced systemic resistance (ISR), are related but distinctversions of a single systemic plant response. Both responses enhance defense mechanisms, so that on subsequent infection by the same or a different phytopathogen, the defense responses are quickly activated, and as a consequence the plant shows an enhanced resistance to a broad spectrum of pathogens [32, 36]. SAR is mediated by the phytohormone salicylic acid (SA), and activated by biotrophic microorganisms, its characteristics are an increase in pathogenesis related proteins (PRP) and the early establishment of oxidative stress by reactive oxygen species (ROS). ISR is mediated by the phytohormones jasmonate (JA) and ethylene, and is stimulated after root colonization by non-pathogenic micro-organisms such as *Pseudomonas fluorescens* [36].

The third well studied, class of resistance mechanisms is that based on the interaction of R-gene (resistance) products with Avr-gene (avirulence) products [37]. Most R-genes encode proteins with NBS-LRR domains (nucleotide binding site - leucine rich region), the NBS domain is involved in signal transduction, while the LRR domain is involved in modulation of signal transduction and recognition of specific phytopathogens [38]. When an Avr protein (from fungi, bacteria, virus or nematodes) is recognized by an R protein present in the plant, specific plant defense responses are trigged. Such specific responses are strong enough to inhibit infection normally by producing a hypersensitive response (HR), which provokes cell death around the infection site [38, 39, 40].

However, despite the high specificity of R-gene mediated resistance, this is not a durable mechanism. Phytopathogens diversify and overcome specific R-gene resistance, while novel R-genes capable of detecting different avirulence factors are also produced [40]. R-gene mediated resistance can be lost, due to Avr gene mutations, without loss of phytopathogen fitness [41].

MOLECULAR BREEDING METHODS EMPLOYED TO ENHANCE PLANT DISEASE RESISTANCE

Plant-pathogen interactions are specific, complex, dynamic and difficult to monitor. The identification of genes involved in disease resistance by conventional breeding requires several years of work, and the offspring produced can lose important commercial traits such as taste, fruit quality and even disease resistance to a specific pathogen.

However, new strategies with a molecular focus have been developed, for the identification, characterization and functional analysis of genes for the molecular breeding of resistance [42].

An advantage of the emerging technologies, such as gene expression analysis, is the ability to identify differentially expressed transcripts and to propose a possible defense function based on the differential expression profile [32]. Additionally, the gene expression profiles provide information about gene networks.

Microarrays have been used successfully in many studies, however, their use and potential are restricted to organisms whose genomes or transcriptomes are sequenced and characterized [32, 43].

Alternative methods to analyze differential gene expression based on amplification by PCR (polymerase chain reaction) of transcript fragments have been used successfully in many organisms [42, 44, 45, 46]. Among these techniques, cDNA-AFLP (complementary DNA-Amplified fragment length polymorphisms), is a technology that allows the generation of quantitative and qualitative gene expression data from any organism. This technology is robust, reliable and sensitive enough to discover genes differentially expressed under different growth conditions or stages.

The major advantage is that no previous knowledge of any DNA sequence of the organism under study is required; therefore this technology is not restricted to a specific organism or species, in contrast to other expression analysis systems that require the design of specific primers and arrays [47, 48, 49, 50]. The cDNA-AFLP technique has previously been used successfully in the study of several plant-pathogen systems [49, 51, 52, 53, 54, 55].

The overall objective of determining gene expression profiles during plant-pathogen interactions is the understanding of defense responses in order to improve disease resistance in economically important crops, like tomato.

Recently, gene expression analysis based on microarrays data obtained from samples taken during tomato-*Cmm* interactions [6] revealed that the phytohormone ethylene plays an important role in symptom development, although its effective contribution to disease is unclear.

However, it has been reported that ethylene can either promote or prevent the advance of infection [56].

To develop an effective *Cmm* control strategy in tomato it is necessary to characterize any effective disease resistance mechanism, such as, for example, those found in wild relative *S. peruvianum* LA2157 and *S. habrochaites* [22, 23].

The analysis of gene expression profiles in combination with genomic and metabolomic tools can have a substantial impact on the understanding of plant-pathogen interactions that will allow the development of new control strategies for plant diseases.

Comparison of Transcriptome Data Obtained during the *Solanum-Clavibacter* Interaction: Applications in Molecular Breeding for Plant Disease Resistance

This section details an example of the application of molecular methods, to support breeding for the enhancement of disease resistance in tomato. The use of gene expression profiles obtained by cDNA-AFLP during the interaction of *Cmm* with tomato (compatible interaction) and three wild relatives (incompatible interactions), allowed the identification of several genes with a contrasting gene expression profile between susceptible and resistant species [57]. The results obtained suggest a possible manipulation of plant defenses by *Cmm* at the early stages of infection. This finding is important since early responses could be definitive for the development of resistance to pathogen infection [58]. The contrasting gene expression profiles observed for several genes between resistant and susceptible solanum species, at the early stages of *Cmm* infection, are similar to those reported during the watermelon-*Fusarium* interaction [55]. The hypothesis is that molecular defense mechanisms could be inhibited and/or delayed as a result of tomato metabolism manipulation by *Cmm*.

The symptomatology produced by *Cmm* infection on three wild tomato relatives showed differences in degree of severity at 25 days post inoculation (dpi): in *S. peruvianum* LA2157, the symptoms were reduced [22], whereas in *S. peruvianum* LA2172 and *S. habrochaites* LA2128 [23] the symptoms were localized, close to the infection site. However, for these species the inoculated plants survived without any problem. In tomato (*S. lycopersicum*), symptoms were generalized subsequently leading to plant death. This facilitated a comparison of gene expression profiles during a weak (in *S. peruvianum* LA2157), an intermediate (in *S. peruvianum* LA2172 and *S. habrochaites* LA2128) and strong (in tomato) *Cmm* infection [57].

cDNA-AFLP gene expression profiling at 0, 8, 24 and 32 hours post infection in tomato, *S. peruvianum* LA2157, LA2172 and *S. habrochaites* LA2128 allowed the identification of differentially expressed genes involved in several pathways such as photosynthesis, general metabolism, plant defense, transport, signaling and gene regulation. Most of the identified differentially expressed genes showed homology with uncharacterized plant sequences or unknown sequences, which could represent new transcripts associated specifically with the *Solanum-Clavibacter* interaction and/or *Solanum*-actinomycetes interaction. Most of the differentially expressed genes with homology to previously characterized plant sequences have a clear role during disease resistance responses and in addition, some genes such as trealose-6-phosphate synthetase, have been associated with plant responses to abiotic stress suggesting crosstalk between *Cmm*-resistance and abiotic stress responses. In this respect the durability and effectiveness of non-host resistance has also been based on several overlapping molecular mechanisms [35].

Photosynthesis Associated Genes

Several genes involved in photosynthetic pathways were down-regulated in both the resistant and susceptible species highlighting a relationship between photosynthesis and plant defense responses. With a decrease in photosynthesis, the non-essential cellular processes are also decreased, whereas the synthesis of antimicrobial metabolites, such as phytoalexins, becomes a priority in order to stop the advance of the phytopathogen at least close to the infection site [59]. On the other hand, a decreased local photosynthetic rate, may lead to carbohydrate accumulation (in particular hexoses) at the infection site, as a result of increased invertase activity [54, 59]. In this scenario, *Cmm* could have an advantage during the compatible interaction, since the 129 kb pathogenicity island *chpC/tomA*, contains several genes involved in carbohydrate metabolism [7].

Senescence and Ethylene

A transcript with homology to a protein associated with senescence was found to be differentially expressed in all species. In *S. peruvianum* LA2172 and *S. habrochaites* LA2128 it was up-regulated, whereas in *S. peruvianum* LA2157 and the commercial tomato cultivar it was down-regulated. Premature senescence can be induced by phytopathogens and decreased photosynthetic rate [60, 61, 62]. Cell death is also a link between plant defense responses, senescence and the effects of ethylene. Therefore ethylene has a role in both plant resistance [60] and plant susceptibility [6]. The transcription factor ERF/AP2 involved in ethylene signaling was induced at 4 and 8 dpi during the tomato-*Cmm* interaction [6], when the *Cmm* infection is already well established whereas in resistant *S. peruvianum* LA2172 it was down-regulated at the initial stages of infection, again suggesting that early ethylene signaling could be a determining factor in the development of disease or resistance. However more detailed work in this area is necessary in order to confirm this hypothesis [63].

Are *Cmm*-Susceptibility Genes Present in Tomato?

Interestingly, differentially expressed transcripts with homology to the Per1 protein family, including the calmodulin-binding protein and the MBF1c transcriptional coactivator multiprotein bridging factor were identified in the tomato analysis. Per1 is a LEA protein found in the Mp420 maize genotype, which has been shown to be involved in abiotic stress tolerance and response to *Aspergillus* infection [64]. In contrast, the calmodulin-binding proteins are involved in Ca^{2+} signaling and are required for SA accumulation and HR establishment [65]. The transcriptional coactivator multiprotein bridging factors can interconnect with other transcription factors through the basic transcription machinery and enhance defense responses [66]. These transcripts were down-regulated during tomato-*Cmm* interaction at the early stages of infection therefore the defense responses mediated by these genes could be suppressed as a result of *Cmm* infection.

Genes of Wild Relatives Involved in *Cmm*-Resistance

To validate the differential gene expression patterns obtained by cDNA-AFLP during *Solanum-Clavibacter* interactions, the gene expression profiles of five transcripts were confirmed by quantitative PCR in both resistant and susceptible species [57]. The differentially expressed transcripts were selected based on available bibliographic information concerning their roles in plant resistance. The selected transcripts showed homology to: secretory peroxidase, 30 kDa endochitinase, MRP1 putative multidrug resistance-associated protein, PBC1 (20S proteasome beta subunit C1) and SUMO E2 conjugating enzyme SCE1.

The secretory peroxidase, 30 kDa endochitinase, SCE1 and PBC1, were induced in resistant species as a result of *Cmm* infection at early stages and repressed in the susceptible, commercial tomato species. This suggests that defense responses are stronger and/or faster in wild relatives that in tomato. The gene expression profile of secretory peroxidase suggests a possible role for the apoplastic oxidative metabolism in the basic response mechanism of wild *Solanum* species to *Cmm* infection [67] and/or as a signaling pathway mediated by reactive oxygen species (ROS), since H_2O_2 can act as an intracellular or intercellular secondary messenger for systemic induction of defense mechanisms during abiotic and biotic stress [68, 69].

The induction of the 30 kDa endochitinase in wild relatives and its repression in the cultivated species highlights the importance of chitinases against fungal and bacterial infection. Chitinase activity induces phytoalexin synthesis thereby enhancing non-host resistance [70].

Recent evidence suggests that the ubiquitin/proteasome pathway plays an important role during defense responses [71], due to the involvement of this pathway in R-gene signaling, HR establishment, SAR and ISR [72]. A transcript with homology to PBC1 (20S proteasome beta subunit C1) was induced in wild species but repressed in cultivated tomato, whereas a transcript with homology to the SUMO E2 conjugating enzyme SCE1 showed no significant changes in gene expression levels in either wild relatives or cultivated tomato. Recently, the gene silencing of ubiquitin activating enzyme in *Nicotiana benthamiana* was shown to enhance susceptibility to *Cmm*, suggesting an important role for the ubiquitin/proteasome pathway in the *Cmm* resistance of wild tomato relatives [73].

A transcript with homology to MRP1, an ABC (ATP binding cassette) transporter, showed a similar gene expression profile between wild relatives and the commercial tomato cultivar; however, quantitative differences were observed with higher expression levels found in wild relatives. It is possible that MRP1 lacks a definitive function during defense responses against *Cmm*, or that the quantitative differences in gene expression levels between wild relatives and tomato, are sufficient to produce the resistant response. On the other hand, several ABC transporters, such as MRP1 and PDR1 (both differentially expressed in wild relatives) are essential for resistance responses in other systems [74, 74, 76].

Although there is no direct evidence that R-gene resistance is involved in the response of wild species to *Cmm* attack [77], R-gene mediated resistance cannot be completely ruled out. Constitutively expressed genes and/or variation in R gene homologs (RGH) are abundant in plant genomes, including domesticated species such as tobacco with low levels of genetic diversity [78]. During the tomato-*Cladosporium* interaction an RGH transcript was induced, an exceptional occurrence since it is accepted that R genes are normally constitutively expressed [51]. Therefore, the probability of discovering new RGHs with new specificities and expression patterns in wild relatives with a high degree of genetic diversity, such as *S. peruvianum* and *S. habrochaites* is probably higher than expected. Additionally, wild relatives have marked morphological differences that could act in passive defense, such as the structure, density and size of the vascular bundles, which in combination could have an influence on *Cmm* pathogenesis and/or enhance the defense mechanisms.

CONCLUSION

Cmm is a highly adapted tomato phytopathogen, able to overcome the early defense responses of the plant but in contrast is not well adapted to produce infection in the wild relatives *S. peruvianum* and *S. habrochaites,* where the successful defense mechanisms initiate at the early stages of *Cmm* infection and are stronger and faster than in the commercial tomato species. The high degree of genetic variability of *Cmm* contrasts with the low genetic diversity found in commercial tomato cultivars, this fact could be a determining factor to consider for molecular breeding of *Cmm* resistant tomato cultivars- In order to develop an effective control strategy the first step is to understand how *Cmm* overcomes the basic defense responses in commercial tomato cultivars. In addition, the identification of the pathogen associated molecular patterns (PAMPs), perceived by wild relatives and of the mechanisms involved in their recognition, could enhance the development of an innovative and durable *Cmm* control strategy, while the support of genetic engineering applied to crop protection, will allow molecular breeding with the use of genetic resources from wild relatives.

REFERENCES

[1] J.B. Jones, J.P. Jones, R.E. Stall, T.A. Zitter (1991). Compendium of Tomato Diseases. APS Press, St. Paul, MN, USA

[2] X. Xu, S.A. Miller, F. Baysal-Gurel, K.H. Gartemann, R. Eichenlaub, G. Rajashekara, *Appl Environ Microbiol,* 12, 3978 (2010).

[3] H. Jahr, R. Bahro, A. Burger, J. Ahlemeyer, R. Eichenlaub, *Environ Microbiol,* 1, (1999).

[4] K.H. Gartemann, O. Kirchner, J. Engemann, I. Gräfen, R. Eichenlaub, A. Burger, *J Biotechnol,* 106, 179 (2003).

[5] W.M. Carlton, E.J. Braun, M.L. Gleason, *Phytopathol,* 88, 525 (1998).

[6] V. Balaji, M. Mayrose, O. Sherf, J. Jacob-Hirsch, R. Eichenlaub, N. Iraki, S. Manulis-Sasson, G. RechavI, I. Barash, G. Sessa, *Plant Physiol,* 146, 1797 (2008).

[7] K.H. Gartemann, B. Abt, T. Bekel, A. Burger, J. Engemann, M. Flügel, L. Gaigalat, A. Goesmann, I. Gräfen, J. Kalinowski, O. Kaup, O. Kirchner, L. Krause, B. Linke, A. McHardy, F. Meyer, S. Pohle, C. Rückert, S. Schneiker, E.M. Zellermann, A. Pühler, R. Eichenlaub, O. Kaiser, D. Bartels, *J Bacteriol,* 190, 2138 (2008).

[8] J. Wittmann, R. Eichenlaub, B. Dreiseikelmann, *Microbiol,* 156, 2366 (2010).

[9] H. Jahr, J. Dreier, D. Meletzus, R. Bahro, R. Eichenlaub R, *Mol Plant Microbe Interac,* 13, 703 (2000).

[10] J. Dreier, D. Meletzus, R. Eichenlaub, *Mol Plant Microbe Interact* 10, 195 (1997).

[11] O. Kaup, I. Gräfen, E.M. Zellermann, R. Eichenlaub, K.H. Gartemann, *Mol Plant Microbe Interact,* 18, 1090 (2005).

[12] I. Stork, K.H. Gartemann, A. Burger, R. Eichenlaub, *Mol Plant Pathol,* 51, 149 (2008).

[13] A. Savidor, D. Teper, K.H. Gatermann, R. Eichenlaub, L. Chalupowicz, S. Manulis-Sasson, I. Barash, H. Tews, K. Mayer, R.J. Giannone, R.L. Hettich, G. Sessa, *J. Proteome Res.,*11, 736 (**2012).**

[14] L. Chalupowicz, M. Cohen-Kandli, O. Dror, R. Eichenlaub, K.H. Gartemann, G. Sessa, I. Barash, S. Manulis-Sasson, *Phytopathol,* 3, 252 (2010).
[15] D. Meletzus, A. Bermphol, J. Dreier, R. Eichenlaub, *J Bacteriol,* 175, 2131 (1993).
[16] Y. Bai and P. Lindhout, *Annals of Botanny*, 100, 1085 (2007).
[17] S.D. Tanksley, *The Plant Cell,* 16, 181 (2004).
[18] A. Barone, A Di-Matteo, D. Carputo, L. Frusciante, *Current Genomics*, 10, 1 (2009).
[19] T. Arie, H. Takahashi, M. Kodama, T. Teraoka, *Plant Biotechnol,* 24, 135 (2007).
[20] D.R. Pantee and F. Chen, *Current Genomics*, 11, 30 (2010).
[21] R.E. Stall, J.B. Jones, G.V. Minsavage, *Annu. Rev. Phytopathol*, 47, 265 (2009).
[22] A.W. van Heusden, M. Koornneef, R.E. Voorrips, W. Brüggemann, G. Pet, R. Vrielink-van Ginkel, X. Chen, P. Lindhout, *Theor Appl Genet,* 99, 1068 (1999).
[23] D.M. Francis, E. Kabelka, J. Bell, B. Franchino, D. St. Clair, *Plant Dis,* 85, 1171 (2001).
[24] F. Kleitman, I. Barash, A. Burger, N. Iraki, Y. Falah, G. Sessa, D. Weinthal, L. Chalupowicz, K.H. Gartemann, R. Eichenlaub, S. Manulis-Sasso, *Eur J Plant Pathol,* 121, 463 (2008).
[25] R.B. Abramovitch and G.B. Martin, *Curr Opin Plant Biol,* 7, 356 (2004).
[26] R.B. Abramovitch, J.C. Anderson, G.B. Martin, *Nature,* 7, 601 (2006).
[27] J.D.G. Jones and J.L. Dangl, *Nature,* 444, 323 (2006).
[28] B.M. Mole, D.A. Baltrus, J.L. Dangl, S.R. Grant, *Trends Microbiol,* 15, 363 (2007).
[29] S.A. Hogenhout and R. Loria, *Curr Opin Plant Biol,* 11, 449 (2008).
[30] L. Chalupowicz, M. Cohen-Kandli, O. Dror, R. Eichenlaub, K.H. Gartemann, G. Sessa, I. Barash, S. Manulis-Sasson, *Phytopathol,* 3, 252 (2010).
[31] J. Setubal, L.M. Moreira, A.C.R. da Silva, *Curr Opin Microbiol,* 8, 595 (2005).
[32] J, Wan, F.M. Dunning, A.F. Bent, *Funct Integr Genomics,* 2, 259 (2002).
[33] B. Baker, P. Zambryski, B. Staskawicz, S.P. Dinesh-Kumar, *Science*, 276, 726 (1997).
[34] J.G. Bishop, A.M. Dean, T. Mitchell-Olds, *PNAS,* 97, 5322 (2000).
[35] K.S. Mysore and C.M. Ryu, *Trends Plant Sci,* 9, 97 (2004).
[36] G.E. Vallad and R.M. Goodman, *Crop Sci,* 44: 1920 (2004).
[37] H.H. Flor, *Phytopathol*, 9, 275 (1971).
[38] Y. Belkhadir, R. Subramaniam, J.L. Dangl, *Curr Opin Plant Biol,* 7, 391 (2004).
[39] J.L. DangI and J.M. McDowell, *PNAS*, 103, 8575 (2006).
[40] B.J. DeYoung and R.W. Innes, *Nature Immunol* 7, 1243 (2006).
[41] J.M. McDowell and B.J. Woffenden, *Trends Biotechnol,* 21, 178 (2003).
[42] K.E. Hammond-Kosack and J.E. Parker, *Curr Opin Biotechnol,* 14, 177 (2003).
[43] P. Breyne, R. Dreesen, B. Cannoot, D. Rombaut, K. Vandepoele, S. Rombauts, R. Vanderhaeghen, D. Inze, M. Zabeau, *Mol Gen Genomics,* 269, 173 (2003).
[44] P. Breyne and M. Zabeau, *Curr Opin Plant Biol,* 4, 136 (2001).
[45] P. Breyne, R. Dreesen, K. Vandepoele, L. De Veylder, F. Van Breusegem, L. Callewaert, S. Rombauts, J. Raes, B. Cannoot, G. Engler, D. Inzé, M. Zabeau, *PNAS,* 23, 14825 (2002).
[46] J. Donson, Y. Fang, G. Espiritu-Santo, W. Xing, A. Salazar, S. Miyamoto, V. Armendarez, W. Volkmuth, *Plant Mol Biol,* 48, 75 (2002).
[47] C.W.B. Bachem, R.S. van der Hoeven, S.M. de Bruijn, D. Vreugdenhil, M. Zabeau, R.G.F. Visser, *Plant J,* 9,745 1996;
[48] C.W.B. Bachem, R.J.F.J. Oomen, R.G.F. Visser, *Plant Mol Biol Rep,* 16, 157 (1998).

[49] W.E. Durrant, O. Rowland, P. Piedras, K.E. Hammond-Kosack, J.D. Johnes, *Plant Cell,* 12, 963 (2000).

[50] M. Vuylsteke, J.D. Peleman, M.J.T. van Eijk, *Nat Protoc,* 2, 1399 (2007).

[51] S.H.E.J. Gabriëls, F.L.W. Takken, J.H. Vossen, C.F. de Jong, Q. Liu, S.C.H.J. Turk, L.K. Wachowski, J. Peters, H.M.A. Witsenboer, P.J.G.M. de Wit, M.H.A.J. Joosten, *Mol Plant Microbe Interact,* 19, 567 (2005).

[52] M. Polesani, F. Desario, A. Ferrarini, A. Zamboni, M. Pezzotti, A. Kortekamp, A. Polverari, *BMC Genomics,* 9, 142 (2008).

[53] M. Ventelon-Debout, C. Tranchant-Dubreuil, T.T.H. Nguyen, M. Bangratz, C. Siré, M. Delseny, C. Brugidou, *BMC Genomics,* 8, 26 (2008).

[54] L. Wang, K. Tsuda, M. Sato, J.D. Cohen, F. Katagiri, J. Glazebrook, *PLoS Pathogens,* 5, 1 (2009).

[55] S. Sestili, A. Polverari, L. Luongo, A. Ferrarini, M. Scotton, J. Hussain, M. Delledonne, N. Ficcadenti, A. Belisario, *BMC Genomics*, 12, 122 (2011).

[56] L.C. Van Loon, B.P.J. Geraats, H.J.M. Linthorst, *Trends Plant,* 11, 184 (2006).

[57] J.P. Lara-Avila, M.I. Isordia-Jasso, R. Castillo-Collazo, J. Simpson, A.G. Alpuche-Solis. *Plant Mol. Biol Rep.* 30, 498 (2012).

[58] C. Muellenborn, J.H. Krause, C. Cerboncini, *Plant Mol Biol Rep,* 29, 597 (2011).

[59] M.D. Bolton, *Mol Plant Microbe Interact,* 22, 487 (2009).

[60] C. Espinoza, C. Medina, S. Somerville, P. Arce-Johnson, *J Exp Bot,* 58, 3197 (2007).

[61] B.F. Quirino, J. Normanly, R.M. Amasino, *Plant Mol Biol,* 40, 267 (1999).

[62] B.F. Quirino, Y.S. Noh, E. Himelblau, R.M. Amasino, *Trends Plant Sci,* 5, 278 (2000).

[63] X. Meng, F. Li, C. Liu, C. Zhang, Z. Wu, Y. Chen, *Plant Mol Biol Rep,* 28, 176 (2010).

[64] Z.Y. Chen, R.L. Brown, K.E. Damann, T.E. Cleveland, *Phytopathol* 97, 1094 (2007).

[65] L. Du, G.S. Ali, K.A. Simons, J. Hou, T. Yang, A.S.N. Reddy, B.W. Poovaiah, *Nature,* 457, 1154 (2009).

[66] N. Suzuki, L. Rizhsky, H. Liang, J. Shuman, V. Shulaev, R. Mittler, *Plant Physiol,* 139, 1313 (2005).

[67] S.O. Kotchoni and E.W. Gachomo, *J Biosci,* 31, 389 (2006).

[68] R. Mittler, S. Vanderauwera, M. Gollery, F. Van Breusegem, *Trends Plant Sci,* 9, 490 (2004).

[69] A.K. Nanda, E. Andrio, D. Marino, N. Pauly, C. Dunand, *J Integr Plant Biol,* 52, 195 (2010).

[70] T. Yamaguchi, A. Yamada, N. Hong, T. Ogawa, T. Ishii, N. Shibuya, *Plant Cell,* 12, 817 (2000).

[71] W. Dong, D. Nowara, P. Schweizer, *Plant Cell,* 18, 3321 (2006).

[72] G. Shirsekar, L. Dai, Y. Hu, X. Wang, L. Zeng, G.L. Wang, *J Plant Biol,* 53, 10 (2010).

[73] V. Balaji, G. Sessa, C.D. Smart, *Phytopathol,* 101, 349 (2010).

[74] R. Hückelhoven, *Curr Opin Plant Biol,* 10, 573 (2007).

[75] C. Kwon, P. Bednarek, P. Schulze-Lefert, *Plant Physiol,* 147, 1575 (2008).

[76] Y. Stukkens, A. Bultreys, S. Grec, T. Trombik, D. Vanham, M. Boutry, *Plant Physiol,* 139, 341 (2005).

[77] E. Kabelka, B. Franchino, D.M. Francis, *Phytopathol* 92, 504 (2002).

[78] X. Leng, B. Xiao, S. Wang, Y. Gui, Y. Wang, X. Lu, J. Xie, Y. Li, L. Fan, *Plant Mol Biol Rep,* 28,152 (2010).

In: Tomatoes: Cultivation, Varieties and Nutrition
Editor: Tadahisa Higashide
ISBN: 978-1-62417-915-0

Chapter 6

BIOCONTROL OF FUNGAL DISEASES OF TOMATO: CONTRIBUTION OF *PSEUDOMONAS* SP. PCI2

Nicolás A. Pastor, Susana B. Rosas, Javier A. Andrés, Marina C. Niederhauser and Marisa Rovera

Laboratorio de Interacción Microorganismo–Planta. Facultad de Ciencias Exactas, Físico-Químicas y Naturales, Universidad Nacional de Río Cuarto. Río Cuarto, Córdoba, Argentina

ABSTRACT

Tomato is the most diffused vegetable plant worldwide. Thus, demand for tomato cultivation, production and commerce is continually increasing. Production of tomato fruits is destined for direct consumption and for processing into industrialized products, such as puree, pulp, juice and sauces.

Fungal diseases are one of the major factors limiting tomato production. These diseases are managed by the application of chemical fungicides. However, chemical fungicides are costly, can cause adverse environmental effects, can result in health hazards to humans and other organisms, and can induce pathogen resistance. Currently, it is widely accepted that biological control of plant diseases, involving the use of antagonistic microorganisms, is an alternative or a supplemental way to reduce the use of chemicals in agriculture for disease management.

A large number of microorganisms from the rhizosphere have been considered to play a key role in sustainable agriculture because of their biocontrol potentials and plant growth-promoting activities. Among these, the soil-borne fluorescent *Pseudomonas* have received particular attention because of their catabolic versatility, their excellent root-colonising abilities, their adaptation to environmental stresses and their capacity to aggressively compete with other microorganisms. The identification of antagonistic strains represents only the first step towards the development of effective biological control strategies. After the identification of organisms as potential antagonists, it is important to determine the specific mechanisms, interactions and conditions responsible for biological control. The understanding of these characteristics will allow scientists to establish both the limitations and the full potential of biocontrol for developing effective strategies for disease management.

Tomato is susceptible, among other fungi, to *Phytophthora infestans*, the causal agent of late blight, *Fusarium oxysporum* f.sp. *lycopersici* (Fol), which causes a severe wilt disease, *F. oxysporum* f.sp. *radicis-lycopersici* (Forl), the causal agent of crown and root rot, *Alternaria solani*, the causal agent of early blight, *Verticillium dahliae* and *V. albo-atrum*, which cause vascular wilt, and to *Sclerotium rolfsii* and *Alternaria alternata*. These two later fungi cause heavy losses in the yield of tomato. *S. rolfsii* is a major problem for the cultivation of tomato in the tropical regions of the world, causing damping-off in nursery seedlings as well as stem rot, wilting and blight in adult plants. *A. alternata* is frequently isolated from diseased tomato plants and causes stem canker and leaf spot.

In this chapter, we will review some studies on biological control of tomato fungal diseases by fluorescent *Pseudomonas*. We will also contribute data from our own work on the antagonistic activity of strain *Pseudomonas* sp. PCI2, isolated from tomato rhizosphere, against *S. rolfsii* and *A. alternata*.

INTRODUCTION

Tomato (*Solanum lycopersicon*) is the second most important vegetable crop next to potato. Present world production is about 100 million tons of fresh fruits from 3.7 million ha [1]. Tomato plants are economically attractive because of their relatively rapid growth and high yields. Tomato belongs to the *Solanaceae* family, which includes other known species such as *Solanum tuberosum* (potato), *Nicotiana tabacum* (tobacco) and *Capsicum annuum* (pepper) [2]. The optimum daily temperature for growing tomato plants ranges from 18 to 25 ºC, with night temperatures between 10 and 20 ºC. Larger differences between day and night temperatures adversely affect yield. Tomato is a frost-sensitive crop plant. Temperatures above 25 ºC, together with high humidity and strong wind, result in reduced yields. Night temperatures above 20 ºC, together with high humidity, lead to an excessive vegetative growth and to a poor production of fruits. In addition, a high humidity leads to a high incidence of pests and diseases [1]. Tomato plants have a vigorous root system that grows to a depth of 50 cm or more. The main root develops dense adventitious and lateral roots. The growth habit of the stem ranges from vertical to prostrate. The stem is solid, rough and shaggy, and grows to a height of 2 to 4 meters. Leaves arrange spirally. They are between 15 and 50 cm long and between 10 and 30 cm wide. Leaflets are between oval and oblong in shape and are covered with glandular hairs. The inflorescence is clustered and produces between 6 and 12 flowers. Fruits are succulent, globular-shaped and have between 2 and 15 cm in diameter. Immature fruits are green and hairy whereas mature fruits can be yellow, orange or red. Fruits can be rounded and smooth or ridged. The first harvest can be carried out between 45 and 55 days after flowering (or between 90 and 120 days after planting). Tomato fruits are rich in minerals, vitamins, essential amino acids, sugars and fibers [2]. Fructose and glucose represent nearly 50% of the dry matter and more than 95% of the sugars in the tomato fruit. In addition, tomatoes are an important source of vitamins A and C. A medium-sized tomato fruit (148 grams) can supply between 20 and 40% of the daily requirement for these vitamins [3]. Furthermore, fruits contain high concentrations of iron (Fe) and phosphorus (P). Tomato fruits can be consumed fresh in salads or cooked as sauces or soup. Fruits can also be processed as puree, juice and ketchup. Canned and dry tomatoes are processed products of economic importance. Typically, tomato seedlings are transplanted because plants develop better when they are previously grown under greenhouse conditions. With this procedure, lower amounts of seeds are required for planting and seedlings can be selected based on health and growth before planting in the field. Moreover, seedlings can be better protected [2].

In Argentina, just over 1,185 ha are destined for producing tomatoes under cover. There is concentration of the activity. An 87.7% of Argentina's planting surface is concentrated between the provinces of Corrientes and Buenos Aires. The province of Corrientes, with approximately 630 ha, has the highest concentration of greenhouses destined for tomato cultivation. All of the fruits obtained under cover are destined for direct fresh consumption. Approximately 14,389 ha (6.34% of Argentina's surface) are used for production of tomato in the field. Production of tomato in the field is not as concentrated as production under cover. About 87% of the total production surface is distributed among eight provinces. Mendoza is the most relevant province, with 5,201 ha destined for the production of tomatoes. Approximately 370,000 tons of processed tomatoes are consumed each year in Argentina. The industry transforms tomatoes into three different products: tomato puree, natural peeled tomato and triple tomato extract [4].

Fungal Diseases that Affect Tomato Plants

Diseases are the main biological repression for production of vegetables and fruit crops, and several pathogens, including fungi, bacteria, viruses and nematodes can infect these crops. Of these, fungal diseases are one of the major factors limiting crop production and causing the greatest proportion of overall economic losses. A fungal infection is caused by spores that germinate and penetrate plant tissues through stomata or wounds. The fungal filaments grow at an increasing rate in the affected tissue and, there, they extract nutrients and secrete substances that are toxic to the plant. As a consequence, the plant tissue usually dies. The harmful effects of the fungus are generally limited to the affected area. However, some fungal species invade the vascular tissues of the plant and spread throughout the entire organism [2, 5]. Due to increasing demand, tomato has a great potential for increased commercialization. Therefore, a more efficient tomato production requires better knowledge of its pathogens and control methods.

Late Blight, Caused by *Phytophthora infestans*

Late blight can devastate tomato crops during periods of cold, rainy weather. This disease can infect both young and old leaves. First, gray spots appear which give the plant an appearance of damage by frost. Usually, the spots begin to form at the edge of leaves and spread inwards. In addition, spots may spread from the center of the leaves outwards. Spots are white on the bottom of leaves. The infection of green or mature fruits produces large, irregularly shaped brown spots. *P. infestans* also causes similar symptoms in potato plants. Therefore, this fungus can spread from potatoes to tomatoes [2, 6]. During the disease cycle, zoospores are essential propagules for the pre-infection process. Zoospores are a potential target for controlling *P. infestans* [7].

Fusarium Wilt, Caused by *Fusarium oxysporum* f. sp. *lycopersici* (Fol)

Fusarium oxysporum is an important pathogen of many agricultural and ornamental crops [8]. *F. oxysporum* is known to infect a large number of host plants, causing wilt, root rot and crown rot. Besides decreases in yield, *F. oxysporum* produces toxins which can accumulate in the final products and, therefore, can be dangerous to people and animals [9]. *F. oxysporum* f. sp. *lycopersici* only attacks certain tomato cultivars. Infected plants exhibit wilting of leaves, which progresses from the base of the stem towards the upper part of the plant. Initially, only one side of the leaves, branches, or the plant is affected. Then, the symptoms quickly spread to the rest of the plant. Wilted leaves fall prematurely. A pink fungal fluff can be observed on dead plants [2, 6]. The vascular wilt is favored by root and stem wounds. *F. oxysporum* f. sp. *lycopersici* is transmitted by seeds and contaminated soil. The disease is more common in acidic, poorly drained soils. Affected plants left in the field are the main source of inoculum. The pathogen sporulates easily in diseased plants and spreads towards healthy plants by water and wind. The fungus survives in the soil and in crop residues as chlamydospores. Individual bagging of plants growing in pots containing previously disinfected soil is recommended in order to control the disease [10]. Application of fungicides is a common practice. Nevertheless, it may not be very effective since the disease appears late in the life cycle of the crop and the persistence of fungicides throughout this cycle is doubtful [11].

Foot and Root Rot, *Fusarium oxysporum* f. sp. *radicis-lycopersici* (Forl)

F. oxysporum radicis-lycopersici (Forl) is a special form of *Fusarium oxysporum* that was first described in 1978 [12]. Foot and root rot (TFRR) affects tomato plants from both greenhouses and field, causes significant losses in production and has been reported in 32 countries [13]. TFRR cannot be efficiently prevented by chemical pesticides [14]. Forl causes symptoms that are different from those caused by Fol. Symptoms on mature crops are root and basal stem rot [15]. TFRR occurs at cool temperatures (18 °C). In addition, Forl has a broader host range than that of Fol [16, 17].

Early Blight, Caused by *Alternaria solani*

Among the different fungal diseases affecting tomato plants, early blight caused by *A. solani* is one of the most destructive. Early blight causes heavy losses, ranging from 35 to 78%, in yield [18]. The premature loss of lower leaves is the most obvious symptom of the disease. First, brown or black spots, with dark edges, appear on lower leaves. Later, these spots fuse and leaves turn yellow and dry. *A. solani* also attacks fruits, causing large sunken areas with concentric rings and a black, velvety appearance. Infection of plants can result in a complete loss of production as yields are reduced by the destruction of foliage. Fruits are directly damaged by the pathogen or by sunblotch if plants are defoliated [19]. A warm, humid weather favors the rapid spread of early blight. *A. solani* can infect plants at any growth stage, but progresses more rapidly after the plants produce fruits. This phytopathogen spreads by seeds, wind, rain and debris from infected plants [2, 6]. The ineffectiveness of fungicides due to frequent heavy rainfall increases the severity of the disease. Susceptible hybrids can be severely damaged by early blight [20].

Verticillium Wilt, Caused by *Verticillium albo-atrum* and *Verticillium dahliae*

V. albo-atrum and *V. dahliae* can attack over 200 plant species, including potato, pepper, eggplant, strawberry, watermelon and radish. The economic losses caused by these pathogens can reach up to 50%, especially in tomato cultivars severely infected by *V. dahliae* [21]. Once symptoms have appeared, it is difficult to efficiently control the disease. These fungi spread through plant debris, especially in slightly acidic soils. As *Fusarium* wilt, this disease appears first on lower leaves and progresses upward. Yellow spots develop on lower leaves. Then, leaves rapidly turn wilted and fall. Infected plants can survive throughout the growth cycle, but they show little development and, therefore, decreases in yield. *Verticillium* wilt causes an inner browning of the water-carrying tissue inside the stem. Furthermore, several lateral roots may appear at the base of the plant. This disease is more common during cold weather [2, 6].

Damping-off and Stem Rot, Caused by *Sclerotium rolfsii* Sacc.

S. rolfsii is a soil pathogen that attacks over 500 different plant species [22]. *S. rolfsii* is a major problem for cultivation of tomato in the warm, moist tropical regions of the world [23], causing damping-off in nursery seedlings as well as stem rot, wilting and blight in adult plants [24, 25, 26], with consistent loss of production [27]. Rot development is favored by high temperatures and humidity. Generally, the first above-ground symptoms are yellowing of leaves and wilting of infected plants. The stem often appears soft and sunken in the soil line and develops a brownish discoloration both internally and externally. Under humid conditions, a white fungal growth can be observed on the lower part of the stem, near the soil surface. Usually, brown spherical sclerotia form in the mycelium. Under dry conditions, the mycelium and the sclerotia may not be visible. *S. rolfsii* spreads as mycelium in infested organic matter or as sclerotia in infested soil. Infection usually occurs at the soil surface but may also occur below the soil line. *S. rolfsii* can spread over a meter through the soil and from plant to plant. Sclerotia produced on crop debris and on infected plants serve as inoculum for the next crop [28].

Leaf Spot, Caused by *Alternaria alternata* f. sp. *lycopersici*

Tomato is also susceptible to *A. alternata*, which infects vegetable crops belonging to several families and reduces their yield both qualitatively and quantitatively. *A. alternata* is frequently isolated from diseased tomato plants and causes stem canker [29] and leaf spot [30]. The primary sources of inoculum are wintering structures, in the form of mycelium from infected plants and conidia or mycelium from seeds [31]. When a tomato plant is infected with *A. alternata*, circular chlorotic spots that are either brown or black, and which later on necrose, will normally appear on the lower leaves of the plant. These spots will eventually migrate to the stem and even to the fruits. Such spots are actually cankers and can eventually overtake a plant and kill it. The predisposing factors include temperature and humidity. Humidity (rain) promotes the germination and spread of *A. alternata* conidia, whereas temperature favors the growth of this fungus. Production of spores is favored by temperatures between 21 and 30 °C and is impaired by temperatures below 15 and above 32 ºC [32]. *A. alternata* is usually controlled by spraying tomato plants with chemical fungicides, such as chlorothalonil [33].

BIOCONTROL OF TOMATO FUNGAL DISEASES WITH FLUORESCENT *PSEUDOMONAS* SPP.

Currently, several organizations and people see the increased application of chemical pesticides as a threat to public health, the environment and to the maintenance of agriculture itself. On the other hand, agronomists have growing problems due to increased pesticide resistance in populations of target pests [34]. In addition, soil-borne diseases are difficult to manage by the application of chemicals since the causal agents are in the soil, where chemicals cannot be easily applied [5]. Supplementary pest control schemes and new alternatives to the currently dominant pesticide applications are seen by many people as an urgent requirement to continue with the successful story of agriculture and to feed the growing human population in a healthy environment. Therefore, biological control is considered as an alternative or as a supplemental way for reducing the use of chemicals in agriculture [34]. In fact, biological control may provide control of diseases which cannot be, or only partially, handled by other strategies [35]. The terms "biological control" and its abbreviated synonym "biocontrol" are used in different fields of biology. In entomology, these terms are used to describe the use of living predatory insects, entomopathogenic nematodes, or microbial pathogens to suppress populations of different harmful insects. In plant pathology, the terms apply to the use of microbial antagonists to suppress diseases and to the use of host-specific pathogens to control populations of weeds. The organism that suppresses a pest or pathogen is called biological control agent (BCA). Specifically, "biological control" refers to the beneficial use of resident or introduced living organisms, different from the resistant host plants, to suppress the activities and populations of plant pathogens [36].

Fluorescent *Pseudomonas* are ubiquitous soil microorganisms, prominent members of the microflora in the rhizosphere of many crop plants and usually the first to colonize young roots [5]. The main saprophytic *Pseudomonas* species associated with plants include *P. fluorescens*, *P. putida* and *P. aeruginosa* [37]. Certain strains suppress plant diseases by protecting seeds or roots against infection by pathogenic bacteria and fungi [5]. Fluorescent *Pseudomonas* have been proven to posses many traits that make them well suited as biological control agents [reviewed in references 5, 38, 39, 40, 41]. These traits include their catabolic versatility, their excellent root-colonising abilities, their adaptation to environmental stresses and their capacity to aggressively compete with other microorganisms [42, 40]. Indeed, it has been demonstrated that fluorescent *Pseudomonas* induce defence mechanisms in plants [43] and produce a wide range of antifungal metabolites such as siderophores [44], hydrolytic enzymes [45, 46], volatile compounds [47], 2,4-diacetylphloroglucinol (DAPG), pyoluteorin (PLT), pyrrolnitrin (PRN) [48, 49] and phenazine 1-carboxylic acid (PCA) [50].

Fluorescent *Pseudomonas* spp. have emerged as the most promising group of rhizobacteria involved in the biocontrol of plant diseases. For example, fluorescent *Pseudomonas* were used for controlling *Fusarium* wilt [51] and *Verticillium* wilt [52] in tomato. In the first study, several of the tested *P. fluorescens* strains showed significant reductions (30-40%) in disease incidence. In the second study, significant differences in growth (27% for shoot height) between plants infested with *V. dahliae* and inoculated with *Pseudomonas* sp. PsJN were observed. Thomas *et al.* [53] screened bacteria from tomato rhizosphere for antagonistic activity *in vitro* against the fungal phytopathogen *F. oxysporum* f. sp. *radicis-lycopersici.* These authors found that the most active isolate was *P. chlororaphis* PCL1391. During biocontrol tests under greenhouse conditions, strain PCL1391 was able to reduce the number of diseased tomato plants by 45%, as compared to the untreated control. Further characterization carried out by these authors showed that strain PCL1391 has a broad-spectrum antifungal activity and that it produces a variety of potential antifungal metabolites. Dekkers *et al.* [54] showed that tomato foot and root rot caused by *F. oxysporum* f. sp. *radicis-lycopersici* was controlled by inoculation of seeds with cells of *P. fluorescens* WCS365. Their biocontrol experiments, performed in a *F. oxysporum* f. sp. *radicis-lycopersici*-tomato system, showed that *P. fluorescens* WCS365 significantly suppressed the disease. The mean percentage of diseased plants was 53.3 in the untreated controls, whereas the mean percentage was 27.6 in the inoculation treatment. El-Rafai *et al.* [55] evaluated the efficiency of different application treatments (soil inoculation, seed soaking, seed coating) of a *P. fluorescens* strain in reducing the incidence of *Fusarium* wilt, *Verticillium* wilt and early blight of tomato, and the effect on plant growth. In the infested control treatments, the percentage of infested plants were 100, 92 and 89% for *F. oxysporum* f. sp. *lycopersici*, *V. dahliae* and *A. solani*, respectively, after 9 weeks of experimentation. By comparison, the percentage of infested plants in the treatment of infestation with *F. oxysporum* f. sp. *lycopersici*, *V. dahliae* or *A. solani* and seed coating with *P. fluorescens* were 32.4, 42.6 and 0.0%, respectively. In addition, the authors reported that plant growth improved after coating tomato seeds with the *P. fluorescens* strains. Tran *et al.* [56] applied cell suspensions of *P. fluorescens* SS101 to leaves of tomato plants 1 day before inoculation with zoospores of *P. Infestans*. They observed a substantially reduced incidence of tomato late blight. These authors informed that the area of the few lesions observed on leaves treated with strain SS101 was significantly smaller than that of leaves inoculated only with zoospores of *P. infestans* (infestation treatment). In the infestation treatment, disease severity increased exponentially over a period of 9 days after zoospore inoculation, whereas lesion area remained very low for leaves treated with strain SS101. This study also showed that *P. fluorescens* strain SS101 significantly reduced the expansion of existing late blight infections and the formation of sporangia. Srinivasan *et al.* [57] evaluated the efficacy of *P. putida* FC-6B, *Pseudomonas* sp. FC-7B, *P. putida* FC-8B, *Pseudomonas* sp. FC-9B and *Pseudomonas* sp. against wilt of tomato caused by *F. oxysporum* f. sp. *lycopersici*. They reported that the lowest disease incidence (3.3 in a scale of 0-100) was obtained with a single application to soil of *P. putida* FC-6B at 10^8 CFU ml^{-1} and two applications of *P. putida* FC-6B and FC-8B (10^7 CFU ml^{-1}). They also observed that bacterial treatment, especially with *P. putida* FC-6B and FC-8B, significantly increased plant growth when compared with untreated and healthy controls. Validov *et al.* [58] tested *P. putida* strain PCL1760 for effective biocontrol against tomato foot and root rot caused by *F. oxysporum* f. sp. *radicis-lycopersici*. These authors reported that *P. putida* strain PCL1760, under laboratory conditions, reduced TFRR symptoms by 45%

to 83% in the experiments with statistically significant biological control. Under greenhouse conditions, *P. putida* strain PCL1760 also showed significant biological control. In these experiments, the disease incidence decreased from 1.5 in the negative control to 0.8 (according to the disease severity index used by these authors, healthy plants were given a value of 0, plants with small lesions were given a value of 1, and plants with developed lesions were given a value of 2. The value for dead plants was 3). In addition, the percentage of germinated seeds was 57% in the presence of Forl, but the addition of both Forl and *P. putida* PCL1760 resulted in a value of 76% germinated seeds. Finally, these authors proved that addition of strain *P. putida* PCL1760 decreased the amount of Forl DNA in plant material approximately 5-fold in comparison with the amount found in the negative control. More recently, De Curtis *et al.* [27] reported that *Pseudomonas* sp. T4B-2A was capable of markedly inhibiting the growth of the causal agents of tomato damping-off, *S. rolfsii* and *R. solani in vitro*. In growth chamber experiments, *Pseudomonas* sp. T4B-2A was ineffective against *S. rolfsii*, whereas it significantly reduced the incidence of *R. solani* by 47.4%, compared to the untreated controls. They also observed that strain T4B-2A was more effective against *R. solani* than a commercial biofungicide based on *Bacillus subtilis* (BSF4). In field experiments, *Pseudomonas* sp. T4B-2A reduced the incidence of *R. solani* by 62.7% and 59.2% and that of *S. rolfsii* by 59% and 73.2% in 2006 and 2007, respectively.

Contribution of *Pseudomonas* sp. PCI2 to the Biocontrol of *S. rolfsii* and *A. alternata* in Tomato Plants

We isolated a native strain from tomato rhizosphere, *Pseudomonas* sp. PCI2. Strain PCI2 was identified as a member of *P. putida* by means of amplification and sequencing of a 730 bp fragment from the 16S rRNA, which was aligned with sequences retrieved from the GenBank database during a phylogenetical study. Strain PCI2 exhibited inhibition of the fungal phytopathogens *S. rolfsii* and *A. alternata in vitro*. In addition, *Pseudomonas* sp. PCI2 was found to solubilize P, to synthesize indole-3-acetic acid (IAA), to produce siderophores and to exhibit chitinolytic activity. Finally, we evaluated the potential of *Pseudomonas* sp. PCI2 to control tomato damping-off caused by *S. rolfsii*. Plastic pots were filled with 600 g of sterile mixture (soil:sand:perlite at 2:1:1 w/w/w). Each pot was then moistened with sterilized distilled water and infested in the mixture surface with 30 mg of sclerotia. Pots were kept for 8 days in a growth chamber under controlled conditions: 16 h light at 28 ± 2 °C, 8 h dark at 20 ± 2 °C (light intensity of 220 $\mu E\ m^{-2}\ s^{-1}$). After incubation, tomato seeds were soaked for 30 min in a 10^9 CFU ml^{-1} aqueous cell suspension of *Pseudomonas* sp. PCI2 and sown into the infested pots at eight seeds per pot. Damping-off was determined by counting the total healthy stand after 40 days, compared to non-infested control plants. Shoot and root dry weights (72 h at 70°C) were also determined (Table 1). In *S. rolfsii*-infested soil mix, inoculating tomato seeds with *Pseudomonas* sp. PCI2 improved seedling stand by 29% and significantly increased shoot and root dry weight of plants over the untreated pathogen controls by 84.7 and 59.9 mg, respectively. No evident differences between bacterized and control seeds were observed in non-infested potting mixture when recording plant stand [59].

Table 1. Biocontrol activity of strain PCI2 against *S. rolfsii*

Treatment	Root dry weight (mg)[a]	Shoot dry weight (mg)[a]
Non-infested, non-bacterized healthy control	123.2 ± 17b	207.3 ± 11b
Infested with *S. rolfsi,* non-bacterized control	86.7 ± 15.2c	133.5 ± 14.8c
Infested with *S. rolfsii* and bacterized with PCI2	146.6 ± 10.1b	218.2 ± 18.3b
Bacterized with PCI2 alone	195 ± 11a	240.9 ± 12.8ab

[a] Data represent the average of two experiments ± standard deviation. Values in each column with different letters are significantly different according to the LSD test ($P<0.05$).

Results from this study were similar to those reported by Lewis *et al.* [60] and Jetiyanon [61]. Lewis *et al.* [60] reported that, depending on the formulation used, the biocontrol fungus *Gliocladium virens* increased the plant stand of bean by 42% after 35 days of growth in soil infested with *S. rolfsii* and inoculated with *G. virens*, as compared to plants growing in infested soil. On the other hand, Jetiyanon [61] reported that inoculation of tomato and pepper plants with a mixture of *Bacillus* strains reduced the number of plants affected by *S. rolfsii* by 46%. The tested mixture of *Bacillus* strains suppressed the disease by activating an increase in the activity of the enzymes superoxide dismutase (SOD) and peroxidase (PO), which are associated with Induced Systemic Resistance (ISR).

We also investigated the ability of *Pseudomonas* sp. PCI2 to control leaf spot of tomato plants caused by *A. alternata.* A number of studies demonstrated biological control of this phytopathogen in tomato at a postharvest stage [62, 63]. To our knowledge, few studies have adressed the effect of a biocontrol agent against *A. alternata* at previous growth stages of tomato plants. In our study, the aerial parts of the plants were atomized with a suspension of strain PCI2. The plants were incubated for 24 h under controlled conditions (as described above). Then, leaves were inoculated on their abaxial surfaces with two droplets of conidial suspension, at both sides of the midrib. We observed that previous inoculation of plants with *Pseudomonas* sp. PCI2 decreased the percentage of infected leaves by 24%, as compared to non-bacterized/infested controls (Figure 1).

Figure 1. Tomato leaf showing leaf spot disease by *A. alternata*.

Sid *et al.* [30] reported on the selection of four *Bacillus* sp. strains for the biocontrol of leaf spot of pepper caused by *A. alternata*. These authors observed that the selected strains reduced leaf spot *in vivo* between 53 and 74%. In our study, we observed that the percentage of infected leaves was reduced by 24% compared with non-inoculated plants infested with *A. alternata*. Sid *et al.* [30] did not investigate the precise mechanism responsible for the bacterial biocontrol against *A. alternata*.

On the other hand, Kishore *et al.* [64] applied the chitinolytic strains *B. circulans* GRS 243 and *Serratia marcescens* GPS 5 by foliar spray to control late leaf blight (LLS) of peanut caused by *Phaeoisariopsis personata*. These authors observed that the chitin-supplemented foliar application of chitinolytic strains resulted in an optimized biological control of the disease, compared with the application of the strains alone.

The chitin-supplemented foliar application was also effective for control of leaf blight of peanut when using a chitinolytic strain of *B. cereus* [65]. Finally, Lanna Filho *et al.* [66] reported on the effective control of early blight caused by *A. solani* by previous spraying of tomato leaves with strains of *Paenibacillus macerans* or *Bacillus pumilus*. These bacteria reduced disease severity by almost 50%, compared to control plants. These authors hypothesized that antagonists may have acted through multiple mechanisms such as competition for space and nutrients, antibiosis, and, possibly, ISR.

CONCLUSION

The results presented in this chapter suggest that *Pseudomonas* sp. PCI2 could contribute to control tomato damping-off caused by *S. rolfsii* and tomato leaf spot disease caused by *A. alternata*. However, further studies are needed to analyze the effect of varying the initial concentration of the bacterial inoculum. On the other hand, more knowledge on the mechanism of action of PCI2 and the best form of introduction into the host is needed to improve the efficacy of the strain.

In addition, further work is underway in order to investigate the effect of PCI2 in the system tomato-Forl and to study other characteristics of the strain such as production of the phytohormones salicylic acid (SA), jasmonic acid (JA) and ethylene. These studies will contribute to the knowledge on the behavior of PCI2 and on the potential practical application of this strain for an integrated management of tomato diseases.

ACKNOWLEDGMENTS

This work was supported by grants from Secretaría de Ciencia y Técnica, Universidad Nacional de Río Cuarto (Córdoba, Argentina), Agencia Nacional de Promoción Científica y Tecnológica (Secretaría de Ciencia y Técnica de la Nación) and Consejo Nacional de Investigaciones Científicas y Técnicas (CONICET, Argentina).

REFERENCES

[1] FAO, 2003. *World Agriculture Information Center Database.* Rome, Italy. Available from: http://www. Fao.org/waicent/portal/ statistics_en.asp.

[2] Naika, S., van Lidt de Jeude, J., de Goffau, M. Hilmi, M. & van Dam, B. Agromisa Foundation and CTA, Wageningen, Netherlands. Available from: http:// journeytoforever.org/farm_library/AD17.pdf (2005).

[3] Nakama, M. & Fernández Lozano, J. INTA. Available from: http://www. mercadocentral.com.ar/site2006/publicaciones/boletin/pdf/Tomate13.pdf (2008).

[4] Corvo Dolcet, S. Available from: http://www.seedquest.com/News/release/2005/ pdf/13528.pdf (2003).

[5] Anjaiah, V. *Kluwer Academic Publishers*. The Netherlands pp. 453-500 (2004).

[6] Gleason, M. & Edmunds, B.A. Available from: http://www.extension. iastate.edu/ publications/pm1266 pdf (2006).

[7] van West, P., Morris, B.M., Reid, B., Appiah, A.A., Osborne, M.C., Campbell, T.A., Shepherd, S.J. & Gow, N.A.R. *MPMI 15*, 790–798 (2002).

[8] Nelson, P.E., Toussoun, T.A. & Cook, R.J. *Pennsylvania State University Press*, University Park, P.A. (1981).

[9] Pitt, J.I. *Med. Mycol. 38*, 17-22 (2000).

[10] Jaramillo, J., Rodríguez, V.P., Guzmán, M., Zapata, M. & Rengifo, T. FAO. Available from: http://www.fao.org/docrep/010/a1374s/a1374s00.htm (2007).

[11] Srivastava, R., Khalid, A., Singh, U.S. & Sharma, A.K. *Biol. Control 53*, 24–31 (2010).

[12] Jarvis, W.R. & Shoemaker, R.A. *Phytopathology 68*, 1679-1680 (1978).

[13] Jones, J.B., Jones, J.P., Stall, R.E. & Zitter, T.A. *The American Phytopathological Society*, St. Paul, MN, USA (1991).

[14] Kamilova, F., Kravchenko, L.V., Shaposhnikov, A.I., Makarova, N. & Lugtenberg, B. *MPMI 19*, 1121–1126 (2006).

[15] Kim, J.T., Park, I.H., Hahm, Y.I. & Yu, S.H. *Plant Pathol.* J. *17*, 290-294 (2001).

[16] Rowe, R.C. *Phytopathology 70*, 1143-1148 (1980).

[17] Menzies, J.G., Koch, C. & Seywerd, F. *Plant Dis. 74*, 569-572 (1990).

[18] Datar, V.V. & Mayee, C.D. *Indian Phytopathol. 34*, 191-195 (1981).

[19] Rotem, J. The olcani center Bet Dagan. Israel. *The american Phytopathological Society*, St. Paul, Minnesota. 336p. (1994).

[20] Mathur, K. & Shekhawat, K.S. *Indian J. Mycol. Plant Pathol. 16*, 235-236 (1986).

[21] Jabnoun-Khiareddine, H., Daami-Remadi, M., Ayed, F., Jebari, H. & El Mahjoub, M. *The Afr. J. Plant Sci. Biotechnol. 1*, 10-15 (2007).

[22] Wydra, K. Annual Report IITA, Ibadan, 1995:68 (1996).

[23] Aycock, R. North Carolina Agricultural Experiment Station Technical Bulletin No. 174. Florida, USA (1966).

[24] Tindall, H.D. London. *Macmillan Press*, pp. 506-7 (1983).

[25] Chamswarng, C., Sangkaha, K. & Kateprasard, N. *Kasetsart J. (Nat. Sci. Suppl.) 26*, 25-29 (1992).

[26] Flores-Moctezuma, H.E., Montes-Belmont, R., Jiménez-Pérez, A. & Nava-Juárez, R. *Crop Prot. 25*, 95-201 (2006).

[27] De Curtis, F., Lima, G., Vitullo, D. & De Cicco, V. *Crop Prot. 29*, 663–670 (2010).

[28] Sikora, E.J. Available from: http://www.aces.edu/pubs/docs/A/ANR-0797/ (2009).

[29] Gilchrist, D.G. & Grogan, R.G. *Phytopathology 66*, 165-177 (1975).

[30] Sid, A., Ezziyyzni, M., Egea-Gilabert, C. & Candela, M.E. *Biologia Plantarum 47*, 569-574 (2003).

[31] Agrios, G. Fitopatología. *Editorial Limusa*, 756p. (1991).

[32] Galaz Delgado, J.M. Available from: http://www.bionativa.cl/pdf/tesis/trichonativa/t14.pdf (2001).

[33] Davis, R.M., Miyao, E.M., Mullen, R.J., Valencia, J., May, D.M. & Gwynne, B.J. *Plant Dis. 81*, 601-603 (1997).

[34] Gerhardson, B. *Trends Biotechnol. 20*, 338-343 (2002).

[35] Cook, R.J. *Annu. rev. Phytopathol. 31*, 53-80 (1993).

[36] Pal, K.K. & McSpadden Gardener, B. *The Plant Health Instructor* DOI: 10.1094/PHI-A-2006-1117-02 (2006).

[37] Ramamoorthy, V., Raguchander, T. & Samiyappan, R. *Eur. J. Plant Pathol. 108*, 429-441 (2002).

[38] Haas, D. & Défago, G. *Nature Rev. Microbiol. 3*, 307-319 (2005).

[39] Botelho, G.R. & Hagler, L.C.M. *Braz. J. Microbiol. 37*, 401-416 (2006).

[40] Weller, D.M. *Phytopathology 97*, 250-256 (2007).

[41] Couillerot, O., Prigent-Combaret, C., Caballero-Mellado, J. & Moënne-Loccoz, Y. *Lett. Appl. Microbiol. 48*, 505-512 (2009).

[42] Walsh, U.F., Morrisey, J.P. & O´Gara, F. *Curr. Opin. Biotechnol. 12*, 289-295 (2001).

[43] Ongena, M., Duby, F., Rossignol, F., Fauconnier, M.L., Dommes, J. & Thonart, P. *MPMI 17*, 1009-1018 (2004).

[44] Loper, J.E. & Buyer, J.S. *MPMI 4*, 5-13 (1991).

[45] Friedlander, M., Inbar, J. & Chet, I. *Soil Biol. Biochem. 25*, 1211–1221 (1993).

[46] Commare, R.R., Nandakumar, R., Kandan, A., Suresh, S., Bharathi, M., Raguchander, T. & Samiyappan, R. *Crop Prot. 21*, 671-677 (2002).

[47] Fernando, W.G.D. & Linderman, R. *Biol. Agric. Hort. 12*, 1-14 (1994).

[48] Raaijmakers, J.M., Weller, D.M. & Thomashow, L.S. *Appl. Environ. Microbiol. 63*, 881–887 (1997).

[49] Ramette, A., Frapolli, M., Défago, G. & Möenne-Loccoz, Y. *MPMI 16*, 525–535 (2003).

[50] Timms-Wilson, T.M., Ellis, R.J., Renwick, A., Rhodes, D.J., Mavrodi, D.V., Weller, D.M., Thomashow, L.S. & Bailey, M.J. *MPMI 13*, 1293-1300 (2000).

[51] Larkin, R.P. & Fravel, D.R. *Plant Dis. 82*, 1022-1028 (1998).

[52] Sharma, V.K. & Nowak, J. *Can. J. Microbiol. 44*, 528-536 (1998).

[53] Chin-A-Woeng, T.F.C., Bloemberg, G.V., van der Bij, A.J., van der Drift, K.M.G.M., Schripsema, J., Kroon, B., Scheffer, R.J., Keel, C., Bakker, P.A.H.M., Tichy, H.-V., de Bruijn, F.J., Thomas-Oates, J.E. & Lugtenberg, B.J.J. *MPMI 11*, 1069-1077 (1998).

[54] Dekkers, L.C., Mulders, I.H.M., Phoelich, C.C., Chin-A-Woeng, T.F.C., Wijfjes, A.H.M., & Lugtenberg B.J.J. *MPMI 11*, 1177-1183 (2000).

[55] El-Rafai, I.M., Asswah, S.M.W. & Awdalla. O.A. *Pakistan J. Biol. Sci. 6*, 399-406 (2003).

[56] Tran, H., Ficke, A., Asiimwe, T., Höfte, M. & Raaijmakers, J.M. *New Phytol. 175*, 731-742 (2007).

[57] Srinivasan, K., Gilardi, G., Garibaldi, A. & Gullino, M.L. *J. Plant Pathol. 91*, 147-154 (2009).

[58] Validov, S.Z., Kamilova F. & Lugtenberg B.J.J. *Biol. Control 48*, 6–11 (2009).

[59] Pastor, N.A., Reynoso, M.M., Tonelli, M.L., Masciarelli, O., Rosas, S.B. & Rovera, M. *J. Plant Pathol. 92*, 737-745 (2010).

[60] Lewis, J., Papavizas, G.C. & Hollenbeck, M. *Biol. Control 3*, 109-115 (1993).

[61] Jetiyanon, K. *Biol. Control 42*, 178-185 (2007).

[62] Chalutz, E., Ben-Arie, R., Droby, S., Cohen, L., Weiss, B., Wilson, C.L. *Phytoparasitica 16*, 69-75 (1988).

[63] Wang, Y., Bao, Y., Shen, D., Feng, W., Yu, T., Zhang, J. & Zheng, X.D. *Int. J. Food Microbiol. 123*, 234-239 (2008).

[64] Kishore, G.K., Pande, S., & Podile, A.R. *Biol. control 95*, 1157-1165 (2005).

[65] Kokalis-Burelle, N., Backman, P.A., Rodriguez-Kabana, R. & Ploper, L.D. *Biol. Control 2*, 321-328 (1992).

[66] Lanna Filho, R., da Silva, R.R. & Alves, E. *Pesq. Agrop. Bras. 12*, 1381-1387 (2010).

In: Tomatoes: Cultivation, Varieties and Nutrition
Editor: Tadahisa Higashide

ISBN: 978-1-62417-915-0

Chapter 7

BEGOMOVIRUS DISEASE: THE MOST DEVASTATING DISEASE OF CULTIVATED TOMATO

Walter N. Leke[1*], Abwe N. M. Dione[2], Sali A. Ndindeng[1], and Noé Woin[1]

[1]Institute of Agricultural Research for Development (IRAD), Yaounde, CR, Cameroon

[2]Institute of Agricultural Research for Development (IRAD), Ekona, SWR Cameroon

ABSTRACT

Recently, several begomoviruses (genus *Begomovirus*; family *Geminiviridae*) have been associated with leaf curl and yellow leaf curl symptoms in tomato crops worldwide. These viruses are transmitted by the whitefly *Bemisia tabaci* (Genn.) sibling species group, and they have a genome of single stranded (ss) DNA (about 2.7-2.8 kb) with an organization typical for Old World monopartite begomoviruses. In addition, many monopartite begomoviruses are associated with alphasatellite and betasatellite molecules that may influence symptom development, pathogenicity, and virulence. Wild, uncultivated plant species are commonly infected with viruses, and may serve as reservoirs of viruses also pathogenic to cultivated plants, and can exert dynamic forces on the ecology of natural systems.

INTRODUCTION

Tomato (*Solanum lycopersicum*) is an important vegetable crop worldwide. Recently, several begomoviruses (genus Begomovirus; family Geminiviridae) have been associated with tomato plants of all known cultivars showing symptoms such as but not limited to leaf curling and yellow leaf curling [1, 2, 3, 4, 5, 6]. A review by [7] emphasized the urgent need for holistic research efforts to ease the control of the diseases caused by geminiviruses. This

* Corresponding author: Email: lekwat@yahoo.com.

notwithstanding, there has been a real increase in the emergence of whitefly-transmitted geminiviruses, (Family *Geminiviridae*, Genus *Begomovirus*) and their vector *Bemisia tabaci* sibling species group [8, 9, 10]. Although the diseases caused by geminiviruses represent a serious constraint to agriculture, little was known about the causal agents of the diseases until the isolation of virus particles with unique twinned quasi-isomeric morphology associated with maize streak and beet curly top diseases [11]. This attribute provided the name geminivirus, from Gemini, the sign of the zodiac symbolized by twins [12], and has remained a unifying feature of this family of viruses. Structural analysis demonstrated that the 22 x 38 nm particles associated with MSV consist of two incomplete $T = 1$ icosahedra [13] and a similar structure was subsequently observed for *African cassava mosaic virus* (ACMV) [14]. In groundbreaking research, [12] and [15] demonstrated that the so-called geminate particles associated with ACMV, MSV and *Bean golden mosaic virus* (BGMV) contained circular ssDNA, and this genomic DNA was infectious when re-introduced to plants by mechanical inoculation [16], setting geminiviruses apart from all other plant viruses that had been characterized at that time. Evidence was provided for BGMV and *Tomato golden mosaic virus* (TGMV) to suggest that at least some genimiviruses had divided genomes [17, 18, 19]. Shortly afterwards, the nucleotide sequence of cassava latent virus (subsequently renamed ACMV) was established, and infectious clones were used to demonstrate a bipartite genomic structure [20, 21]. Subsequently, the monopartite geminiviruses MSV, BCTV and *Tomato pseudo-curly top virus* (TPCTV) were similarly characterized [22, 23], resulting in the present-day recognition of four genera (*Mastrevirus*, *Begomovirus*, *Curtovirus* and *Topocuvirus*) in the family *Geminiviridae* by International committee on the taxonomy of viruses, ICTV [24] Viruses of the *Geminiviridae* are characterized by having either one or two small, circular, ssDNA components of ~2.6-3.0 kb encapsidated in paired icosahedral or 'geminate' particles (22 x 38 nm) [13]. The viruses in the family *Geminiviridae* replicate in the host cell nucleus and are transmitted in a non-propagative, persistent and circular manner by insect vectors and have the propensity to infect phloem cells [7, 25]. The components share a common region that contains motifs required for the control of gene expression and replication, notably conserved reiterated motifs and a putative stem-loop structure containing the highly conserved nonanucleotide sequence, TAATATTAC that functions in the initiation of rolling circle replication [26]. There are now more than 200 officially recognized geminivirus species [27]. Members of the family *Geminiviridae* are classified into four genera based on their genome organization, hosts and insect vectors [24, 27]. Mastreviruses (type species MSV) are transmitted by leafhoppers and have one genome component. They have been found only in the Eastern Hemisphere (Europe, Africa, Asia and Australia) and have been associated with many diseases of monocot plants though some mastreviruses have also been associated with diseases of dicot plants [24]. Their genome has four genes: *V1* and *V2* encode the coat protein (CP) and the movement protein (MP), respectively, on the virion strand. The replication-associated proteins: Rep (C1:C2) and RepA (C1) are encoded by the complementary sense strand. The genes on the virion and complementary strands are separated by the long intergenic region (LIR) and the short intergenic region (SIR). Topocuviruses (type species TPCTV) have a single genome component of ~3 kb and are transmitted by treehoppers. Topocuviruses have only been reported in the Americas (New World, NW) and are thought to have emerged as a recombination between a mastrevirus and a begomovirus [28]. Curtoviruses (type species *Beet curly top virus*, BCTV) are similar to topocuviruses in terms of the genome size of ~3 kb, but are transmitted by leafhoppers and infect exclusively dicot plants [24]. The economically most important, geographically widespread and numerous geminiviruses are within the genus *Begomovirus* (type species BGMV), which contains more than 200 species [27]. Begomoviruses are vectored by whitefly *B. tabaci* [7, 8, 9, 25, 29] and infect only dicot plants. Begomoviruses contain a circular, single-stranded (ss) DNA genome comprising either one (~2.8 kb) (monopartite) or two components (~2.6 kb each) (bipartite) [24, 30]. The single DNA component of monopartite begomoviruses contains six open reading frames (ORFs). The coat protein gene (cp or V1) and the V2 ORF are expressed from the viral sense strand, and the rep or C1 ORF, which encodes the Rep-initiator protein, and the C2, C3, and C4 ORFs, which are involved in movement and pathogenicity, are expressed from the complementary strand [30, 31]. The genome organization of begomoviruses is illustrated below (Figure 1). The earliest known record of a plant virus disease in 752 AD is considered to have been caused by a begomovirus [32]. However, in the past two decades, begomoviruses have become one of the most important group of plant viruses in tropical and subtropical regions, causing devastating yield losses of many crops, including bean, cassava, cotton, cucurbits, and tomato [25, 33] (Figures 2). Agricultural intensification has been proposed as one of the main causes for begomovirus disease outbreaks [25, 33, 34]. Agricultural intensification has also been responsible for the large crop yield increases observed since the 1960s and has involved the application of fertilizers, pesticides, and/or irrigation to cropping systems based increasingly on extensive monocultures [35, 36]. Such practices have led to increases in vector populations worldwide and so to associated begomovirus disease epidemics. Other underlying factors influencing the emergence of begomovirus epidemics have been the evolution of more aggressive virus variants, increased movement of infected planting material, and the introduction of more vulnerable crop plant varieties [25].

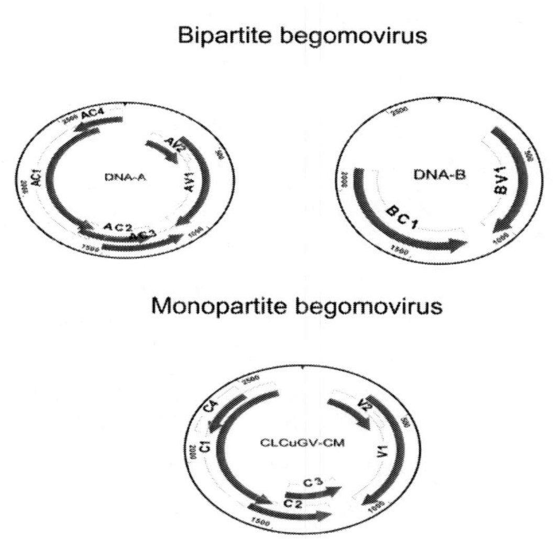

Figure 1. Genome organisation of members of the genus Begomovirus. Arrows represent open reading frames. *V1/(AV1)* coat protein gene, *V2/AV2* precoat prorein gene, *C1* (*AC1*) replication initiator protein gene, *C2*(*AC2*) transcription activator protein gene, *C3* (*AC3*) replication enhancer protein gene, *C4* (*AC4*) symptom determinant protein gene, *BV1* nuclear shuttle protein gene and *BC1* movement protein gene.

Figure 2. Tomato plants in a farmers' fields in southwestern Cameroon with characteristic yellow leaf curl (A) and leaf curl symptoms (B).

THE VECTOR OF BEGOMOVIRUSES

Whitefly *B. tabaci* (Figure 3) is a widespread phloem-feeding pest of dicotyledonous crops in the tropical and subtropical regions of Africa, America, Asia, Europe and Oceania and can transmit more than 200 different plant viruses responsible for the loss of millions of hectares of valuable food and industrial crops [31, 33, 37]. The *B. tabaci* complex colonizes more than 600 host plant species, belonging to 77 families [38]. Almost all of the plant viruses transmitted by *B. tabaci* are begomoviruses. *B. tabaci* is the only known vector species for begomoviruses, and hence the global distribution of begomoviruses is closely related to that of this species. The prevailing *B. tabaci* biotypes present in a region will affect the begomoviruses present. Most biotypes can transmit a range of begomoviruses, but they do so with very differing efficiencies depending on both virus species and bioype [39, 40]. Increase in populations of *B. tabaci* is a general characteristic of begomovirus disease epidemics, and facilitates the spread of begomoviruses into and within crops, and their transmission to and from weed hosts. Increased populations of *B. tabaci* are associated with a range of factors including conducive climatic conditions [41], the spread of the more fecund B-biotype [42, 43], the cultivation of particular crops or varieties [25, 36, 44], and virus infection of the host [45]. Begomoviruses are transmitted by *B. tabaci* in a circulative and persistent manner. Except for *Tomato yellow leaf curl Sardinia virus* (TYLCSV) whose DNA was detected in eggs, nymphs, and adults, of the first generation progeny [46], no replication of begomoviruses has been found in *B. tabaci.* Apart from causing damage to plants by the begomoviruses that they transmit, *B. tabaci* also causes direct damage to crops by inducing noticeable physiological disorders (irregular ripening, silver leaf, and severe chlorosis) [47, 48]. Also, the large populations of *B. tabaci* observed on suitable hosts induce the growth of superficial fungi (sooty molds) on the honeydew excreted by immature and adult whiteflies.

These fungi greatly reduce the photosynthetic capacity of the plants and may eventually lead to the death of the plants and total crop failure.

Figure 3. The infestation of a tomato plant in Cameroon by Bemisia tabaci which is the insect vector of begomoviruses.

DNA Satellites Associated with Begomovirus

Satellites are defined as viruses or nucleic acids that depend on the helper virus for their replication but lack extensive nucleotide sequence homology to the helper virus and are dispensable for its proliferation [49]. Satellite viruses encode a structural protein that encapsidates its own nucleic acid while satellite nucleic acids rely on the helper virus structural protein for encapsidation and do not necessarily encode additional non-structural proteins. A third type of agent, referred to as satellite-like nucleic acid, also depends on the helper virus for its replication but provides a function that is necessary for the biological success of the helper virus and is therefore considered as part of the helper virus genome. The first satellite RNA was identified in 1969 in association with the nepovirus *Tobacco ringspot virus* [6], and since then a large number of satellite RNAs, associated with several groups of plant viruses have been reported [49], with the majority of the satellites interfering with the replication of the helper viruses, resulting in attenuated symptoms. Some satellites exacerbate disease symptoms induced by the helper virus or produce novel symptoms which are usually not associated with the helper virus infections [50].

The first begomovirus satellite DNA to be discovered was found to be associated with the monopartite begomovirus *Tomato leaf curl virus* (ToLCV) from Australia [51]. This 682 nt circular ssDNA depends on ToLCV for its replication and encapsidation, but its replication can also be supported by other begomoviruses. It has no proven effects on the viral replication or on symptoms caused by ToLCV. It has no extensive ORFs and has little sequence similarity to its helper virus (ToLCV) except for the nonanucleotide TAATATTAC sequence present in the stem loop of all geminiviruses [52]. Failure to reproduce yellow vein symptoms in *Ageratum conyzoides* (goat weed) by re-introduction of *Ageratum yellow vein virus* (AYVV) [53, 54], another monopartite begomovirus isolated from goat weed showing yellow vein symptoms, which was otherwise shown to be infectious in *Nicotiana benthamiana* [55], suggested that another factor was required to restore pathogenicity in the natural host. In a search for additional viral components, a number of small circular recombinant components, each containing the AYVV origin of replication together with sequences of unknown origin, were isolated from infected goat weed [56]. Similar recombinants were also identified for the begomoviruses associated with cotton leaf curl disease (CLCuD) [57, 58, 59]. The significance of the unidentified sequences within the recombinants was not appreciated at the time, but they were to provide a vital clue in the discovery of a new class of satellites. As was observed for AYVV, the cloned genomic component of Cotton leaf curl virus (renamed *Cotton leaf curl Multan virus* (CLCuMV)) failed to produce typical CLCuD symptoms, suggesting the presence of another factor [58], whose search resulted in the isolation of a small circular ssDNA, referred to as DNA-1 [60], that is a representative of a new class of components associated with monopartite begomoviruses [61].

Recently, many monopartite begomoviruses have been identified that associate with a type of satellite molecule referred to as betasatellite, composed of ssDNA, ~1.3 kb in size and approximately half the size of the helper begomoviruses. Many of the betasatellites are required for typical disease symptom development [32, 54, 59, 62, 63, 64]. Despite their recent discovery, betasatellites may have existed for many centuries, e.g., EpYVB in association with EpYVV were later on found to cause EpYVD that was described about 1250 years ago [32]. Betasatellites all require a helper begomovirus for replication, local and systemic spread, and whitefly vector-mediated transmission, and some have been shown to modulate symptom severity [64]. The betasatellite molecule contains one ORF (*βC1*) (Figure 4), an A-rich region ~240 nts long and a satellite-conserved region (SCR), of ~220 nts. The *βC1* has in one instance been shown to be responsible for the suppression of jasmonic acid signaling involved in at least one gene silencing pathway [65]. The transgenic *Arabidopsis thaliana* expressing *β*C1 of *Tomato yellow leaf curl China virus* (TYLCCNV) was shown to develop disease symptoms like that observed in begomovirus-infected tobacco plants, in that plants exhibited upward leaf curling, foliar enations, and sterile flowers [66]. Apart from the nonanucleotide sequence, betasatellites do not share any significant sequence similarity with the helper begomoviruses.

The begomovirus/betasatellite complexes are often associated with a second type of circular ssDNA satellite, initially referred to as DNA-1 [53, 60, 61, 67], but now called alphasatellites [68]. Alphasatellites encode a single protein that shares high nt identity with the replication-associated (activator) protein (Rep) (Figure 4), a rolling-circle replication initiator protein encoded by viruses in the genus *Nanovirus*, family *Nanoviridae* that also have a genome of circular ssDNA [69]. Consequently, alphasatellites are capable of autonomous replication, but require a helper begomovirus for spread in plants and for whitefly vector transmission. In addition to Rep, alphasatellites also have an A-rich region, ~200 nts long, downstream of the Rep-encoding region. Recently, it has been demonstrated that the Rep of the alphasatellite associated with *Tobacco curly shoot virus* (TbCSV) can be used as a virus-induced gene silencing vector (VIGS) [70]. In contrast to betasatellites, alphasatellites possess in their stem loop the nonanucleotide sequence, TAGTATTTAC also found in the stem loop of viruses in the family *Nanoviridae*. Alphasatellites can affect both begomovirus titer and symptom development in host plants [53, 71]. Initially it was thought satellite molecules were limited to the OW, but recently, alphasatellites have been found associated with NW begomoviruses, thus expanding the geographical distribution of satellite molecules associated with begomoviruses.

THE ROLE OF WEEDS (UNCULTIVATED PLANT SPECIES) IN THE SPREAD OF BEGOMOVIRUS DISEASE

Wild, uncultivated plant species are commonly infected with viruses, and may serve as reservoirs of viruses also pathogenic to cultivated plants [72], and can exert dynamic forces on the ecology of natural systems [73]. Ageratum conyzoides (family; Asteraceae) for example, is widely distributed in tropical and sub-tropical regions throughout the world. In Southeast Asia, yellow vein symptoms are commonly observed in A. conyzoides, which have been associated with a complex consisting of the begomovirus Ageratum yellow vein virus (AYVV), the betasatellite Ageratum yellow vein betasatellite (AYVB) and several different types of alphasatellite molecules [53, 74]. In China and Indonesia, additional begomoviruses and DNA satellites have been identified in A. conyzoides plants showing yellow vein or leaf curl symptoms [3, 70, 75, 76, 77, 78].

CONTROL OF BEGOMOVIRUSES

Attempted control measures to reduce losses caused by begomoviruses, their satellite DNAs and their vector have been described [79, 80]. Control measures seek to reduce virus inoculums and vector population levels by biological, chemical, cultural and physical means, and by growing crop varieties with at least some degree of resistance or tolerance to virus or vector.

Studies on plant virus evolution support the use of control measures to reduce vector and/or virus populations. Epidemiological models suggest a high density of vectors can favor the evolution of higher virulence in virus populations [81]. Higher vector populations will increase transmission rates and decrease the effects of genetic drift, while increasing mixed virus infections and effects thereof.

A reduction in virus inoculums in host plants will assist control of begomovirus diseases by limiting the diversity of the virus populations [82], because large virus populations have more evolutionary potential than small populations. Genetic drift will also have less effect on large rather than small populations. Similarly, viruses that are able to maintain high titers and incidence year-round may show a great evolutionary potential, as favorable mutations will have time to accumulate before a genetic bottleneck being encountered.

1. Continuous versus Discontinuous Cropping Systems

Successful and sustained control of begomoviruses has been achieved only rarely, and in general only in cropping systems in which there is a temporal discontinuity in the cultivation of host plants supporting virus infection. In this chapter these are called discontinuous cropping systems. *Tomato mottle virus* (ToMoV) has been controlled in Florida assisted by this virus having no significant weed hosts and tomato production not occurring year-round [47]. Production is synchronized and growers are able to maintain a tomato-free period before the growing season. Moreover, transplants are obtained from commercial areas distant from sources of infection. Such systems are not feasible for most tropical agroecosystems, because even if reservoirs of infection are known, their removal will be difficult. For example, in southern India, it is impossible to achieve a whitefly host-free period, because of the occurrence of at least 173 plant species reported as *B. tabaci* hosts [83]. Similarly in Pakistan, 229 *B. tabaci* host plant species, belonging to 48 families, are known [84]. Although not all hosts are present year-round, and different hosts may select for different *B. tabaci* genotypes, the sheer number of hosts results in the ability for some *B. tabaci* populations to survive well throughout the year.

Further challenges faced in continuous cropping systems are that their prevailing climates are those which favor the vector, *B. tabaci* [42]. Moreover, the ease with which begomoviruses can adapt to new hosts also contributes to the control of begomovirus diseases being most difficult in tropical climates, which support a greater diversity of potential plant hosts [85]. For example, *Tomato yellow leaf curl virus*-Isreal (TYLCV-Is) has persisted in Florida following its introduction in the 1990s despite a campaign to eradicate it [86]. In contrast to ToMoV, TYLCV-Is has a wider host range, and this may be why the use of the insecticide imidacloprid in Florida has had a much greater impact on ToMoV than TYLCV [2].

Continuous cropping systems in the tropics are also characterized by asynchronized planting, and hence crops are often vulnerable to infection at the seedling stage from neighboring older crops. [87] found tomato leave curl disease (ToLCVD) infection most rapid when new plantings were adjacent to older tomato crops. The incidence also increased more rapidly where tomato was grown continuously compared to where tomato was grown only once a year.

Discontinuous cultivation of tomatoes will provide more epidemiological and evolutionary bottlenecks than encountered in continuous cropping systems. These constraints are expected to occur on vector transmission to, and infection of, a new host or following seasonal population dynamics of the virus hosts and vectors. A reduction in viral population size will also reduce the chances for recombination events. They reported greater intra-species diversity and predicted recombination events in tomato begomoviruses grown under continuous (southern India) versus discontinuous cropping systems (northern India). Intra-species diversity of introduced TYLCV populations in Spain was also found to increase over time.

It is unlikely that smallholders in the tropics can introduce much disruption to the begomovirus and *B. tabaci* annual cycles, due to the many alternative wild and crop hosts for both virus and vector, and the difficulty in coordinating plant hosts where these are cultivated in neighboring small holdings of less than 1 ha. Farmers with sufficient resources, often use insecticides. Such treatments usually have only limited efficacy due to lack of control in neighboring plots, and the development of insecticide resistance in *B. tabaci*, which has been reported to all major chemical groups [88]. [89] highlight that for TYLCV a viruliferous adult can transmit the virus efficiently with a single phloem feeding of less than two minutes. Control measures must therefore include measures to disrupt virus acquisition and transmission as well as vector numbers. Consequently, the deployment of begomovirus-resistant lines has been viewed as the most desirable control option for many farmers, particularly those with limited resources.

2. Host Resistance

"A host plant is resistant if it can suppress the multiplication of a virus, and consequently suppress the development of disease symptoms." Resistance levels range from low to immunity, when the plant could be termed a non-host. In contrast, "Tolerance is a unique instance where in response to virus infection, the host expresses negligible or mild disease symptoms, but supports normal levels of virus multiplication" [90]. Susceptible plants are those in which symptoms are pronounced and yield losses high.

There are almost no non-transgenic sources of immunity, and only a few sources of resistance to begomoviruses, or *B. tabaci*, available in most host crops [90]. Nevertheless, there are many examples where resistance as part of an IPM strategy has brought epidemics under control, as in the cassava mosaic epidemics in Madagascar in the 1930s [91, 92], in Uganda in the 1990s [93, 94], and the current use of tomato lines resistant to tomato leaf curl viruses in India [95].

Cropping practices should incorporate available host resistances as part of an integrated approach so that the resistance remains as durable as possible. This applies to resistance genes introduced by traditional breeding and those introduced by transgenic means. A potential advantage of transgenic resistances are that this technology offers the ability to pyramid genes, and providing monocultures are avoided, may present a useful tool as part of an integrated control approach [82]. Also, some transgenic breeding lines resistant to begomovirus(es) might be less attractive to whitefly feeding as in the case of a transgenic luteovirus-resistant breeding line affected in its volatile production and with reduced attractiveness to aphids [96].

[82] proposed that quantitative resistance would be more effective against begomoviruses, as a breakdown in resistance would be more likely to be in "the gentler form of erosion." Their pest risk model indicates that pathogen populations exposed to strong directional selection evolve faster than those exposed to weaker selection factors such as quantitative resistance, or to disruptive selection that can be introduced through spatial or temporal patterning of the selective force. Gene rotation and cultivar mixtures need much more research, despite these practices having been known to offer promise for a long time [39].

Intercropping with other crops non host to *B. tabaci* is also known to be useful for reducing the disease incidence in crops such as tomatoes which are usually grown as monocultures, but only little progress has been made in incorporating it in a rational way in cropping systems [39, 97]. Intercropping decreased whitefly numbers and the spread of tomato yellow leaf curl disease [80]. Its effect on the genetic diversity of virus populations remains, however, unknown and studies are needed to determine whether it increases the chances of mixed infections and hence recombinations. Intercropping is known to influence the feeding behavior of vectors and the time spent on different hosts. Tomato is not a preferred host for long-term adult feeding, and so there is significant adult whitefly movement into and out of tomato fields [87]. This process may encourage mixed infections, and subsequent recombinations, and could be one reason for the enormous diversity of begomoviruses in tomato. To date, there are more than 23 recognized species, as well as more than 15 provisional species, of begomoviruses that infect tomato, more of which are emerging from West and Central Africa and recombination has been shown to have contributed significantly to their evolution [6, 98, 99, 100].

3. Quarantine

Epidemics are often the result of a new host or pathogen having been introduced into an area and causing "disruptions of the mutual adaptation to be expected between hosts and pathogens as they coevolved over long periods in a particular environment" [35]. International trade in plants has led to the severe losses incurred by the spread of the *B. tabaci* B-biotype worldwide, and the viruses it vectors; TYLCV-Is has been transported in plants from the Middle East to the Americas and Mediterranean [1, 101], Japan [102], Reunion Island [103], and northern Africa [104, 105]. The economic losses caused by the B-biotype in the Americas were estimated at US $500 million in 1992 alone, and control of TYLCV-Is is particularly difficult [86]. In addition to the movement of vector and virus populations, the exchange of plant germplasm may also lead to the increase in begomovirus diseases through introduction of genotypes highly susceptible to virus, or that support large indigenous vector populations [25, 36, 79, 106].

Symptomless tomato fruit can provide a source of TYLCV for *B. tabaci* under experimental conditions [107]. This should be considered as another possible, although inefficient, mechanism for long range dispersal of viruses. Such transmission poses a potential threat to crops grown in tropical and subtropical regions, and also to those in more temperate countries which import large quantities of tomatoes and where *B. tabaci* has become a major pest in greenhouses. Produce will also present a route for betasatellites/alphasatellites-containing disease complexes to reach the Americas and Australia in a similar manner to the previous introductions of both TYLCV and the B-biotype [43]. Such complexes could have a serious impact, even in the Americas where bipartite begomoviruses predominate.

Quarantine at borders and other points of entry is thus paramount, particularly between geographically isolated regions. Models can be used to predict countries that need to pay particular attention to the import of viruses or vectors in vegetative planting material, or even produce. For example, the model of [42] highlights that Chile is a country that needs particularly vigilant quarantine services as its Central Valley has an ideal climate for B-biotype. Such models will also help to predict the impact of global warming on vector distribution, and hence countries which need to introduce even stricter quarantine measures.

Conclusion

A study of the genetic diversity within the genus *Begomovirus* and factors affecting it illustrates that most cropping systems employed can affect the size and genetic structure of both vector and virus populations. The potential of begomoviruses to evolve rapidly by acquiring genes from other begomoviruses or from viruses of other genera adds further complexity. Clearly there will be no single solution to controlling these epidemics, and human impact on begomovirus evolution can only be minimized by limiting or changing some of the practices that have led to the emergence of these viruses and their vector populations.

REFERENCES

[1] H. Czosnek, H. Laterrot, *Arch Virol.* 142, 1391-1406, (1997).

[2] L.R. Schneider, *Science* 166: 1627-1629,(1969).

[3] S. E.Seal, F. vandenBosch, M. J. Jeger, *Crit. Rev. Plant Sci*, 25: 1, 23-46 (2006).

[4] L i – Fang. Chen, R. Maria, T. Kon, K. Gamby, B. Xoconostle-Cazares, R. L. Gilbertson, *Mol. Plant Pathol.* 10, 415– 430, (2009).

[5] P. Pandey, N. R. Choudhury, S. K. Mukherjee, *Virol. J.* 6, 152, (2009).

[6] W. N. Leke, A. Kvarnheden, E. B. Ngane, V. P. K. Titanji, J. K. Brown, *Arch Virol.* 156:925–928, (2011).

[7] B.D. Harrison, *Annu. Rev. Phytopathol.* 23, 55-82, (1985).

[8] J.K. Brown, *Plant Dis. Feature* (January), (2007). *http://www.apsnet.org/online /feature/btabaci/*

[9] J.K. Brown, 2010. Phylogenetic biology of the *Bemisia tabaci* sibling species group. Chapter 2 In: P. A. Stansly and S. E. Naranjo (eds.), Bionomics and Management of a Global Pest. Springer Science, The Netherlands, pp. 31-67.

[10] J.K. Brown, D. Frohlich, R.C. Rosell, *Annu. Rev. Entomol.* 40, 511–534, (1995).

[11] *D.L. Mumford, Phytopathology 64: 136-139, (1974).*

[12] B.D. Harrison, H. Barker, K.R. Bock, E.J. Guthrie, G. Meredith, & M. Atkinson, *Nature* 270: 760-762 (1977).

[13] W. Zhang, N.H. Olson, T.S. Baker, L. Faulkner, Agbandje-McKenna., M., Boulton., M. I., Davies., J.W. & McKenna, R. *Virology* 279: 471-477, (2001).

[14] B. Böttcher, S. Unseld, H. Ceulemans, R.B. Rusell, & H. Jeske, *J. Virol.* 78: 6758-6765 (2004).

[15] R.M. Goodman, *Virology* 83: 171-179, (1977a).

[16] R.M. Goodman, *Nature* 266: 54-55, (1977b).

[17] S. Haber, M. Ikegami, N.B. Bajet, & R.M. Goodman, *Nature* 289: 324-326, (1981).

[18] D.M. Bisaro, W.D.O. Hamilton, R.H.A. Coutts & K.W. Buck, *Nucleic Acids Res.* 10: 4913-4922, (1982).

[19] W.D.O. Hamilton, D.M. Bisaro & K.W. Buck, *Nucleic Acids Res.* 10: 4901-4912, (1982).

[20] *J. Stanley, Nature 305: 643-645, (1983).*

[21] *J. Stanley, & M.R. Gay, Nature 301: 260-262, (1983).*

[22] *S.H. Howell, Nucleic Acids Res. 12: 7359-7375, (1984).*

[23] *P.M. Mullineaux, J. Donson, B.A.M. Morris-Krsinich, M.I. Boulton & J.W. Davies, EMBO J. 3: 3063-3068, (1984).*

[24] J. Stanley, D.M. Bisaro, R.W. Briddon, J.K. Brown, C.M. Fauquet, B.D. Harrison, E.P. Rybicki & D.C. Stenger, (2005). *Geminiviridea* In: Fauquet, C.M., Mayo, M.A., Maniloff, J., Desselberger, U., Ball, L.A. (Eds), VIIIth Report of the International

Committee on Taxonomy of Viruses. Virus Taxonomy. Elsevier/Academic Press, London, pp. 1163-1169.

[25] A., Varma and V. G. Malathi, *Ann. Appl. Biol.* 142: 145–164, (2003).

[26] L. Hanley-Bowdoin, S.B. Settlage, B.M. Orozco, S. Nagar & D. Robertson, Crit. Rev. Plant Sci. 18: 71-106, (1999).

[27] C.M. Fauquet, R.W. Briddon, J.K. Brown, E. Moriones, J. Stanley, M. Zerbini, & X. Zhou, *Arch. Virol.* 153: 783-821, (2008).

[28] R.W. Briddon, I.D. Bedford, J.H. Tsai, & P.G. Markham, *Virology* 219: 387-394, (1996).

[29] D.R. Jones, *Eur. J. Plant Pathol.* 109: 195-219, (2003).

[30] M.R. Rojas, C. Hagen, W.J. Lucas & R.L. Gilbertson, *Annu. Rev. Phytopathol.* 43: 361-394, (2005).

[31] *N. Navot, E. Pichersky, M. Zeidian, D. Zamir & H. Czosnek, Virology 185: 151-161, (1991).*

[32] K. Saunders, I.D. Bedford, T.Yahara & J. Stanley, *Nature* 422: 831, (2003).

[33] F. J. Morales and P. K. Anderson, *Arch. Virol.* 146: 415–441, (2001).

[34] Y. Xie and X.-P. Zhou, *Arch. Virol.* 148: 2047–2054, (2003).

[35] J. M. Thresh, *Annu. Rev. Phytopathol.* 20: 193–218, (1982).

[36] P. A. Matson, A. G. Parton Power and M. J. Swift, *Science* 277: 504–509, (1997).

[37] J.P. Legg & C.M. Fauquet, *Plant Mol. Biol.* 56: 585-599, (2004).

[38] A.N. Basu, (1995). Bemisia tabaci (Gennadius): Crop pest and principal whitefly vector of plant viruses. Oxford and IBH Publishing, New Delhi, 183pp.

[39] I.D. Bedford, R.W. Briddon, J.K. Brown, R.C. Rossel & P.G. Markham, *Ann. Appl. Biol.* 125: 311-325, (1994).

[40] M.N. Maruthi, J. Colvin, S.E. Seal, G. Gibson & J. Cooper, *Virus Res.* 86. 71-85, (2002).

[41] *F.J. Morales & P.G. Jones, Virus Res. 100: 57-65, (2004).*

[42] J. E. Polston and P. L. Anderson, *Plant Dis.* 81: 1358–1369, (1997).

[43] T.M. Perring, *Crop Protect.* 20: 725-737, (2001).

[44] A.S. Costa, (1975). Increase in the population density of Bemisia tabaci, a threat of widespread virus infection of legume crops in Brazil. In: J. Bird and K. Maramorosch (eds). Tropical Disease of Legumes, Academic Press, New York. pp 27-49.

[45] J. Colvin, C.A. Omongo, M.N. Maruthi, G.W. Otim-Nape & J.M. Thresh, *Plant Pathol.* 53: 577-584, (2004).

[46] D. Bosco, G. Manon & G.P. Accotto, *Virology* 323: 276-283, (2004).

[47] A.S. Costa & J.K. Brown, *Entomol. Exp. Appl.* 61: 211-219, (1991).

[48] J.P. Legg, P. Sseruwagi & J.K. Brown, (2004). In: Sixth International Scientific Meeting of the Cassava Biotechnology Network, p 78. CIAT, Cali, Colombia

[49] M.A. Mayo, M.J. Leibowitz, P. Palukaitis, K.B.G. Scholthof, A.E. Simon, J. Stanley & Taliansky, M. (2005). VIIIth Report of the International Committee on Taxonomy of Viruses. Virus Taxonomy. Elsevier/Academic Press, London, pp. 1163-1169.

[50] M.J. Roossinck, D. Sleat & P. Palukaitis, *Microbiol. Rev.* 56: 265-279, (1992).

[51] I. Dry, L.R. Krake, J.E. Rigden & M.A. Rezaian, *Proc. Natl. Acad. Sci.* U.S.A. 94: 7088-7093, (1997).

[52] S.A.A. Behjatnia, I.B. Dry & M.A. Rezaian, *Nucleic Acids Res.* 26: 925-931, (1998).

[53] K. Saunders, J. Stanley, *Virology* 264, 142–152, (1999).

[54] *K. Saunders, I.D. Bedford, R.W. Briddon, P.G. Markham, S.M. Wong & J. Stanley, Proc. Natl. Acad. Sci. USA 97: 6890-6895, (2000).*

[55] *H.N.P. Tan, Wu, M.Wong, I.D. Bedford, K. Saunders & J. Stanley, J. Gen. Virol. 76: 2915-2922, (1995).*

[56] J. Stanley, K. Saunders, M.S. Pinner, S.M. Wong, *Virology* 239, 87–96, (1997).

[57] Y. Liu, D.J. Robinson & B.D. Harrison, *J. Gen. Virol* 79: 1501-1508, (1998).

[58] R.W. Briddon, S. Mansoor, I.D. Bedford, M.S. Pinner & P.G. Markham, *Virus Genes* 20: 17-24, (2000).

[59] R.W. Briddon, S. Mansoor, I.D. Bedford, M.S. Pinner, K. Saunders, J. Stanley, Y. Zafar, K. Malik & P.G. Markham, *Virology* 285: 234-243, (2001).

[60] S. Mansoor, S.H. Khan, A. Bashir, M. Saeed, Y. Zafar, K.A. Malik, R.W. Briddon, J. Stanley & P.G. Markham, *Virology* 259: 190-199 1999.

[61] R.W. Briddon, S.E. Bull, I. Amin, *S. Mansoor,* I.D. Bedford, N. Rishi, S.S. Siwatch, M.Y. Zafar, A.M. Abdel-Salam & *P.G. Markham, Virology* 324: 462-474, (2004).

[62] J. Jose & R. Usha, *Virology* 305: 310-317, (2003).

[63] X. Zhou, Y. Xie, X. Tao, Z. Zhang, Z. Li & C.M. Fauquet, *J. Gen. Virol.* 84: 237-247, (2003).

[64] *R.W. Briddon, S. E. Bull, I. Amin, A. M. Idris, S. Mansoor, I. D. Bedford, P. Dhawan, N. Rishi, S. S. Siwatch, A. M. Abdel-Salam, J. K. Brown, Y. Zafar & P.G. Markham, Virology 312: 106-121, (2003).*

[65] J-Y. Yang, M. Iwasaki, C. Machinda, Y. Machinda, X. Zhou & N-H. Chua, *Genes Dev.* 22: 2564-2577, (2008).

[66] *X. Tao & X. Zhou, Plant J. 38: 850-860, (2004).*

[67] S. Mansoor, I. Amin, M. Hussain, Y. Zafar, S. Bull, R.W. Briddon & P.G. Markham, *Plant Dis.* 85: 922, (2001).

[68] *M. Mubin, R.W. Briddon & S. Mansoor, Arch. Virol. 154: 365-368, (2009b).*

[69] G. Gronenborn, *Vet. Microbiol.* 98: 103-110, (2004).

[70] J.F., Huang, X.P. Zhou, *J. Phytopathol.* 154, 648–653, (2006).

[71] B.L. Patil & C.M. Fauquet, *J. Gen. Virol.* 91: 1871-1882, (2010).

[72] R.A.C. Jones, *Virus Res.* 141, 113–130, (2009).

[73] J.D. Wren, M.J. Roossinck, R.S. Nelson, K. Scheets, M.W. Palmer, U. Melcher, *PLoS Biol.* 4, e80, (2006).

[74] R.W. Briddon, J. Stanley, *Virology* 344, 198–210, (2006).

[75] T., Jiang, X.P. Zhou, *Plant Pathol.* 53, 799, (2004).

[76] Y. Shibuya, J. Sakata, N. Sukamto, T. Kon, P. Sharma, M. Ikegami, *Plant Dis.* 91, 1998, (2007).

[77] Y. Xie, P. Wu, P. Liu, H. Gong, X. Zhou, *Virol. J.* 7, 178, (2010).

[78] Q. Xiong, S.F. Fan, J. Wu, X. Zhou, *Virology* 97, 405–411, (2007).

[79] J. K. Brown and J. Bird, *Plant Dis.* 76: 220–225, (1992).

[80] B. Pic´o, M. J. Diez and F. *Nuez, Sci. Hortic.* 67: 151–196, (1996).

[81] F. Escriu, A. Fraile and F. Garc´ıa-Arenal, *Evolution* 57: 755–765, (2003).

[82] B. A. McDonald and C. Linde, *Annu. Rev. Phytopathol.* 40: 349–379, (2002).

[83] A. K. Saikia and V. Muniyappa, *Trop. Agric.* 66: 350–354, (1989).

[84] M. R. Attique, M. Rafiq, A. Ghaffar, Z. Ahmad and A. I. Mohyuddin, *Crop Prot.* 22: 715–720, (2003).

[85] S. L. Pimm and J. K. Brown, *Science* 304: 831– 833, (2004).

[86] J. E. Polston, R. J. McGovern and L. G. Brown, *Plant Dis.* 83: 984–988, (1999).
[87] H. K. Ramappa, V. Muniyappa and J. Colvin, *Ann. Appl. Biol.* 133:187–198, (1998).
[88] I. Denholm, M. Cahill, F. J. Byrne and A. L. Devonshire, (1996). Progress with documenting and combating insecticide resistance in *Bemisia*. In: *Bemisia*: 1995, *Taxonomy, Biology, Damage, Control and Management*. pp. 577–603.
[89] Y. X. Jiang, C. De Blas, L. Barrios and A. Fereres, *Ann. Entomol. Soc. Am.* 93: 573–579 (2000).
[90] M. Lapidot and M. Friedman, *Ann. Appl. Biol.* 140: 109–127, (2002).
[91] G. Cours, D. Fargette, G.W. Otim-Nape and J. M. Thresh, *Trop. Sci.* 37: 238–248, (1997).
[92] S. Ranomenjanahary, R. Rabindran and D. J. Robinson, *Ann. Appl. Biol.* 140:315–318, (2002).
[93] G.W. Otim-Nape, J. M. Thresh and M.W. Shaw, *Ann. Appl. Biol.* 130: 503–521, (1997).
[94] J. P. Legg, *Crop Prot.* 18:627–637, (1999).
[95] M. N. Maruthi, H. Czosnek, F. Vidavski, T. Shlomo-Yedidia, J. Milo, S. Leviatov, V. Mallithimmaiah, Hagalawadi, Seetharam, P., Attiganal, Subbappa, K., Rudrappa, and Muniyappa, V. *J. Plant Pathol.* 109: 1–11, (2003).
[96] E. S. Jim´enez-Mart´ınez, N. A. Bosque-P´erez, P. H. Berger, R. S. Zemetra, H. Ding and S. Eigenbrode, D. *Environ. Entomol.* 33:1207–1216, (2004).
[97] V. N. Fondong, J. M. Thresh and S. Zok, *J. Phytopathol.* 150: 365–374, (2002).
[98] M. Padidam, S. Sawyer and C. M. Fauquet, *Virology* 265: 218–224, (1999).
[99] C. M. Fauquet, S. Sawyer, A. M. Idris and J. K. Brown, *Phytopathology* 95: 549–555, (2005).
[100] W. N. Leke, J. K.Brown, M. E. Ligthart, N. Sattar, D. K. Njualem, A. Kvarnheden, *Virus Res.* 163: 229–237, (2012).
[101] G. P. Accotto, J. Navas-Castillo, E. Noris, E. Moriones and D. Louro, *Eur. J. Plant Pathol.* 106:179–186, (2000).
[102] K. Kato, M. Onuki, S. Fuji and K. Hanada, *Annu. Phytopathol. Soc. Jpn.* 64: 552–559, (1998).
[103] M. Peterschmitt, M. Granier, R. Mekdoud, A. Dalmon, O. Gambin, J. F. Vayssieres and B. Reynaud, *Plant Dis.* 83: 303 1999b.
[104] M. Peterschmitt, M. Granier and S. Abvoulama, *Plant Dis.* 83:1074, (1999a).
[105] I. Fekih-Hassan, F. Gorsane, F. Djilani, H. Fakhfakh, M. Nakhla, D. Maxwell and M. Marrakchi, *EPPO Bull.* 33: 347–350, (2003).
[106] R.W. Briddon and P. G. Markham, *Virus Res.*71: 151–159, (2000).
[107] H. Delatte, A. Dalmon, D. Rist, I. Soustrade, G. Wuster, J. M. Lett, R.W. Goldbach, M. Peterschmitt and B. Reynaud, *Plant Dis.* 87: 1297–1300, (2003).

In: Tomatoes: Cultivation, Varieties and Nutrition
Editor: Tadahisa Higashide

ISBN: 978-1-62417-915-0

Chapter 8

TOMATO ANTIOXIDANT CONTENT AS INFLUENCED BY CONVENTIONAL AND ORGANIC PRODUCTION METHODS AND POST HARVEST MANAGEMENT

Ruplal Choudhary* and S. Alan Walters
Department of Plant, Soil and Agricultural Systems, Southern Illinois University, Carbondale, Illinois, US

ABSTRACT

Consumers are becoming more aware of the nutritional benefits that vegetable provide to their diets and are interested in purchasing those that have increased nutrient or antioxidant content. Many studies have focused on the nutritional value of vegetables produced in organic and conventional (or non-organic) production systems; and, the nutritional value of organic vegetables compared to those produced using conventional production methods is widely debated. However, many consumers often perceive organically produced vegetables as being safer (reduced pesticide residues), produced using more environmental friendly production methods, and more nutritious compared to those produced using conventional production practices. This chapter will focus on the comparison of organic and conventional production systems for the production of two antioxidants in tomato fruit, ascorbic acid and lycopene. In addition, a brief review of post-harvest management system for tomato packaging, transportation and storage will be discussed in relation to antioxidant content of fruit.

INTRODUCTION

Tomato (*Solanum lycopersicum*) is one of the most widely consumed vegetables in the world. It is an important source of antioxidants, such as β-carotene (Vitamin A), ascorbic acid (Vitamin C), and lycopene [1; 2; 3; 4]; and, although ascorbic acid and β-carotene levels in tomato fruit are lower than many other vegetables, this vegetable ranks high as a source of

* E-mail address: choudhry@siu.edu.

vitamin A and C in human diets of many countries because of high consumption rates in diets [5; 6]. One medium-sized fresh tomato (148 g) generally provides about 40% and 20% of the recommended daily allowance (RDA) of vitamin C and vitamin A, respectively [7]. Furthermore, Chassy et al. [1] indicated that tomatoes provide about 19 and 3 mg/100g fresh fruit weight of vitamin A and lycopene, respectively. Several studies have indicated that since tomatoes are major sources of antioxidants, particularly lycopene and ascorbic acid, regular consumption of tomatoes will decrease the risk of chronic degenerative diseases [8]. Many epidemiological studies have suggested that regular consumption of tomato and their products may lead to decreased incidence of various forms of cancer and heart diseases [9; 10; 11].

Ascorbic Acid and Lycopene as Antioxidants in Tomato Fruit

Ascorbic acid and lycopene (Figure 1) are both antioxidants which mean they have the ability to quench free radicals of oxygen. Free radicals are highly aggressive, electrochemically imbalanced molecules with unpaired electrons that are reactive with cell components and can cause permanent damage, which can lead to mutations and possibly cancer in human cells. Oxygen-derived free radicals are the most reactive species, and are formed naturally as by-products during oxidative cellular metabolism. Ascorbic acid is the major antioxidant found in tomato fruit when consumed fresh. Ascorbic acid provides an excellent source of electrons and can donate electrons to free radicals such as hydrogen peroxide, superoxide and singlet oxygen and quench their reactivity [12]. Lycopene is the primary carotenoid found in tomato and provides the red pigment for the color of the fruit [13]. The molecular configuration of lycopene also enables it to inactivate free radicals. As an antioxidant, lycopene has a singlet-oxygen-quenching ability that is twice as high as beta-carotene (pro-vitamin A) and ten times higher than alpha-tocopherol (pro-vitamin E) [14]. Tomato and tomato-derived products are the main source of lycopene obtained from food in Western countries [13].

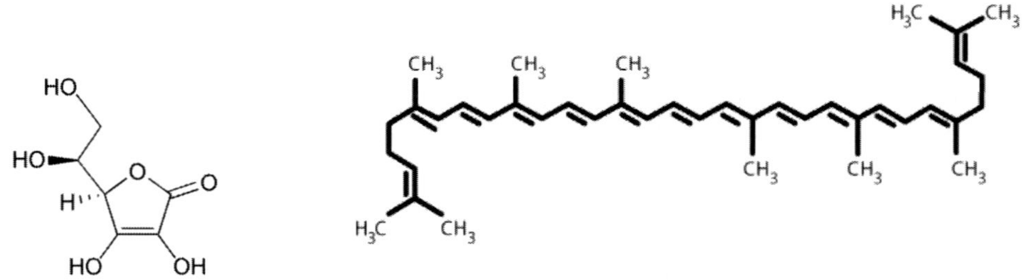

Ascorbic acid Lycopene

Figure 1. Molecular stuctures of the antioxidants, Ascorbic Acid (vitamin C) and Lycopene.

Although the amount of ascorbic acid in tomato is diminished when cooked, the processing of tomato fruits into juice, sauce, paste, or ketchup increases the concentration of bio-available lycopene. Lycopene in tomato paste is four times more bio-available than in fresh tomatoes. Processing tomato fruit makes the lycopene more bio-available by increasing the surface area available for digestion; and more importantly, the chemical form of lycopene is altered by the temperature changes involved in processing to make it more easily absorbed by the body. Lycopene has garnered much attention during the decade due to the association of lycopene-rich diets and lower risks of certain cancers, heart disease, and age-related diseases [15; 16].

CONSUMER AWARENESS OF NUTRITION

Consumers are becoming more aware of the nutritional benefits that fruits and vegetables provide to their diets and are interested in purchasing those that have increased nutrient or antioxidant content [17]. This has led directly to an increased demand for organic produce in the marketplace. Many consumers are willing to pay significantly more for organic produce, as they associate these foods as being healthier, more nutritious, better tasting, safer, and produced through more environmentally friendly methods than those produced through conventional agricultural practices [18]. Marketing tomatoes as pesticide-free or organic and being high in lycopene or other antioxidants is an effective method to increase consumer sales and consumption [19]. It is widely thought that any benefit derived from organic vegetables is due primarily to an absence of pesticide residues [20]; and, pesticide residues in vegetables concern consumers since these products are often eaten fresh and are not processed [4]. However, prior to the widespread use of pesticides, those that advocated organic food consumption claimed that these foods were more nutritious compared to conventional (or non-organic) practices based on the soil and fertilizer management practices used by organic growers [20].

Many studies have focused on comparing the nutritional value of vegetables produced in organic and conventional production systems to determine which system consistently produces the most nutritious products. However, the nutritional content of organic vegetables compared to those produced using conventional production methods is still widely debated due to the inconsistency among studies that have tried to compare the two production systems. Even with the lack of sufficient scientific data, many consumers still perceive organically produced vegetables as being more nutritious compared to those produced using conventional production practices. Therefore, this chapter will focus on providing an unbiased perspective comparing two important antioxidants in tomato fruit that were produced under organic and conventional production systems. In addition, a brief description of post-harvest management practices of tomato that help in retaining nutrition and quality has also been provided.

Comparison of Organic and Conventional Production Practices

Although the nutritional content of fruits and vegetables is much more complicated than just comparing conventional and organic production methods, there are fundamental differences between organic and conventional agricultural production practices that can definitely influence the nutritional content of fruits and vegetables (Table 1). In organic systems, careful attention is placed on improving soil characters using various methods that are focused to increase soil organic matter, improve soil water conservation practices, reduce excessive nutrient pollution and enhance soil microbial activity; and, this is achieved by using alternative methods such as applying composts, manures, organic mulches or the use of cover crops and crop rotations to maintain soil productivity. It has been well documented that soil composition has a major influence on determining the nutritional quality of fruits and vegetables [21]. Organic systems rely on creating a diverse soil ecosystem to provide nitrogen and other nutrients to plants, while conventional farm methods use synthetic petroleum-based fertilizers to provide nitrogen and other nutrients to plants. Furthermore, the high availability and constant uptake of inorganic nitrogen directly influences the synthesis of secondary metabolites, proteins, and soluble solids in plants and plant products [1].

More importantly, to maximize plant product quality and yields, conventional systems use a wide variety of different pesticides to control plant pests while organic systems do not typically use pesticides; and if pesticides are used, naturally occurring products, such as the bacterium *Bacillus thuringiensis* (for insect control) or unrefined sulfur (for disease control) are used. However, many conventional vegetable growers are transitioning to more sustainable methods that often include the use of soil building practices (such as no-tillage or cover cropping systems) and integrated pest management (which reduces pesticide use and application).

Although these basic differences between the two production methods could potentially lead to differences in the final nutrition content of vegetables, there are several field production practices that will affect the nutritional content of tomato or other vegetables regardless of whether organic or conventional production systems are used. Research has indicated that the nutritional quality of tomato fruit depends upon many factors such as the tomato cultivar used, specific growing conditions at production site (especially the soil management/fertility system), cultural practices, maturity at harvest (on- or off-vine ripening), and postharvest conditions.

Soil management System

The methods used for soil and fertility management in the production scheme have a definite influence on the nutritional composition of tomato fruit. First, the soil type is an important factor in determining the antioxidant content of fresh produce. Even under very similar production practices, the soil can play a major role in vegetable nutritional content [21]. Thus, farms that are in close proximity, but having differing soil types, can produce fruits and vegetables with differing nutritional and antioxidant content, although similar production methods were used.

According to the carbon balance theory and growth differentiation balance hypothesis, high nutrient availability (especially nitrogen) will lead to increases in plant growth and development rates and biomass, with decreased allocation of resources toward the production of carbon-containing compounds such as starch, cellulose, and the non-nitrogen containing phenolic antioxidants [1; 22; 23]. Thus, when plants are fertilized with high analysis inorganic nitrogen sources, the resulting high availability of nitrogen in the soil leads to more uptake and higher amounts of nitrogen-containing compounds in plants. Wilkins et al. [24] indicate that phenolic antioxidants decrease in tomato fruit with increasing soil nutrient availability. The use of over-fertilization is a common occurrence in conventional vegetable production to ensure high yields and this management practice tends to result in soils with high nitrogen availability and less antioxidants in tomato fruit or other vegetables.

Table 1. Comparison of practices used in conventional (non-organic) and organic farming systems for specific production issues

Production Issue	Conventional	Organic
Fertility Enhancement of Soils	Apply chemical, inorganic fertilizers to promote plant growth In some instances, practices such as no-tillage or cover cropping are used	Apply organic fertilizers such as manure or compost to improve soil and provide nutrients to plants Some other 'alternative' method (e.g., cover cropping) is generally used to improve soil structure, microbial activity and organic matter content
Insect Control	Use Integrated Pest Management (IPM) practices for insect pest control, with labeled insecticides used to control specific insect pest problem if populations are high Calendar spraying (e.g., on a 7 day schedule) of pesticides is often used to prevent disease buildup in crops Insect diversity is generally not maintained as broad-spectrum insecticides kill both pests and beneficials	Use Integrated Pest Management (IPM) practices to control insects, with minimal pesticide use, if there is an organically certified insecticide available Maintain insect diversity, through using beneficial insects, mating disruption, trap crops, or other low environmental impact methods
Disease Control	Use Integrated Pest Management (IPM) practices for disease pest control, with labeled pesticides used to control specific disease problem Calendar spraying (e.g., on a 7 day schedule) of insecticides is often used to prevent insect buildup in crops Although most use cultural practices, such as crops rotations, resistant cultivars and many other cultural practices, the focal point of most control programs is through the use of fungicides, bactericides, and nematicides	Use Integrated Pest Management (IPM) practices for disease pest control, although limited labeled pesticides for use in controlling specific diseases Use of crop rotations, resistant cultivars, field sanitation practices and other cultural management practices are generally the focus of disease control programs on organic farms
Weed Control	Use herbicides primarily to manage weed populations, although many try to use tillage, crop rotation, hand weeding or mulches to manage weed populations	No herbicides of any sort generally used, as alternative methods, such as flame weeding, tillage, crop rotation, hand weeding or mulches are used to manage weed populations

Climate

The climatic conditions at the field production site have a definite influence on tomato antioxidant content. Environmental conditions, such as sun exposure, can have a direct influence on ascorbic acid levels in tomato fruit. Furthermore, the quality and duration of sunlight must also be favorable for tomato fruit to ripen and reach maximum lycopene levels. Drought conditions can also improve the nutritional composition of tomato fruit due to reduced allocations of water which increases the nutrient concentration in the fruit on a dry weight basis [4].

Cultural Practices

The cultural practices used in the field for tomato production influences the amount of antioxidants present in fruit. Field cultural practices such as pruning and thinning of foliage and fruit will determine the tomato crop load and fruit size, which directly influences the nutritional composition of the fruit [4]. Other cultural practices such as the use of organic or synthetic plastic mulches and drip irrigation will impact the amount of moisture in the soil, which will have an effect on the nutritional content of fruit.

Tomato Cultivar

The specific tomato genotype grown has a great influence on the fruit nutritional composition. Tomato cultivars differ for ascorbic acid and lycopene content [1; 9; 25; 26; 27] and, the cultivar used in production plays a significant role in the amounts of these nutritional compounds found in harvested tomato fruit [4]. Woese et al. [28] indicated that fruit and vegetable variety selection played a much larger role in determining the final content than the specific production method. In an evaluation of tomato genotypes, the highest ascorbic acid content was found in the wild small-fruited species, S. *pimpinellifolium*, and in the smaller *S. lycopersicum* entries [3]. Furthermore, small fruited tomatoes, such as the wild species (*S. pimpinellifolium*), generally have three or more times the lycopene content than the larger domesticated *S. lycopersicum* fruit [3]. The smaller-sized tomatoes, due to their higher surface area/volume ratio, can provide 30 to 40% higher antioxidant activity than the larger-sized tomatoes. Similarly, cultivars having a brighter red color can provide up to 50% more lycopene than the tomatoes with less red coloration [29]. Thus, it generally appears that the selection of small, brightly colored tomato cultivars optimize ascorbic acid and lycopene fruit levels.

Organic and Conventional Production Systems for Tomato Nutritional Content

Although there have been multiple studies comparing the nutritional quality of conventionally and organically produced tomatoes and other vegetables, most have provided inconsistent results and are difficult to interpret because of differing cultivar selections and agronomic conditions in which the studies were conducted [22; 28]. There is really insufficient information available to resolve the issue based on the few, unbiased high-quality studies that have been conducted [22]. Most studies represent only one seasonal harvest, provide little information on pre-planting conditions (e.g., tillage, cover crop, previous crop, etc.) and provide almost no information on the length of time the organic fields were under organic cultivation prior to study initiation [1].

Based on the current literature, it is quite confusing when trying to understand if an organically produced tomato fruit generally has more ascorbic acid and lycopene content compared to one produced in a non-organic system. Some studies indicate that organic production systems improve both tomato ascorbic acid and lycopene content compared to conventional systems, while others state that no differences exist between the two production methods or that conventionally produced tomato contain more phytochemicals. For example, Toor et al. [30], Caris-Veyrat et al. [13] and Chassey et al. [1] indicated in that tomatoes grown using organic methods produced more ascorbic acid in fruit compared to plants grown with inorganic nitrogen sources in conventional production systems. Furthermore, Caris-Veyrat et al. [13] also found that organic cultivation practices provide tomatoes and tomato-derived products with significantly higher lycopene content. Whereas, in contrast, two-year field studies using identical tomato cultivars conducted by Juroszek [4] and Pieper and Barrett [31] found no significant difference between organic and conventional farming systems for lycopene and ascorbic acid content; and, Lumpkin [32] could not identify significant differences in nutritional composition between tomatoes grown under conventional or organic farming systems. With such contradictory results, it is really not possible to distinguish which farming system is superior in terms of consistent effects on antioxidant compounds in tomato fruit [32].

Although there are fundamental differences between organic and conventional production systems (particularly regarding soil fertility management which has the potential to drastically influence the nutritive composition of plants) [1; 22], both systems include combinations of farming practices that can vary greatly depending upon the climate, soil quality, pest and disease incidence, and farm management practices [1]. Results from Lumpkin [32] indicated that the development of secondary plant metabolites in tomatoes is influenced by many factors, with some controllable and others that are not. Due to this lack of control, it may not be possible to identify all factors in the farming systems used on organic or conventional commercial farms that influence tomato fruit phytochemical content.

Overall, most research findings indicate that a straight comparison of organic and non-organic foods do not differ significantly for nutritional content. However, most have indicated a tendency for higher ascorbic acid levels and less nitrate content in organic products than non-organic [20; 22]. The lower amounts of nitrates in organic products compared to those produced conventionally confirm that organic vegetables are grown with lower nutrient inputs than their conventionally produced counterparts, since the nitrogen source used as the fertilizer source plays a major role in determining nitrate content [22; 28].

Although many studies indicate that tomato fruit lycopene content does not differ between the organic and conventional farming systems [4], comparisons among production systems for tomato fruit quality characters (or any other vegetable) should only be made using the same cultivar in similar soil and environment conditions [4]. The climatic conditions of a particular field site and year also play a major role in determining the nutritional content of a tomato fruit. Chassy et al. [1] found significant year-to-year variability when comparing quality and nutritional characteristics for organic and conventional fresh market tomatoes over a three-year period at one growing location.

Based on the previous research, we think that there is no direct evidence that indicates that organically grown tomatoes consistently produce more antioxidants than those from conventionally produced fields. We feel that the tomato cultivar chosen is more important than the production scheme to improve the ascorbic acid and lycopene content in fruit. However, it is important to remember that both the ascorbic acid and lycopene content in tomato fruit varies considerably not only among cultivars, but also between fertility systems, environmental growing conditions, and stage of maturity at harvest.

Post Harvest Management (Transport, Packaging, Storage)

Post-harvest management of tomatoes should maintain or improve the quality of tomatoes for desirable end use. The major indicators of tomato quality are color, texture, firmness and flavor [33]. Color is a major quality characteristic and uniformity of color is a major requirement of the US quality standards. The characteristic color of ripe tomatoes is imparted to it upon synthesis of carotenoids such as alpha and beta carotene in early stage of ripening and lycopene in later stage of ripening. Apart from being a source of color and flavor, tomato is a valuable source of antioxidants coming from lycopene, ascorbic acid, phenolics and flavonoids [34]. Antioxidants should be considered as one of the important qualities attributes of tomatoes and the post-harvest management practices should consider maximizing retention of antioxidants during transport packaging, distribution and storage. Vine-ripened tomatoes have more ascorbic acid than post-harvest-ripened fruit [35]. Lycopene and total antioxidant contents increase as the fruit ripens and color turns from pink to red [36]. Synthesis of lycopene is inhibited at temperatures below 12 °C and prohibited at temperatures higher than 30 °C [37].

Tomatoes are harvested at different stages of ripeness and maturity. As shown in Table 2, there are nine stages of maturity and within the mature green stages there are four stages. The vineripe tomatoes are harvested at breaker stage or later. After harvesting the tomatoes at mature green or later, tomatoes should be precooled to a temperature of 12-13 °C to delay ripening and prevent microbial growth. Tomatoes are also very sensitive to chilling. The chilling injury causes non-uniform ripening, decreased flavor, softness and mealiness. Table 3 indicates the number of days to reach full red color at a specified storage temperature.

Table 2. Ripening stages for tomatoes (with permission from Chakraverty and Paul 2001)

Class	Description
Mature green 1	Seeds cut by a sharp knife on slicing the fruit; no jelly-like material in any of the locules; fruit is more than 10 days at 20 °C from breaker stage
Mature green 2	Seeds fully developed and not easily cut by of slicing fruit; jelly-like material in a least on locule; fruit is six to ten days at 20 °C from breaker stage; minimum harvest maturity.

Class	Description
Mature green 3	Jelly-like material well developed in locules but still completely green; fruit is two to five days at 20 °C from breaker stage.
Mature green 4	Internal red coloration at the blossom end, but no external color change; fruit is one to two days at 20 °C from breaker stage.
Breaker	First external pink, red or yellow color at the blossom end (USDA color stage 2)
Turning	More than 10 percent but not more than 60 percent of the surface, in the aggregate, shows a definite change in color from green to tannish yellow, pink, red, or a combination thereof (USDA color stage 3)
Pink	More than 30 percent but not more than 60 percent of the surface, in the aggregate, shows pink or red color (USDA color stage 4)
Light red	More than 60 percent of the surface, in the aggregate, shows pinkish-red or red, but less than 90% of the surface shows red color (USDA color stage 5)
Red	More than 90% of the surface, in aggregate, shows red color (USDA color stage 6)

Table 3. Days to full red color in tomatoes after harvest at different ripeness stages (with permission from Chakraverty and Paul 2001)

Ripeness stage	Days to full red color at indicated temperature					
	12.5 °C	15°C	15.5°C	20°C	22.5°C	25°C
Mature green	18	15	12	10	8	7
Breaker	16	13	10	8	6	5
Turning	13	10	8	6	4	3
Pink	10	8	6	4	3	2

Packaging and Transportation

Tomatoes are packed in a variety of packaging types and materials depending on whether the package is for retail sale or for transportation and distribution. In the past, wooden crates were used with some cushioning materials such as scraps of paper. Cardboard boxes have replaced wooden boxes in tomato industry for bulk distribution packaging. Several types of distribution boxes are available for holding capacities ranging from 15 to 25 pounds. Loose tomatoes in these corrugated boxes are transported with a general purpose polystyrene (GPPS) liner. The consumer packages are also transported inside the above corrugated paperboard boxes. Popular consumer packagings are expanded polystyrene trays (PS trays) wrapped with polyethylene or polyethylene tetrapthalate clamshell (PET Clamshell). In each of these consumer packagings, 4 or 6 tomatoes are packaged [38].

Post-harvest Treatment and Storage Conditions

It is essential to provide rapid precooling soon after harvest of tomato for optimal postharvest keeping quality. Forced-air cooling is considered as the most effective but room cooling is more common in practice. Vacuum cooling is not recommended for tomatoes. The precooling endpoint is typically recommended as 12.5°C (55°F) [34]. Tomato is susceptible to chilling injury when cooled below 12°C, which may prevent subsequent ripening and quality. Tomatoes may be stored under controlled atmospheric conditions to enhance storage life. The controlled atmosphere should retard respiration rate of tomatoes by reducing oxygen and increasing carbon dioxide concentrations. The recommended modified atmosphere conditions for mature green tomatoes are 3-5% oxygen and 2-3 % carbon dioxide while stored at 10-13 °C at 90-95% relative humidity (RH). The firm ripe tomatoes can be stored at the same gaseous composition at 10 °C and 85-90% RH for 2 weeks [33].

Modified atmospheres created by packaging in polyethylene films may delay ripening and prolong shelf life [37]. The quality characteristics (vitamin C and carotenoids) of tomatoes can be preserved only when the oxygen permeability of the film is high enough to prevent anaerobic respiration [39]. Modified atmosphere storage recommendation for tomatoes harvested at breaker or light pink conditions are 0% carbon dioxide with 3-5% oxygen and 1-3% carbon dioxide with 3-5% oxygen for mature green tomatoes. Some additives such as ethylene absorbers or scavengers may delay ripening of tomatoes. Bailen et al [40] reported that the inclusion of granular activated carbon within polypropylene bags for storage of tomatoes at 8 °C reduced level of ethylene for 21 days. As a result, the changes in color, and weight loss were reduced and quality was improved.

CONCLUSION

Many consumers are becoming more aware of the nutritional benefits that vegetable provide to their diets and are interested in purchasing those that have increased nutrient or antioxidant content. A greater realization of the importance of phytochemicals in promotion of human health has prompted the broadening of dietary recommendations to include greater daily intake of a wider range of vegetables and fruit [41; 42]. This has led to an increasing demand for organic foods obtained from alternative organic cultural practices that limit the use of mineral soluble fertilizers and synthetic pesticides [13]. However, unlike consumers that regularly eat organic produce, non-consumers of organic food are not convinced that organic produce tastes better and has more nutrition than conventionally grown produce [18].

Although many studies have focused on determining the nutritional benefits of vegetables produced in organic compared to conventional (or non-organic) production systems, the nutritional value of organic vegetables compared to those produced using conventional production methods is widely debated. Some studies report higher levels of ascorbic acid and lycopene in organically produced tomatoes, while other indicate no difference between production systems or even higher levels of ascorbic acid and lycopene in conventionally produced fruits and vegetables. There is really a lack of sufficient, unbiased information that either confirms or denies that organic foods are more nutritious [13]. It is important to remember that several factors play a role in determining the final nutritional value of a tomato fruit including differences in soil types and nutrients applied, tomato cultivar, irrigation practices, environmental conditions under which they were grown, and postharvest handling [31]. Thus, large-scale, multi-year comparison with carefully defined inputs and variables are still needed to evaluate whether specific agronomic practices influence antioxidant phytochemical contents in tomato and other vegetables as well as determining those specific cultivars that have the highest antioxidant content [1]. Future comparisons among production systems for tomato fruit quality characters (or any other vegetable) should only be made using the same cultivar in similar soil and environment conditions [4].

The ability to manage and control antioxidant levels in plant products through production and post-harvest practices has the potential to enhance the nutritive content of foods. Specific cultural and fertility management practices as well as cultivar choice has the potential to improve the nutritional content of fresh fruits and vegetables, although currently, there is insufficient scientific data to either confirm or deny that organic foods are more nutritious that those produced through conventional means. However, we agree with Lester [43] that increased consumer nutrition and health through regular consumption of 5 to 9 fruits and vegetables per day is far more important than convincing consumers that organic produce is better than non-organic.

REFERENCES

[1] A.W. Chassy, L. Bui, E.N.C. Renaud, M. Van Horn, and A.E. Mitchell, *J. Agric. Food Chem. 54*, 8244-8252 (2006).

[2] G. Giovanelli, V. Lavelli, C. Peri, and S. Nobili, J. *Sci. Food. Agric. 79*, 1583-1588 (1999).

[3] P.M. Hanson, R.Y. Yang, J. Wu, J.T. Chen, D. Ledesma, S.C.S. Tsou, and C. Lee, *J. Amer. Soc. Hort. Sci. 129*, 704-711 (2004).

[4] P. Juroszek, H.M. Lumpkin, R.Y. Yang, D.R. Ledesma, and C.H. Ma, J. *Agric. Food Chem. 57(4),*1188-94 (2009).

[5] C.M. Rick, *Hybridization of Crop Plants*, pp. 669–680, American Society of Agronomy/Crop Science Society of America, Madison,Wisconsin (1980).

[6] M.A. Stevens, In J.G. Atherton and J. Rudich (eds.) p. 35-109. *The tomato crop*. Univ. Press, Cambridge, U.K. (1986).

[7] USFDA, Vegetables – Nutrition facts. *<http://www.fda.gov/downloads/Food/LabelingNutrition/FoodLabelingGuidanceRegulatoryInformation/InformationforRestaurantsRetailEstablishments/UCM169237.pdf>* accessed 21 Dec, 2011.

[8] L. Frusciante, P. Carli, M.R. Ercolano, R. Pernice, A. Di Matteo, V. Fogliano, and N. Pellegrini, *Mol Nutr Food Res. 51(5)*:609-17 (2007).

[9] V. Lavelli, C. Peri, and A. Rizzolo, *J. Agr. Food Chem. 48*:1422-1448 (2000).

[10] A.V. Rao, and S. Agarwal, *Nutr. Res. 19*, 305 -323 (1999).

[11] World Cancer Research Fund (1997). Food, Nutrition and the Prevention of Cancer: A Global Perspective. Washington, DC: American Institute for Cancer Research.

[12] S.T. Omaye and P. Zhang. p. 53-75. In: W.R. Bidlack, S.T. Omaye, M.S. Meskin, and D. Jahner (eds.). *Phytochemicals- a new paradigm.* Technomic Publishing Co., Lancaster, PA. (1998).

[13] C. Caris-Veyrat, M. J. Amiot, V. Tyssandier, D. Grasselly, M. Buret, M. Mikolajczak, J. C. Guilland, C. Bouteloup-Demange, and P. Borel, *J. Agric. Food Chem. 52*:6503-6509 (2004).

[14] S. Agarwal and A.V. Rao, *Canadian Med. Assoc. J. 163(6)*:739-744 (2000).

[15] P.M. Bramley, *Photochemistry 54*:233-236 (2000).
[16] D. Heber and Q.Y. Lu, *Expt. Biol. Med. 227*:920-923(2002).
[17] K. A. Steinmetz and J. D. Potter, *J Am Diet Assoc. 96(10)*:1027-39 (1996)
[18] A. Saba and F. Messina, *Food Qual. Prefer. 14*:637-645 (2003).
[19] C. Kubota, C.A. Thomson, M. Wu, and J. Javanmardi, *HortSci. 41*:522-525 (2006).
[20] V. Worthington, *J. Altern Complement Med. 7(2)*:161-73 (2001)
[21] G.E. Lester, and K. Crosby. *J. Amer. Soc. Hort. Sci. 127*:843-847 (2002).
[22] K. Brandt and J.P. Mølgaard, *J. Sci. Food Agric. 81*:924-931 (2001).
[23] N.E. Stamp, *Q. Rev. Biol. 78*:23-55 (2003).
[24] R.T. Wilkins, J.M. Spoerke, and N.E. Stamp, *Ecology 77*:247-258 (1996).
[25] B. Akbudak, H. Bolkan, and N. Cohen, *Int. J. Food Sci. Nutr. 60(S1)*:1226-138 (2009).
[26] C. Leonardi, P. Ambrosino, F. Esposito, and V. Fogliano, *J. Agr. Food Chem. 48*:4723-4727 (2000).
[27] J.R. Stommel and K.G. Haynes. *J. Hered 85*:401-404 (1994).
[28] K. Woese, D. Lange, C. Boess, and K. W. Bögl, *J. Sci. Food Agric. 74*:281-293 (1997).
[29] R.K. Toor, C.E. Lister, and G.P. Savage. *Int. J. Food Sci. Nutr. 56(8)*: 597-605 (2005).
[30] R.K. Toor, G.P. Savage, and A. Heeb, *J. Food Comp. Anal. 19*:20-27 (2006).
[31] J.R. Pieper and D.M. Barrett, *J. Sci. Food Agric. 89*:177-194 (2008).
[32] H. Lumpkin, Technical Bulletin No. 34. AVRDC publication number 05-623. Shanhua, Taiwan: AVRDC—The World Vegetable Center. 48 pp. (2005)
[33] A. Chakraverty and S.R. Paul. *Postharvest Technology*. Science Publishers Inc., Enfield, NH, USA (2001).
[34] A.K. Thompson, *Fruit and vegetables: harvesting, handling and storage.* Blackwell Publishing Ltd., Oxford, UK (2003).
[35] G. Giovanelli, V. Lavelli, C. Peri, and S. Nobili, *J. Sci. Food. Agric. 79*:1583-1588 (1999).
[36] S. Brandt, Z. Pek, E. Barna, A. Lugasi, and L. Helyes, *Journal of the Science of Food and Agriculture 86*: 568-572 (2006).
[37] H.C. Passam, I.C. Karapanos, P.J. Bebeli, and D. Savvas, *The European Journal of Plant Science and Biotechnology 1(1)*: 1-21(2007).
[38] US EPA, http://www.epa.gov/wastes/partnerships/stewardship/docs/tomato-packaging-assessment.pdf. (2010). Accessed on February 27, 2012.
[39] G. Muratore, M.A. Del Nobile, G.G. Buonocore, C.M. Lanza, and C.N. Asmundo. *Journal of Food Engineering 67*: 393-399 (2005).
[40] G. Bailen, F. Guillen, S. Castillo, M. Serrano, D. Valero and D. Martínez-Romero, *Journal of Agricultural and Food Chemistry, 54*, 2229–2235 (2006).
[41] C.J. Peters, G.W. Fick and J.L. Wilkins, *Agron. J. 95*:1424-1430 (2003).
[42] J.H. Weisburger, *Food Chem. Toxicol. 37*:943-948 (1999).
[43] G.E. Lester, *HortScience 4*:296-300 (2006).

In: Tomatoes: Cultivation, Varieties and Nutrition
Editor: Tadahisa Higashide

ISBN: 978-1-62417-915-0

Chapter 9

SUSTAINABLE INTEGRATED DISEASE AND PEST MANAGEMENT IN COMMERCIAL TOMATO PRODUCTION IN OPEN FIELDS

Bombiti Nzanza[1], Kgabo Pofu[2] and Phatu Mashela[3]*

[1]Head of Department of Research, ZZ2 Bertie van Zyl, Mooketsi
[2]Researcher, Department of Nematology and Entomology, Agricultural Research Council, Rustenburg
[3]Professor of Plant Protection, University of Limpopo, Sovenga, South Africa

ABSTRACT

A set of research-based environment-friendly approaches are being adopted and applied by tomato (*Solanum lycopersicum*) farmers to respond to increased threats of below- and above-ground virulent pathogen, nematode, mite and insect pests. Primarily, soil health is improved through the application of disease-suppressive and nutrient-enriched compost, cover cropping, effective microorganisms and minimum tillage tactics. Plant-parasitic nematodes are managed using fermented crude extracts of locally available plant materials. Nursery seedlings are inoculated with *Trichoderma harzianum* and arbuscular mycorrhizal fungi for improving tolerance to soil-borne diseases and enhancing overall crop quality and development. Crop protection is achieved through the application of plant extracts, while tomato fields are enriched through augmentation of beneficial insects, coupled with limited application of synthetic chemical pesticides when necessary. A case study for this model is a large-scale commercial tomato farming operation under open fields in South Africa, which has gradually adopted *natuurboerdery* or a hybrid of conventional and natural farming approaches in order to strike a balance between chemical farming and organic farming, with the aim of enhancing sustainable tomato production.

* E-mail: phatu.mashela@ul.ac.za.

INTRODUCTION

Natuurboerdery farming system is an integrated farming approach practiced by ZZ2, the largest open-field tomato (*Solanum lycopersicum*) producer in South Africa. This concept is defined as "an approach to commercial farming that aims to harness the laws and energies of natural ecosystems for health and sustainable crop production without sacrificing the benefits of technology and science [1]. The overall aim of *natuurboerdery* is to maintain ecological balance through minimizing environmental impacts, restoration and conservation of natural resources and biodiversity which lead to overall sustainability and human health promotion [2]. *Natuurboerdery* is characterised by well-researched and developed technical changes and adjustments overtime, which currently represents a completely new, highly innovative and competitive hybrid farming system. This system constitutes environment-friendly conventional and climate-smart agricultural best-practices. Production had since increased, but it is difficult to single out the most important factors that drive the observed improvements. Consequently, all factors are viewed as being equally important, which necessitate good management practices that integrate and optimise soil health management, cultivar selection, correct planting, pest and disease management and any good agricultural practices. Integration of strategies have resulted in 30% increase in marketable yield overtime, with an average of 90-100 t/ha. Production is practiced mainly under open fields all year-round, with farms being at different locations in the northern part of South Africa. *Natuurboerdery* comprises five distinct areas, *viz.* soil health, plant health, environmental health, ecological health and human health [2]. However, in this chapter, we discuss exclusively how soil health and plant health are managed in combination with good agricultural practices to accomplish a sustainable integrated pest and disease management (Figure 1).

Figure 1. A. Composting, B. Mixing compost, C. Scouting, D. Aerial view of selected tomato open fields under *natuurboerdery* at ZZ2.

SOIL HEALTH MANAGEMENT

Soil health is the foundation of *natuurboerdery* system. Farmers constantly strive to achieve soil health status on their farms, which represents a significant paradigm shift in ZZ2 operations for achieving the objectives of climate smart agriculture. In fact, despite high tech chemical farming that characterised ZZ2 operations, consistent gradual decline in crop yield occurred on ZZ2 farms – which were later linked to soil health degradation. Soil health is a function of three components, namely, chemical, physical and biological properties, which can be achieved through deliberate activities intended to achieve a balance (Figure 2).

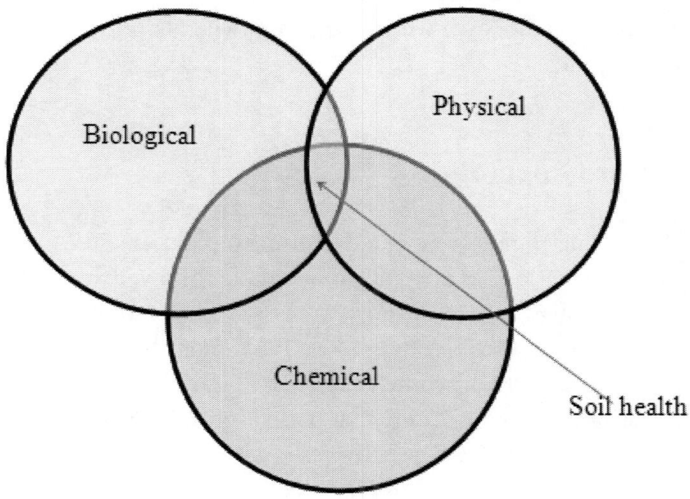

Figure 2. Soil health as a function of biological, chemical, and physical properties of the soil.

Although organic matter is the most important criteria, in a balancing act such as required for achieving soil health status, all three components are equally important. Currently, it is indisputable that to achieve soil health status, management strategies should deliberately be focused to optimise the balancing of the three components of soil health. The starting point in this balancing act is the build-up of the biological component through organic amendment in the form of compost and manure, minimum tillage, fallowing and cover cropping while closely monitoring effective microorganisms and compost tea additions.

Compost and Manure

Compost and manure applications to the soil increases soil organic matter, improves a number of soil properties including soil tilth, water-holding capacity, oxygen content, and soil fertility stimulate beneficial biological activity by increasing biomass and enzymes activities [3,4,5,6,7]. The importance of microbes and understanding of the food-web have since captured the enthusiasm of many producers. Microbes play indispensable roles in the sustainable function of all ecosystems, including recycling of nutrients, regulation of the soil organic matter and soil carbon sequestration, modification of soil physical structure and water regimes, enhancement of the efficiency of nutrient acquisition and plant health, suppression of soil-borne pathogens and detoxification of noxious chemicals [8, 9, 10].

Compost produced by ZZ2 is nutrient-enriched, with disease-suppressive capabilities, and comprises animal manures, grass, invader plant species, *etc*. Animal manures, for instance, contribute to the compost heap an assortment of effective microorganisms, which accelerate the composting process that lasts for several weeks. A curing stage at the end of composting allows the proliferation of beneficial microorganisms. Although this special compost does not eradicate pathogens, the fact that it contains biological control agents and effective microorganisms which multiply and persist in soil, offer advantages to soil health build-up. In fact, since farmers on ZZ2 farms started using this special compost, the prevalence of soil-borne diseases induced by *Verticillium*, *Pythium* and *Fusarium* species are gradually being reduced from the level where they were when conventional chemical farming systems became uneconomic.

Fallowing and Cover Cropping

At the end of harvesting, fields are cleared of tomato plants, poles, twines and any other production materials. Ridges are levelled by ploughing, followed by disking. In the past, fallowing practice lasted for as long as seven years but since the adoption of *natuurboerdery* system, the practice lasts for three to five years, resulting in drastic reductions in the pathogen inoculum and pest cycle life. Generally, the field is left fallow to allow for natural grasses to grow before the next cropping cycle. During that period, grasses are constantly baled to allow for population regeneration, with the hay being used as basal ingredient for composting. In this context, the aim of fallowing is to mitigate the negative effects of monoculture while improving soil health. Basically, under *natuurboerdery*, crop rotation is not being practiced, but fields are predominantly fallowed.

Cover cropping is recognised as one of the best practice for improving soil health. Under *natuurboerdery*, because of the quick regeneration of natural grasses, cover cropping is only used in special circumstances. On-farm-trials were widely conducted to determine the best cover crops for the region. Much attention had been given to evaluating the performance of vetch (*Vicia* spp.) - oat (*Avena sativa*) mixtures for nitrogen fixing and organic matter addition or cowpea *(Vigna unguiculata)*-oat mixture for nitrogen fixing and disease suppression. Considerations have also been given to sunn hemp (*Crotalaria juncea*) and **pearl millet** (*Pennisetum glaucum*) for their individual deep root systems to bring up Ca and K.

Minimum Tillage

Intensive tillage is a standard practice for tomato-producers in South Africa. However, studies have shown that intensive tillage decreases soil C, resulting in the need to adopt new and improved forms of conservation tillage or direct seeding to preserve or increase soil organic matter [11,12]. Strip tillage under *natuurboerdery* focuses on ripping small areas in rows to a depth of approximately 15 cm, followed by direct compost application on strips for transplanting. This practice results in minimum soil disturbance, but is accompanied by increases in quality and quantity of organic matter for conservation purposes [13, 14]. However, others [15] demonstrated that in a strip-grass-till for tomato production competition from Bahia grass (*Paspalum notatum*) for nutrients and water had negative impact on crop yield. Generally, untilled management strategies result in the lowest loss of soil organic cover and N in the topsoil overtime, when compared with tilled soils [10]. Others [16] also reported higher microbial population activities and greater diversity of ammonia-oxidizing bacteria on untilled plots when compared with tilled plots.

In order to convince farmers about the benefits of minimum tillage, on-field experiments were conducted to compare the performance of tomatoes planted on a strip-till in grass and tomatoes grown on ridges or raised beds in an intensive tilled field. Strip tillage resulted in reduced cost production associated with land preparation, reduced soil compaction and water-runoff, with increased water infiltration. The higher active fungal count observed from strip-tilled plots was an indication of improved microbial life [2]. However, the drawback of this system was observed slow growth during the early weeks after transplanting, which necessitated an increase in nitrogen fertilisation. Currently on-farms trials are under way to develop good agricultural practices for tomato production under minimum tillage.

Soil Health Assessment

Soil health assessment is centred on the Cornell soil health principles, which involve integrating and optimising the biological, chemical and physical properties of the soil for farm profitability [17]. Generally, before any cropping, soil samples are collected and sent to the laboratory for soil health assessment, which comprises soil chemical, physical and biological analyses. The soil health analysis consists of 32 potential indicators (Table 1), among which only 12 are selected for soil health assessment [18]. Cornell soil health scoring functions are applied for individual indicators and expressed on a percentage scale [17,19]. The 32 indicators are important in sustainable crop production as they determine soil productivity and respond to management practices in a simple and cost effective manner [3]).

Chemical and physical indicators are primarily used for soil preparation and fertilisation. For instance, irrespective of soil health status, compost at a rate of 10 m^3/ha is applied in the ridges. Chemical analysis is crucial in determining the amount of supplement fertiliser for optimum tomato yield. Basically, in *natuurboerdery* system, healthy plants grown under healthy soils are likely to withstand most abiotic and biotic stresses throughout the production cycle. Biological indicators are used to show progress made towards *natuurboerdery* as they highlight microbial life and bio-diversity in the soil. However, a recent study [18] demonstrated that biological indicators such as active carbon and nematode community were also influenced by certain management practices, for instance, mulching. However, mulching had no significant effect on soil organic carbon, although it influenced active carbon, which was an indicator of soil organic matter that was readily available as an energy source to soil microbes [17, 18]. Incidentally, mulching with grass and *Eucalyptus* woodchips reduced active carbon by 4% and 34%, respectively, while composted *Eucalyptus* woodchips increased the variable by 45%, when compared with untreated controls [18].

Table 1. Thirty-two potential indicators for use in soil health assessment [18]

Biological	Chemical	Physical
1. Active carbon	13.CEC	25. Available water capacity
2. Carbon	14. Potassium	26. Penetration resistance
3. Potential mineralisable nitrogen	15. Magnesium	27. Dry aggregate size (< 0.25 mm)
4. Beneficial nematode population	16. Calcium	28. Dry aggregate size (0.25–2 mm)
5. Parasitic nematode population	17. Phosphorus	29. Dry aggregate size (2–8 mm)
6. Root health assessment	18. Sodium	30. Stone fraction
7. Total fungi	19. Iron	31. Subsurface hardness
8. Total bacteria	20. Zinc	32. Surface hardness
9. Flagellates	21. Copper	
10. Amoeba	22. Boron	
11. Ciliates	23. Manganese	
12. pH	24. Bulk density	

Nematode Community Profiling

The nematode community profiling is often used for research and database building, with some promising results. Worldwide, there has been much interest in using nematodes as soil health indicators, as they are the most abundant soil mesofauna and occur in all ecosystems [20,21]. Nematode populations are very sensitive to environmental disturbances, and thus considered as suitable bio-indicators for assessing the responses of soil to changes in management practices [22,23,24]. In fact, work by others [20] on the influence of mulch types on soil health and nematode community profiling confirmed the need to use nematodes as potential bio-indicators when monitoring soil changes induced by management practices. Also, in the study [20] it was observed that reduced populations of plant-parasitic nematodes were accompanied by increased abundance of fungivores in organically amended plots.

PEST AND DISEASE CONTROL

Scouting for Pests and Diseases

Scouting is intended to identify the types of pest present, quantify the relative levels and determine their spatial distribution in a given field. Scouting is also carried out to evaluate the effectiveness of pest control strategies that have been applied. During tomato production, scouting for pests is conducted twice weekly. *Natuurboerdery* system integrates two scouting methods, namely visual scouting and use of appropriate tools.

Visual pest scouting: Visual scouting of pests is carried out by locally-trained expert scouts on greenhouse whiteflies (*Bemesia tabaci*), leafminer (*Liriomyza trifolii*), thrips (*Thrips tabaci*), aphids (*Myzus persicae*), American bollworm (*Helicoverpa armigera*), tomato semi-looper (*Chrysodeixis acuta*) and Red spider mite *(Tetranychus cinnabarinus*).

Pest scouting is done by counting pests from 24 plots of 2.5 m long in 1 ha. The total number of pests found is recorded and then averaged to estimate the population density of a particular pest/ha.

In case of diseases, scouts record absence or presence of a particular disease on the same plots as for insect pests. Common diseases in the area include powdery mildew (*Leveillula taurica*), bacterial speck (*Pseudomonas syringae* pv. tomato), bacterial spot (*Xanthomonas pruni*) and early blight (*Alternaria solani*). Verticillium (*Verticillium dahliae*) wilt, *Pythium* damping-off, fusarium wilt and bacterial canker (*Clavibacter michiganensis* subsp. Michiganensis) of tomato, all of which used to be prevalent on ZZ2 farms, but are currently viewed as infrequent diseases and therefore, deserving to be scouted for.

Under *natuurboerdery* scouting is done twice a week, before and after spraying. Since *natuurboerdery* aims at the reduction of synthetic pesticides, scout reports are integral parts of decision-making with regard to selection of the pest/disease management option and the material to be applied. The report is divided into three components, namely diseases, insect pests and beneficial insects. Since its inception at ZZ2, *natuurboerdery* has implemented a good record keeping system, where scout reports are linked with the prevailing climatic patterns. For instance, it has been shown that certain diseases and insect pests are directly linked to specific climatic patterns, with the result that each farm has a weather station to ensure that weather variables are constantly available in each ecosystem for use in pest/disease prediction models.

Pheromones and Sticky Traps

The main drawback of visual pest scouting is that the system is prone to human error. The use of pheromones and sticky traps when complemented by visual scouting allow for a better assessment of pests and diseases in the field. Pheromone traps contain sex pheromones, which are male attractant chemicals, with species-specific characteristics that are used by females for attracting males for mating. Potato tuber moth (*Phthorimaea opercullela*) is trapped using the Yellow Bucket Funnel Trap, which contains the PTM Pherolure, baited with a volatile pesticide 2,2 Dichlorovinyl dimethyl phosphate. The bait kills all male potato tuber moths that enter the traps [25,26]. The Green Bucket Funnel Traps, containing TV Pherolure and baited with DDVP, comprise five plant volatiles, which attract mainly American bollworm and tomato semi-looper in a ratio of 70% female and 30% male moths [27]. Whiteflies, thrips, tomato leaf-miner and aphids are attracted to the Yellow Sticky Trap (YST), made from yellow plastic cardboard with standard grid for pest counting [27]. An adhesive paint, referred to as Flytac, is used in combination with YST to ensure that insects making contact with the card get stuck until the cards are cleaned during scouting [26, 7].

Indicator Plants

Indicator plants have been used successfully as monitors for the presence of viruliferous insects among both incoming and resident insect populations. Eggplant (*Solanum melongena*) plants are planted at the beginning of tomato rows produced under *natuurboerdery* for also serving as indicator plants during pest and disease scouting. The plants are highly susceptible to insects and diseases and thus, serve as an early-warning system for managing pests that attack tomatoes under *natuurboerdery* [28].

Use of Economic Threshold Levels

The concept of economic threshold levels (ETL) is important in commercial tomato production. ETL for tomato pests have been developed by ZZ2 scientists and Insect Science (Pty) Ltd over the years through observations and on-farm-trials which are currently underway to refine existing ETL for all pests. ETL for virus-vectors such as whiteflies, thrips and aphids are particularly important in order to minimise viral attacks and the former's implications in reducing visual fruit quality [29]). ETL of selected pests are presented (Table 2). For leafminer flies, no threshold level has been established and chemical control is more leaned towards larva control.

Table 2. Economic threshold levels of selected tomato pests under field production developed by ZZ2 scientists and InsectSA

ARTHROPOD PESTS		STAGE	THRESHOLD
American bollworm	*Helicoverpa armigera*	Egg	1 egg/ 4 plants
		Larva	1 larva/ 8 plants
Tomato semi looper	*Chrysodeixis acuta*	Egg	1 egg/ 4 plants
		Larva	1 larva/ 12 plants
Green peach aphid	*Myzus persicae*	Adult	1 adult/ plant
Leafminer	*Liriomyza trifolii*	Larva	1 larva / 3 plants
		Adult	-
Whitefly	*Bemesia tabaci*	Adult	1 adult/ plant
		Nymph	1 nymph/ 12 plants
Thrips	*Thrips tabaci*	Nymph	1 nymph/ 8 plants
		Adult	1 adult/ 8 plants

Augmentation of Beneficial Insects

The short duration of vegetable crops in rapid successions favours whiteflies and limits the initial numbers of natural enemies on crops, unless the numbers are artificially augmented or local reservoirs are established on indigenous plants or weedy hosts [30]. ZZ2 agronomists and scientists have introduced the neutral zone concept, which consists of small plots of eggplant and tomato plants surrounding the production area. These plots are not sprayed in order to create conducive environment for multiplication of beneficial insects. Because the neutral zones are exempted to any chemical intervention, beneficial insects would multiply and later migrate to the production area to prey or parasitize on insect pests present in the field. The neutral zone also serves as barrier for insect pests from neighbouring fields or natural habitats, since pests would first land on the neutral zone whereby they are controlled by natural enemies before migrating to the field [31]. Under *natuurboerdery* systems, the neutral zone is populated by a large number of predators (Table 3).

Table 3. List of selected predators in the *natuurboerdery* neutral zone [31]

Predator	Insect pest	Life cycle
Parasitic wasp *(Encarsia formosa)*	Whiteflies	Nymphs
Ladybird beetle *(Delphastus pusillis)*	Whiteflies, aphids	Nymphs
Minute pirate bug (*Orius insidiosus*)	All insects	All stages
Lacewing (*Chrysoperla rufilaris*)	All insects	All stages
Chalcid wasp *(Aphytis spp.)*	Aphids	Adults

Plant Extracts

In order to reduce the over-reliance on chemical pesticides under natuurboerdery system, farmers use botanical insecticides derived from plants such as neem (*Azadirachta indica*) and wild garlic (*Tulbaghia violacea*) for pest control. Neem leaf extracts act as a strong antifeedant and repellent, delay and prevent moulting, reduce insect growth, development and oviposition; and therefore have been shown to cause high mortality rates in more than 200 insect species [32, 33, 34, 35, 36, 37]. Under *natuurboerdery* fresh leaves of neem and wild garlic bulbs are fermented with effective microorganisms, molasses and sugar, with the resulting product being called fermented plant extracts (FPEs), for use to control insect pests. Generally, the product is used in the early stages of the crop to keep the pest population below the acceptable threshold values. Additionally, on-farm-trials were conducted using the product to assess its efficacy on whitefly and aphid populations, which have distinct seasonal patterns - with higher counts occurring in summer than in winter. During both seasons, populations of both pests gradually increased under treatment, although the numbers were significantly lower than those on untreated controls [38]. Primarily, the product might have had a repellent effect than an antifeedant activity on both pests. ZZ2 has since established an internationally-recognised herb garden, where different herbs used for research purposes in plant protection are grown. Promising results on suppression of the two pests have been obtained with essential oils of rosemary (*Rosmarinus officinalis*) when used as crude extracts in water [39].

The root-knot nematodes (*Meloidogyne* spp.) are sedentary endo-parasites, which are among the most damaging agricultural pests of tomato, with a wide range of host plants [40, 41, 42]. Before the convention to *natuurboerdery*, ZZ2 farmers relied much on methyl bromide for suppression of *Meloidogyne* spp. However, due to the negative impact of fumigant nematicides on soil health and the environment in general, *natuurboerdery* adopted the use of FPEs from locally available plants such as *Lantana camara*. A product called Nemalan was developed and is currently applied on preventative basis, resulting in sensible reduction of nematode population densities [43].

Use of crude plant extracts to control plant-parasitic nematodes in tomato is an on-going research programme at the University of Limpopo, South Africa. A team of researchers at this university developed bio-nematicide products from two indigenous plants, wild cucumber (*Cucumis myriocarpus*) and wild watermelon (*Cucumis africanus*) - Nemarioc and Nemafrica, respectively. In a recent review of the two products [44], it was demonstrated that crude extracts of the two plants reduced numbers of *Meloidogyne* species by 80-90% in what had been popularised as the ground leaching technology (GLT) among smallholder farmers in marginal communities. The efficacy of Nemarioc bio-nematicide was comparable to that of nemacur and fenamiphos in nematode suppression [44].

In GLT system, both Nemarioc and Nemafrica have had a fertiliser effect on tomato plants [44]. Using the Curve-fitting Allelopathic Response Dosage (CARD) computer model, others [45] demonstrated that tomato plant growth responded to a series of Nemarioc dosages in a density-depended growth pattern, which was characterised by stimulation, saturation and inhibition ranges (Figure 3), where x is dosage and y is g fresh shoot/plant. Using the average dosages generated by CARD model between the stimulation and saturation points, the workers [45] demonstrated that the optimum stimulation dosage for this material on tomato was 2.6 g/plant, which was approximately equivalent to the recommended dosage of 2 g/plant [44]. Currently, both n*emarioc* and *nemafric* bio-nematicides are being developed by the University of Limpopo for use as FPEs in large-scale commercial tomato farming systems through drip irrigation. This concept, being popularised as botinemagation – using botanicals for nematode management through irrigation systems, is more cost-effective than GLT system [44].

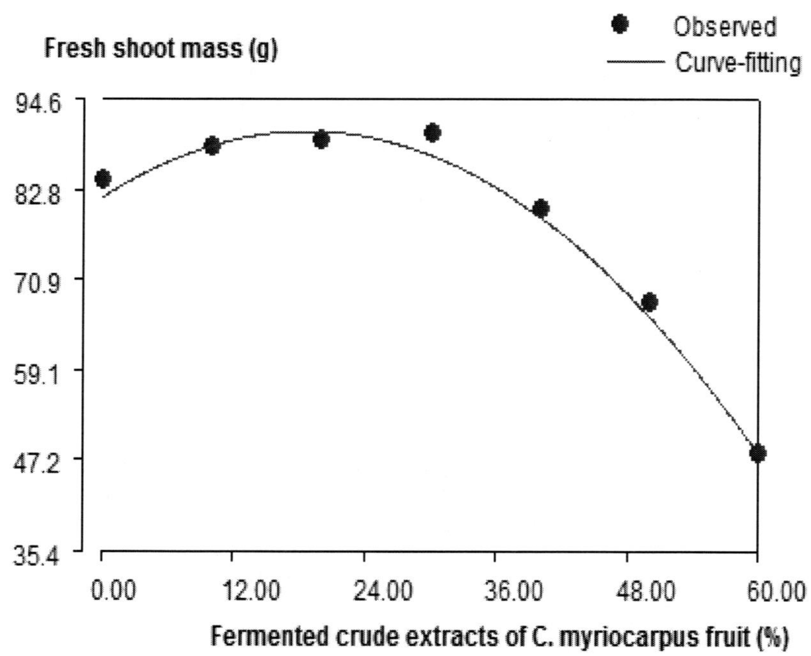

Figure 3. Relation between fresh shoot mass (g) per plant and dosage of *nemarioc* bio-nematicide.

Chemical Control

Farmers in *natuurboerdery* system do not completely preclude the use of synthetic pesticides, but this option is used as a last resort when all other control options are unsuccessful in suppressing population densities of a pest or disease to below ETL. Even when pesticides are needed, farmers are encouraged to consider using the least toxic pesticides. Insecticides are used at the labelled rates and alternated among pesticide groups, ensuring that only the target pest is controlled.

In order to ensure that there is no pesticide-breakdown or precipitation in the tank; water is tested and adjusted primarily for pH using adjuvants. Mixing of pesticides is carefully done to minimise precipitations. ZZ2 scientists and collaborators have developed an IPM coding system which allocates a score to each pesticide on the basis of health and environmental criteria such as acute toxicity. The scores range from 0 (non-toxic) to 5 (highly toxic). At the end of harvesting, a final score for pesticide usage is allocated to the field where it was applied. Trends of pesticides usage on the farm are regularly analysed and farmers are encouraged to use chemical pesticides only when all other means have been deemed unsuccessful. Generally, there had been a significant reduction in the use of synthetic pesticides in terms of volumes and numbers, for the control of pests and diseases on ZZ2 farms. Prior to the adoption of *natuurboerdery*, large volumes of different synthetic pesticides were used on various farms. However, under *natuurboerdery* much less volumes are used during a single tomato growing season, which is from 24 to 28 weeks – depending on environmental conditions.

Insect Control

Phloem-feeding insects such as whiteflies and aphids are virus-vectors and had previous become resistant to a wide range of pesticides, including pyrethroids, organophosphates, carbamates and insect growth regulators due the high frequencies of use [46, 47, 48]. To date, there is no effective chemical control of these insects. To prevent whitefly population build-up during production, seedlings are drenched with with Actara® (thiamethoxam 25%) at label rates of 0.03 ml $plant^{-1}$ 10 days after planting. On-farm-trials demonstrated that Actara® reduced the rate of whitefly infestation during the growing season.

Leafminer is one of the most damaging insect pests in tomato production. Both larvae and adults can cause damage, with larvae primarily mining the palisade mesophyll, while female adults puncture both upper and lower leaf surfaces to feed and lay eggs [49,50,51]. Similarly, leafminers have since developed resistance against an array of pesticides. In our experience, continued control of adult leafminers with available synthetic pesticides appears just impossible to accomplish. Control of larvae is achieved through abamectin, which is a translaminar, but over-reliance on a single chemical product is invariably dangerous in any farming system. Under *natuurboerdery*, to minimise repeated application of abamectin, cyromazine is usually applied as a drench at 7-10 days after transplanting. On-farm-trials conducted by ZZ2 agronomists and researchers demonstrated that when cyromazine was applied after planting, seedlings were less infested by leafminers than untreated plants [52]. Use of any synthetic pesticide for any pest under *natuurboerdery* is orchestrated in such a way that it integrates soil, plant, social, environmental and ecological health systems.

Disease Control

Bacterial speck and bacterial spot are economically important diseases on tomatoes in South Africa. Bacterial speck is more severe under cool and humid conditions, whereas bacterial spot is favoured by warm and rainy weathers. Under ZZ2 farming conditions, the two diseases tend to overlap and strike when temperatures decrease. Both diseases may cause significant reduction in tomato crop yield, especially if the infection appears early in the season [53, 54, 55]. Usually, cases of complete crop failure due to infestation by any of the two pathogens abound in tomato production. Also, infected fruits are of much low quality to the extent that they are not packaged for the market.

Management of any of the two diseases is a challenge in commercial field production due to the limited efficacy of the available disease management strategies [56]. Even the use of high quality seedlings by ZZ2 had not succeeded in curbing the diseases under field conditions. Copper formulations and mancozeb are regularly sprayed and complemented with a plant activator called acibenzolar-S-methyl. Despite all the management strategies in place, the disease always strike and is usually critical at the end of autumn (March-April). Bacterial diseases were more severe in neutral zones (unsprayed plots) when compared to production fields where rigorous management strategies were in place [57]. Severity of these diseases during certain periods have enthused ZZ2 scientists to investigate potential resistance of bacterial diseases to copper formulations. Studies elsewhere had shown that bacterial speck and spot strains are resistant to copper formulations [58].

Powdery mildew (*Leveillula taurica*) is another economic disease in tomato production under open field conditions in South Africa. Worldwide, the control of powdery mildew had been achieved through buprimate, benomyl, bittertanol, carbendazim, triabendazol, trifoline, fenarimol and pyrazophos fungicides, along with various sulphur formulations [59]. Among the listed fungicides, only sulphur formulations are currently being used under *natuurboerdery* system. Incidentally, chemical control under field production for powdery mildew had not been successful in *natuurboerdery* farming systems, and thus, other good agricultural practices, described below, had been necessary for farmers to harvest a good crop.

SEEDLING HEALTH AND DEVELOPMENT

Use of high quality tomato seedlings produced from certified seeds is a prerequisite for the production of high quality seedlings in *natuurboerdery*. In the nursery, seedlings receive extra-care through optimising irrigation, fertigation, pest/disease prevention and applying common sense agricultural health practices. Among other strategies used, pre-sowing inoculation with arbuscular mycorrhizal fungi (AMF) and *Trichoderma harzianum* is practiced [60], while the use of seaweed extract (SWE) and silicon (Si) is still on trial. The main objectives of seed treatment are to protect the developing seedlings from infection by soil-borne pathogens and to improve plant growth through enhanced nutrient uptake [61]. Ultimately any high quality seedling should, invariably, lead to improved fruit yield and quality, although trials have shown mixed results, particularly in fields heavily infested with soil-borne diseases.

Worldwide, use of *Trichoderma* spp. and AMF had been increasing and is a promising alternative for controlling soil-borne diseases in sustainable and organic agriculture [62]. The main purpose of nursery inoculation with AMF is to produce high quality seedlings for improved performance in the field [63]. Both greenhouse and field trials have shown that through its mutualistic relationship with the host plant, AMF could suppress the incidence of soil-borne diseases caused by *Pythium* spp., *Rhizoctonia* spp., *Fusarium* spp., *Phytophthora* spp. and *Verticillium spp.* [64,65,66,67]. In pepper (*Capsicum annum*), AMF reduced the negative effect of *V. dahliae* and improved fruit yield and quality of inoculated plants [68].

The most salient biological control activity of *Trichoderma* spp. has been the suppression of damping-off disease caused by *Pythium* spp., which poses serious threats in greenhouse and field production, particularly in the early stages of seedling growth [69,70]. A review of the biology and systematics of the genus *Trichoderma* had been undertaken, suggesting the presence of effective and non-effective strains in disease suppression [71]. Certain effective strains of *Trichoderma* spp. were shown to provide control of diseases such as damping-off caused by *Rhizoctonia solani* in radish [72], maize (*Zea mays*) and soybean (*Glycine max*) [73], cucumber (*Cucumis sativus*) fruit rot [74], verticillium wilt caused by *V. dahliae* on tomato [75] and grey-mould caused by *Botrytis cinerea* on tomato [76].

Although the application of *Trichoderma* spp. and AMF individually had shown the potential to control soil-borne diseases, others [77] believed that biological control agents applied alone are not likely to perform consistently against soil-borne pathogens under different rhizospheres and therefore, suggested a combination of various inocula. Under *natuurboerdery* system the two biological control agents are simultaneously applied during preparation of seedlings. Initially, the main concern for farmers was that the two biological control agents could negatively interact with each other due to their competition for carbohydrates. However, trials under *natuurboerdery* [60] demonstrated that *T. harzianum* could successfully be inoculated alongside with AMF for improved seedling health and development in tomato production. In a greenhouse trial [78], it was shown that pre-inoculation of seedlings with *T. harzianum* and AMF reduced the incidence of verticillium wilt, but was not able to completely counter infection by *V. dahlia* as all plants expressed verticillium wilt symptoms at the end of the growing season. The observations outlined in this section reinforced the contention that an integrated pest/disease management strategy is indispensable in tomato production.

Bio-Stimulants and Microbial Inoculants

Bio-stimulants are natural or synthetic products of either mineral or organic composition that by their mode of action positively contribute to crop nutrition and the development of healthy plants [79]. Tomato farmers use seaweed extracts (SWE), fluvic and humic acids, plant activators in their fertiliser and spraying programmes. Other techniques used include the application of silicon.

Seaweed Extracts

Seaweed extracts are regularly applied by farmers during the active vegetative growth of tomato production. Crouch and Van Staden [80]) reported that plants treated with SWE often have an improved mineral nutrition and growth due to the more vigorous root system associated with SWE application which provides a larger surface area for nutrient and water uptake. Booth [81] observed that the value of seaweeds as fertilisers was not only due to nitrogen, phosphorus and potash content, but also because of the presence on trace elements and metabolites. Featonby-Smith and Van Staden [82] demonstrated that dilution of 1:500 significantly improved growth of tomato plants irrespective of whether it was applied as a foliar spray at regular intervals, or whether the soil in which tomatoes were planted was flushed once with diluted seaweed concentrate. On-farm-trials confirmed these findings and suggested a combination of drench prior to transplanting and three foliar sprays during early vegetative growth stages [83]. Moreover, Featonby-Smith and Van Staden [82] demonstrated that due to the cytokinin content; SWE might affect the resistance of plants to disease without eliminating the infestation itself while preventing pest infestation. Molecular analysis of SWE-treated plants demonstrated that there was a higher accumulation of defence gene transcripts, with enhanced defence enzyme activities, along with accumulation of phenolics and phytoalexins [84], all of which had been directly linked to plant resistance against pests. This suggests that the use of SWE could be a good preventative tool for pest and disease management under field conditions.

Plant Activators

When plants are attacked by pathogens, they respond by activating a variety of defence mechanisms [85]. Pre-planting treatment of susceptible plants with a biotic inducer can enhance resistance to subsequent attack not only at the site of treatment, but also in tissues distant from the initial injection site [86]. A synthetic compound, which is currently commercially available, ASM, appears to have activities similar to those of salicylic acid, which induces systemic acquired resistance (SAR) [85, 87, 88]. Typically, SAR is effective against various pathogens including viruses, bacteria and fungi [86, 89]. Other studies [86, 90, 91, 92] had shown that the plant activator ASM increased resistance levels to fungal, bacterial and viral agents in tomato plants. Incidentally, under *natuurboerdery* system ASM is applied with copper formulations to control bacterial diseases. However, regular applications of ASM under field conditions stunt tomato growth.

Effective Microorganisms

Effective microorganisms (EM) technology, originally developed in Japan for nature farming, had been adopted through on-farm trials as an integral part of *natuurboerdery* system. Effective microorganisms constitute a mixture of photosynthetic bacteria, *Azotobacter*, *Streptomyces*, *Lactobacillus* spp. and various minor protozoas with different roles. EM improve crop yield by increasing photosynthesis, N fixation, controlling soil-borne diseases and accelerating decomposition of lignin material in the soil [93,94]. Also, EM produce substances acting as antioxidants, such as inositol, ubiquinone, saponin, low-molecual polysaccharides, polyphenols and chelating agents [95,96], which may collectively or individually inhibit harmful microbial species; enhance the proliferation of beneficial microorganisms and detoxify harmful substances.

Under *natuurboerdery* system, EM are primarily used for composting, soil health improvement and plant protection. EM are commercially available as EM stock solution, which is a dormant solution that needs to be activated by the provision of water and sugars. The diluted EM (1:1000) called EM-extended is used for tomato crop production and applied at rate of 10 L/ha every second week through drip to enrich the soil microbial population. Although, the science behind EM is still not fully understood, the benefits under *natuurboerdery* have been noticeable. For instance, when soil is compacted, farmers usually increased the amount of EM, resulting in a breakdown of the compaction layer. Other benefits of EM are that it is used as the main ingredient for the preparation of fermented plant extracts, which are also used for pest and disease control with little evidence as to whether active ingredients are from the plant materials, EM or both.

Compost Tea

Compost tea is a brewed water extract of compost used to enrich soil with beneficial microorganisms and soluble nutrients. The brewing, which last 24-36 h is an aerated process, which was introduced in *natuurboerdery* system by the renowned microbiologist researcher - Dr Ingham from Rodale Institute, USA. The quality of compost tea depends on the quality of compost used, on temperature during brewing, foods added and oxygen concentration. The latter should be greater than 5.5 ppm for active aeration. Because composts produced under *natuurboerdery* contain large numbers of bacteria since the used microorganisms multiply fast, compost tea also has higher bacteria count, although the main focus is to increase fungal population through food additives such as fish meal and seaweed extracts.

The quality of compost tea is regularly assessed at ZZ2 laboratories which use the standard procedures from the Soil Food Web Institute, USA. Standard analyses assess active and total bacteria, active and total fungi, protozoa and nematodes. For human health safety, regular analyses of *Escherichia coli* and salmonella are carried to ensure that the product does not pose any threat to human health. During tomato production, compost tea, with an estimated shelf-life of 6 h, is directly applied in the field through drip irrigation. Assumptions abound that compost teas control diseases through secretion of antibiotic substances, competition for spaces, inhibition of the development of plant diseases, antagonistic effects and parasitism [97]. Physiochemical properties of compost teas, namely, nutrients and other organic molecules such as humic or phenolics compounds may also protect plants against diseases through improved nutritional status, direct toxicity towards pathogens or induced systemic resistance [98, 99].

Silicon

Generally, silicon (Si) is a non-essential nutrient for most plants, while it is required for optimal growth and sustainable production of **rice.** Si is regularly applied in the form of non-toxic silica to confer beneficial effects on plant health when roots are challenged by pathogens [100]. Two mechanisms for Si-enhanced resistance to diseases have been proposed: (1) Si creates a mechanical barrier that prevents penetration of pathogens into the root system [100] and (2) soluble Si acts as a modulator of host resistance to pathogens [101]. In order to benefit from Si, plants must be able to acquire the element in high concentrations regardless of whether they are monocotyledonous or dicotyledonous plants [101]. However, tomato plants belong to the group of excluders, their uptake of Si is quite passive. In one study [102], Si did not inhibit the spread of *Pythium aphanidermatum* in tomato, but in another study [103] a decrease in the incidence of bacterial wilt on tomato was associated with the induction of basal resistance mechanisms at the cell wall level and pit membranes after the plants were exposed to Si. Fauteux et al. [104] observed that continuous exposure to

Si enhanced the ability of plants to mount defences against powdery mildews. Despite the discrepancies in literature on the role of Si in crop protection, tomato farmers believe that Si might have unknown capabilities that play a positive role during tomato production.

Good Agricultural Practices

Plant Nutrition

Plant nutrition plays an important role in tomato production as healthy and vigorous plants tend to be less susceptible to biotic and abiotic stresses than the unhealthy plants. Because *natuurboerdery* is a process characterised by a smooth and gradual conversion to low input agriculture, farmers advertently use both organic and inorganic compounds for plant nutrition, ensuring that the correct amount is applied at the right time, while avoiding over fertilisation. As explained in the soil health management section, soil analysis is mandatory before any planting, with the recommendations being based on the Albrecht model for soil fertility. According to the model, the ideal ratio for tomato production for Ca, Mg, K and Na is 65-70%, 10-15%, 5% and less than 3%, respectively. Basically, the Albrecht model is used to identify plant nutrient shortages and correct them in order to create the right balance for nutrient uptake through the application of nutrient-enriched compost and side-dressing programmes. In some cases, when Ca ratio is very low, gypsum is applied as complementary corrective measure but regardless of the soil status, the amount to be applied is always below 2 t/ha.

Plant sap analysis, which is a simplified method for the extraction and analysis of plant sap, is used to monitor the nutrient status of the tomato plants during the growing season. Results are useful for the prediction of impending nutrient element deficiencies or simply for adjusting a fertiliser program to prevent luxurious feeding and then toxicities [105]. Basically, leaf samples from the most-recently-matured leaves are collected early in the morning and sent to an on-farm-laboratory for sap analysis. A garlic presser is used to squeeze the sap from the petioles, with the aliquot being used for the determination of K and N concentrations, Brix level, pH and EC. Potassium and nitrate ions are determined by the K and N cardy meters, respectively, pH with a pH-meter, EC with EC meter and dissolved solids (% Brix) with a refractometer. Interpretations of results are based on the crop stage of the tomato [105]. Plant sap analysis is used as a temporary measure, with careful considerations of the plant history, field potential, pest/disease incidence, climate and irrigation scheduling before making any changes. The technique does not replace the leaf chemical analysis, which is done only if serious nutrient deficiency or toxicity symptoms are observed.

Irrigation Management

Irrigation scheduling is critical part of integration in *natuurboerdery* system since excessive or inadequate irrigation invariably reduce plant growth, crop yield and fruit quality [106,107]. Under *natuurboerdery* scheduling is manually done, with farmers using drip irrigation exclusively since it allows for precise and controlled water applications to the active root-zone of the crop. The other benefit of drip irrigation is that fertiliser solutions can be injected into the irrigation system so that nutrients can be supplied to the crop throughout the season as needed, thereby reducing the potential loss of nutrients through leaching to outside the active root-zone [108]. Drip irrigation fits well in the integrated pest/disease management tactics under *natuurboerdery* system as overheard irrigation poses serious threats with respect to splashes for spreading bacterial speck and bacterial spot and enhancing foliar and fruit diseases [109]. Blossom-end rot of tomato is also influenced by irrigation as water stress reduced the uptake of Ca.

Various tools, from very sophisticated tools, water software and irrigation modelling, had been tested under field conditions on ZZ2 farms, but it all came back to the feel of soil moisture on the farms. *Natuurboerdery* combines irrigation devices with visual soil moisture inspection for scheduling irrigation. The system consists of making soil profile hole and inspecting water movement in the soil. The amount of water to be applied depends on the soil type, crop growth stage, crop demand using the reference evapotranspiration, weather forecast and other management plans such as compost tea application or bio-nematicide/fertiliser application [109].

Weeding

Control of weeds represents one of the major challenges for tomato producers under *natuurboerdery*, especially with the endeavour to avoid toxic synthetic pesticides. Currently, although costly, weeds are controlled through hand-hoeing. Weeds compete with tomato plants for nutrients and water, and could also harbour pests such as nematodes, whiteflies, bacterial spots and viruses. Common weeds in the area include nightshade (*Atropa belladonna*), morning glory (*Ipomea indica*), crad grass (*Digitaria sanguinalis*), thorn apple (*Datura stramonium*) and couch grass (*Cynodon dactylon*). Generally, these weeds are difficult to control. The research team at ZZ2 had been investigating the possibility of using steam water to control weeds, which offers an ecological and economical means of weed control.

Staking and Pruning

Indeterminate tomatoes grown on ZZ2 farms are characterised by rapid growth, posing a management risk for farmers, if not properly and timely trellised and pruned. Trellising is done by sinking wooden poles of 2 m in the ground every 2.5 m in each row. Tomato plants are transplanted between the two poles and allowed to grow upright between a set of wires, which run from one end to the other. As the plant growths, a new set of twines is gradually tied upward until they reach the top. Basically, this technique is suitable for indeterminate tomato plants, as it allows them to remain off the ground where fruit quality could be compromised by disease infections.

The rapid vegetative growth of indeterminate tomato plants could also compete with the reproductive growth, resulting in poor yield. In order to maintain a balance between vegetative and reproductive growths, pruning is usually done. The technique consists of removing stems and side shoots in order to enhance good air circulation, light penetration and better coverage for spray. Depending on the planting system, the plant densities range from 11,500 to 16,000 plants/ha. Generally, plants that had not been pruned form a bushy canopy, harbouring pests and diseases, which could be difficult to control during spraying. On-farm-trials showed that combined stem and fruit pruning increased yield and improved fruit quality of tomatoes when compared with untreated controls [83].

Cultivar Selection and Planting Date

Cultivar selection is a critical decision for commercial tomato production. On-farm-trials are regularly conducted to determine the right cultivar, which must adapt to the growing seasons, with good yield potential, possessing all the horticultural traits desired in the region. Because some pests and diseases are difficult to control, the use of resistant or tolerant cultivars presents an effective way for pest/disease management. Certain viruses such as tomato yellow leaf curl, transmitted by whiteflies, are no longer a major threat in the region due to the use of resistant cultivars.

Under *natuurboerdery* system, cultivar selection starts with a screening process, where a large number of cultivars are evaluated. The best cultivars will then be planted in a double row system and subjected to semi-commercial production conditions. From this, the best cultivars would then be selected for commercial tomato production. Generally, the possibility of obtaining the right cultivar which combines higher yields, better fruit quality and pest and disease resistant is slim, suggesting necessary compromises. Fruit characteristics such as size, colour, shape and firmness are equally important for the acceptance of the product by consumers.

Selection of planting dates is dictated by several factors including climatic conditions, historical yield data and market predictions. In areas with low and excessively high temperatures which reduce plant growth, planting dates are selected in a way that the extremes will not coincide with the active cropping season. ZZ2 farms are situated in three climatic zones, which allow for winter (January – July), summer (September – May) or all-year-round production.

Conclusion

Natuurboerdery comprises an innovative hybrid farming system from conventional chemical and organic farming systems, with the overall outputs of the system being soil health, plant health, environmental health, ecological health and human health [2]. In this chapter, the inputs, transformation processes and the outputs involved in transforming chemical farming system to *natuurboerdery* system were discussed with specific reference to the management of pests and diseases in tomato production. In ZZ2, where *natuurboerdery* was initiated, the soil health output is being achieved through inputs that include organic amendments, composts, effective microorganisms, crude extracts, minimum tillage, cover cropping, fallowing, regulated fertilisation and regulated irrigation. In contrast, the plant health output is being achieved through cultivar selection, bio-stimulants, microbial inoculants and properly designed irrigation and nutrition practices. Simply stated, pest/disease management using synthetic chemicals is used as the last resort, with scouting, use of plant extracts, effective microorganisms and beneficial insects being actively pursued to maintain pest/disease population build-ups below economic threshold levels. ZZ2 case study provides a successful model of implementing the *natuurboerdery* farming system, which can be described as having the potential benefits of a sustainable farming system in the true sense of climate smart agricultural principles. However, since *natuurboerdery* is a living system, there are still numerous grey areas in the inputs, transformation processes, outcomes and the feedback loop which could be filled through multi-disciplinary intensive research.

REFERENCES

[1] ZZ2, *Natuurboerdery* Production System, Technical Note. ZZ2 Archives, (2006).

[2] B. Nzanza, Introducing *Natuurboerdery*: The ZZ2's integrated farming approach. Technical Note. Research & Development, ZZ2 (2008).

[3] G. Abawi, B. Gugino, O. Idowu, R. Schindelbeck, H., D. van Es, B. Wolfe, B. Moebius and J. Thies, *Cornell Soil Health Training Manual,* University College of Agriculture and Life Sciences, Cornell (2007).

[4] R. Albiach, R. Canet, F. Pomares, F. Ingelmo, *J. Bioresour. Technol.* 75, 48 (2000).

[5] R. Albrecht, R. Joffre, R. Gros, J. Le Petit, G. Terrom and C. Perissol, *Bioresour. Technol* 99, 448-455 (2008).

[6] K. G. Cassman, S. K. de Datta, D. C. Olk, J. Alcantra, M. Sason, J. Descalsota and M. Dizon, *Yield Declined and the Nitrogen Economy of Long-term Experiment on Continuing, Irrigated Rice System in the Tropics*, pp. 181-222 (1995). In: R. Lal and B. A. Stewart (eds.). Soil Management: Experimental basis for sustainability and environmental quality. Lewis Publisher, London, UK (1995).

[7] R. J. Haynes, *Adv. Agron.* 37, 345 (1984).

[8] D. C. Coleman, R. V. Anderson, C. V. Cole, E. T. Elliott, L. Woods and M. K. Campion, *Microb. Ecol.* 4, 380 (1978).

[9] A. C. Kennedy and K. L. Smith, *Plant and Soil* 170, 75-86 (1995).

[10] B. Liu, C. Tu, S. Hu, M. Gumpertz and J. B. Ristaino, *Appl. Soil Ecol.* 37, 202–214 (2007).

[11] R. Lal, R., J. Kimble, R. F. Follett and C. V. Cole. *The Potential for U.S. Cropland to Sequester Carbon and Mitigate the Greenhouse Effect.* Ann Arbor Press, Ann Arbor, MI. 128 pp. (1998).

[12] E. A. Paul, K. Paustian, E. T. Elliott and C. V. Cole, *Soil Organic Matter in Temperate Agroecosystems*, CRC Press, Boca Raton (1997).

[13] R. R. Allen, *App. Eng. Agric.* 4, 191-196 (1988).

[14] M. A. Arshad, M. Schnitzer, D. A. Angers and J. A. Ripmeester, *Soil Biol. Biochem. 22*, 595-599 (1990).

[15] D. O. Chellemi, F. M. Rhoads, S. M. Olson, J. R. Rich, D. Murray and D. M. Sytvia, *Am. J. Alt. Agr.* 14, 59-68.

[16] M. Chiurazzi, *PhD Thesis*. Universita Degli Studi di Napoli "Federico (2008).

[17] B. K. Gugino, O. J. Idowu, R. R. Schindelbeck, H. M Van Es, D. W. Wolfe, B. N. Moebius-Clune, J. E. Thies and G. S. Abawi, *Cornell Soil Health Assessment Training Manual*, Cornell University, New York, USA (2009).
[18] B. Nzanza and P. Pieterse. *South Afrn. Avo. Grower's Assoc.* Yearbook, 34, 78 (2011).
[19] S. S. Andrews, D. L. Karlen and C. A. Cambardella, *Soil Sci. Amer. J.* 68, 1962 (2004).
[20] B. Nzanza and P. Pieterse. *All Afr. Hort. Conf.* 2, 119 (2012).
[21] K. Ritz and D. L. Trudgill, *Plant and Soil* 212, 11 (1999).
[22] D. W. Freckman and C. H. Ettema, *Agric. Ecosyst. Environ.* 45, 261 (1993).
[23] D. A. Neher, J. Nematol. 33, 168 (2001).
[24] G. D. Zheng, L. B. Shi, H. Y. Wu and D. L. Peng, Acta Agric. Scand. DOI:10.1080/09064710.2011.598545 (2011).
[25] P. J. G. Booysen, *Tomato Pest, Fungus & Beneficial Insects:* Scouting Course. Tzaneen, South Africa (2002).
[26] S. Tauranyi, *Masters Dissertation, Univ. Stellenbosch, S. Afr.* (2010).
[27] P. J. G. Booysen and D. Hoets, *Trapping Guidelines for Area-wide Tomato Pest Monitoring Guidelines.* ZZ2, South Africa (2010).
[28] ZZ2, *Technical Note on Natuurboerdery Practices.* ZZ2 Archives, South Africa, (2004).
[29] D. J. Schuster, http://edis.ifas.ufl.edu (2009).
[30] J. Colvin and V. Muniyappa, http://teca.fao.org (1999).
[31] B. Venter, ZZ2 Archives (2010).
[32] D. L. Coudriet, N. Prabhaker and D. E. Meyerdirk, *Environ. Entomol.* 14, 779 (1985).
[33] P. Kumar, H. M. Poehling and C. Borgemister, *J. Appl. Entomol.* 129, 889 (2005).
[34] P. Kumar and H. M. Poehling, *J. Pest. Sci.* 79, 199 (2006).
[35] T. X. Lui and P. A. Stansly, *J. Pest. Sci.* 44, 322 (1995).
[36] P. L. Mitchell, R. Gupta, A. K. Singh and P. Kumar, *J. Econ. Entomol.* 97, 923 (2004).
[37] N. Prabhaker, N. C. Toscano, and D. L.Coudriet, *J. Econ. Entomol.* 82, 988 (1989).
[38] B. Nzanza and P. W. Mashela, Afrn J. Biotech., In Press (2012).
[39] B. Nzanza, *Annual Research Review 2011.* Res. & Dev., ZZ2. S. Afr. (2011).
[40] M. Hashem and K. A. Abo-Elyousr, *Crop Prot.* 30, 292 (2011).
[41] S. Kiewnick and R. A. Sikora, *Biol. Cont.* 38, 187 (2006).
[42] N. Sahebani and N. Hadavi, *Soil Biol. Biochem.* 40, 2020 (2008).
[43] P. Prinsloo, 2010 at http://www.nstf.org.za.www.
[44] P. W. Mashela, D. De Waele and K. M. Pofu, *Sci. Res. and Ess.* 33, 6768 (2011).
[45] T. P. Mafeo, P. W. Mashela and M. S. Mphosi, *Proc. Inter. Hort. Con.* 28, 263 (2010).
[46] M. Cahill, F. J. Byne, I. Denholm, A. Devonshire and K. Gorman, *J. Pest. Sci.* 42, 138 (1994).
[47] A. R. Horowitz, G. Forer and I. Ishaaya, *J. Pest. Sci.* 42, 122 (1994).
[48] C. Jazzar and E. A. F. Hammad, *Bullet. Insect.* 56, 275 (2003).
[49] M. P. Parrella, C. B. Keil and J. G. Morse, *California Agric.* 38, 33 (1984).
[50] H. S. Civelek, *Phytoparasitica* 31, 138 (2003).
[51] M. B. Hossain and H. M. Poehling, *J. Appl. Entmol.* 130, 360-367 (2006).
[52] B. Venter, *Effect of Cyromazine on leafminer infestation in tomato.* Technical report, Natuurboerdery Production, ZZ2, South Africa (2012).
[53] K. Pohronezny and R. B. Volin, *HortScience* 18, 70 (1983).
[54] H. Yunis, H. Bashan and Y. Henis, Plant Dis. 64, 937-939 (1980).

[55] P. Ji, H. L. Campbell, J. W. Kloepper, J. B. Jones, T. V. Suslow and M. Wilson, *Biol. Control* 36, 358-367 (2006).
[56] F. J. Louws, M. Wilson, H. L. Campbell, D. A. Cuppels, J. B. Jones, P.B. Shoemaker, F. Sahin and S. A. Miller, *Plant Dis.* 85, 488 (2001).
[57] B. J. Van Zyl, *ZZ2 Archieves* (2011).
[58] G. W. Sundin and C. L. Bender, *Appl. Environ. Microbiol.* 59, 1024 (1993).
[59] B. Mieslerova and A. Lebeda, *J. Plant Dis. Prot.* 106, 157 (1999).
[60] B. Nzanza, D. Marais and P. Soundy, *Afri. J. Microbiol. Res. 5*, 431 (2011).
[61] A. G. Taylor and G. E. Harman, *Annu. Rev. Phytopathol.* 28, 339 (1990).
[62] O. Erdogan and K. Benlioglu, *Biol. Control* 53, 45 (2009).
[63] S.Gianinazzi, H. Schuepp, J. M. Barea and K. Haselwandter, *Mycorrhizal technology in Agriculture: From genes to bioproducts.* Bikhausser, Bassel, Switzerland (2001).
[64] L. A. Harrier and C. A. Watson, *Advan. Agron.* 79, 225 (2003).
[65] B. Hause and T. Fester, *Planta* 221, 196 (2005).
[66] S. Rosendahl, *Phytopathol. Z.* 114, 41 (1985).
[67] S. Slezack, E. Dumas-Gaudot, S. Rosendahl, R. Kjoller, M. Paynot, J. Negrel and S. Gianinazzi, *New Phytol.* 142, 529 (1999).
[68] I. Garmendia, N. Goicoechea and J. Aguirreolea, *Bio. Control* 31, 305 (2004).
[69] D. Blancard, H. Lecoq and M. Pitrat, *A color atlas of Curcubit diseases: Observation, identification and control,* Manson, London (1994).
[70] H. Rachniyom and T. Jaenaksorn, *J. Agr. Technol.* 4, 71 (2008).
[71] G. J. Samuels, *Mycol. Res.* **100**, 935 (1996).
[72] R. Lifshitz, S. Lifshitz and R. Baker, *Plant Dis.* 69, 4344 (1985).
[73] T. Kommedahl, Windels, C. E., G. Sarbini and H. B. Wiley, *Prot. Ecol.* 3, 61 (1981).
[74] J. A. Lewis and G. C. Papavizas, *Phytopathology,* 70, 89 (1980).
[75] H. Jabnoun-Khiareddine, M. Daami-Remadi, F. Ayed and E. M. Mahjoub, *Afr. J. Plant Sci. Biotechnol.* 3, 36 (2009).
[76] Q. Migheli, A. Herrera-Estrella, M. Avataneo and M. L. Gullino, *Mol. Ecol.* 3, 153-159 (1994).
[77] D. P. Roberts, S. M. Lohrke, S. L. F. Meyer, J. S. Buyer, J. H. Bowers, C. J. Baker, W. LI, J. T. De Souza, J. A. Lewis and S. Chung, *Crop Prot.* 24, 155 (2005).
[78] B. Nzanza, Univ. *Pretoria, PhD thesis* (2012).
[79] G. R. Dixon, *Acta Horticulturae* 635, 103-104 (2004).
[80] I. J. Crouch and J. van Staden, *J. Appl. Phycol.* 4, 296 (1992).
[81] E. Booth, *Proc. Int. Seaweed Symp.* 6, 662 (1969).
[82] B.C. Featonby-Smith and J. Van Staden, *Sci. Hortic.* 20, 146 (1983).
[83] Nzanza, B, D. Robertson-Andersson and D. Marais, Proc. Combined Congr. S. Afr. Soc. of Crop Prod., Soil Sci. Soc. S. Afr., Sn. Afrn. Soc. Hort. Sci. & Sn. Afrn. Weed Sci. Soc., Univ. Stellenbosch, S. Afr. (2009).
[84] J. Jayaraj, A. Wan, M. Rahman and Z. K. Punja, *Crop Prot.* 27, 1366 (2008).
[85] F. R. Cavalcanti, M. L. V. Resende, J. P. M. S. Lima, J. A. G. Silveira and J. T. A. Oliveira, *Physiol. Mol. Plant Pathol.* 68, 198-208 (2006).
[86] S. Soylu, O. Baysal and E. M. Soylu, *Plant Sci.* 165, 1075 (2003).
[87] D. L. Cole, *Crop Prot.* 18, 267-273 (1999).

[88] Resende et al., 2002. M. L. V. Resende, G. B. A. Nojosa, L. S. Cavalcanti, M. A. G. Aguilar, L. H. C. P. Silva, J. O. Perez, G. C. G. Andrade, G. A. Carvalho, R. M. Castro, *Plant Pathol.* 51, 621-628 (2002).

[89] J. A. Ryals, U. H. Neuenschwander, M. G. Willits, A. Molina, H. Y. Steiner and M. D. Hunt, *Plant Cell,* 8, 1819 (1996).

[90] A. H. Anfoka, *Crop Prot.* 19, 401-405 (2000).

[91] N. Benhamou and R. Belanger, *Plant Physiol.* 118, 1203-1212 (1998).

[92] L. Scarponi, R. Buonaurio and L. Martinetti, *Pest. Manage. Sci.* 57, 262-268 (2002).

[93] Hussain T. R. R. Ahmad, G. Jilani and T. Higa, *Applied EM Technology,* Nature Farm. Res. Center, Univ. Agric.; Faisalabad, Pakistan. pp. 1-6 (1994).

[94] A. Javaid, R. Bajwa, I. Siddiqi and U. Bashir, *Int. J. Agri. Biol.,* 2, 2 (2000).

[95] T. Higa, *An earth saving revolution.* Sunmark Publishing, Tokyo (1993).

[96] J. Mayor, M. Rondon, D. Molina, S. J. Riha and J. Lehmann, *Plant Soil* 333, 117-128 (2010).

[97] K. I. Al-Mughrabi, C. Berthélémé, T. Livingston, A. Burgoyne, R. Poirier and A. Vikram, *J. Plant Sci.* 3, 175 (2008).

[98] H. A. J. Hoitink, A. G. Stone and D. Y. Han, *HortScience* 32, 187 (1997).

[99] S. B. Koné A. Dionne, R. J. Tweddell, H. Antoun and T. J. Avis, *Suppressive Effect of Non-aerated Compost Teas on Foliar Fungal Pathogens of Tomato* (2010).

[100] S. G. Kim, K. W. Kim, E. W. Park and D. Choi, *Phytopathology* 92, 1095-1103 (2002).

[101] J. F. Ma and N. Yamaji, Trends Plant Sci. 11, 393-397 (2006).

[102] G. Hein, *J. Plant Nutr. Soil Sc.* 167, 606 (2005).

[103] R. V. C. DIOGO and K. WYDRA. *Physiol. Mole. Plant Pathol.* 70, 120-129 (2007).

[104] F. Fauteux, W. Remus-Borel, J. G. Menzies and R. R. Be ́Langer, *FEMS Microbiol. Lett* .249, 6 (2005).

[105] G. Hochmuth, Fla. Coop. Ext. Serv. Circ. 1144 (1994).

[106] S. J. Locascio and A. G. Smajstria, *J. Am. Soc. Hort. Sci.* 121, 63-68 (1996).

[107] S. M. Sezen, G. Celikel, A. Yazar, S. Tekin and B. Kapur, *Sci. Res. Ess.* 5, 41-48 (2010).

[108] S. Clark, K. Klonsky, P. Livingston and S. Temples, *Am. J. Altern. Agric.* 14, 1009-21 (1999).

[109] M. Le Strange, Univ. Calif., Pub. No. 8017, 2000 at http://anrcatalog.vcdavis.edu.

In: Tomatoes: Cultivation, Varieties and Nutrition
Editor: Tadahisa Higashide

ISBN: 978-1-62417-915-0

Chapter 10

TRENDS IN FOREIGN TRADE COMPETITIVENESS FOR FRESH TOMATO

Jaime de Pablo,[1] Miguel Giancinti[2] and Juan Uribe[3]
[1]University of Almería, Spain
[2]Office MAG, Argentina
[3]University of Almería, Spain

ABSTRACT

Drawing on Porter's competitive forces, we can say that the rivalry of direct competitors is greatest when there is greater concentration, a slow growth market, fixed costs and high stocks, differentiated products, and no absence of rules. Given this situation, the circumstances are perfect for a risk of entry by potential competitors and intelligence systems must be used to reduce uncertainty. It also decreases when there are high barriers to entry or strong retaliatory measures. Possible barriers can be achieved through economies of scale, product differentiation, increased investment, access to distribution channels and government policies. The threat of product substitution is another variable that affects direct competition. In the case of tomatoes, other vegetables can be used in salads. The bargaining power of suppliers and customers depends on relative concentration, producer/client related quality between the product and raw material, product differentiation and the distribution of added value in the final product. This article analyzes the tomato from the perspective of competitiveness in production and consumption in 2004-2010.Thisstudy will analyze the most relevant trends and forecast the main trends for the coming years. The article will first show a structural analysis of the major export countries, after which it will follow the major consumers and trade.

INTRODUCTION

One way to evaluate the comparative advantage is to ask what global trade statistics reveal about the performance of a country. Ballassa (1965) was the first [1] to call this type of analysis ‘comparative advantage’, and coined the term "Revealed Comparative Advantage “in his book *"Trade Liberalization and Revealed Comparative Advantage"*.

Moreover, competitiveness is a complex variable that is affected by a large number of factors at both micro and macro level (Chebil and Briz, 1999) [2]: natural resources, technology, investment, human capital, management skill characteristics of the product, strategy business, distribution channel, infrastructure and environmental regulation.

Some studies related to this work are developed by Sarris [3] which highlights the increase in exports from countries like Spain, Greece and Portugal in the European context, because of its competitive advantages. For agriculture, Chebil and Briz who used Porter's Methodology, came to the conclusion of the importance of innovation, cost reduction, training, and organization of supply and the internationalization of the company.

In 2004, Avendano and Schwentesius studied the factors of competitiveness[4] in the Mexican market for production and export of vegetables, and came to the conclusion that the importance of opening new markets, consolidating export supply that enables negotiation for better prices, diversification supply, product transformation to add value, heavy marketing campaigns that achieve set products in foreign countries and a government committed to defending the interests of exporters, was high thereby supporting the definition of strategies to achieve and maintain their international competitiveness.

Using Balassa’s index, Sánchez [5] found that the theory of comparative advantage states that countries receive trade benefits when they specialize in producing and exporting relatively cheap goods and import relatively more expensive goods. If they do not do this, it incurs a cost which the Revealed Comparative Advantage Index (RCAI) measures by monitoring the comparative advantages from trade flows. Arias and Segura also made an analysis on the same lines [6].

IMPORTING COUNTRIES

The top importing countries of fresh tomato are led by the U.S., with a 20,87% share of world trade (US $ 1.431,40 million),followed by Germany with 19,95% (US $ 1.367,87 million) and the U.K. with 10,91% (US $ 747,88 million). Other relevant countries were Russia, France, the Netherlands, Canada, Poland, the Czech Republic and Italy.

Practically, all the import markets are located in the northern hemisphere, which could be an advantage for off season production.

EXPORTING COUNTRIES

Sometimes we see the competitiveness of countries as a synonym of exportation. This analysis is not accurate because companies are competing and not countries, according to Paul Krugman, a Nobel laureate in economics.

However, the dynamism and composition of exports may help to explain the conditions under which firms operate and the difficulties they may be facing. Thus, the explanatory variables show exports and economic growth or the company size to give indications of the competitiveness of a country.

Table 1. Main importing countries

	2007	2008	Share
U.S.	1.220.452. 309	1.431.405.227	20,87%
Germany	1.304.585.933	1.367.871.700	19,95%
U.K.	773.632.709	747.887.389	10,91%
Russia	528.347.758	628.472.317	9,17%
France	582.935.955	559.699.335	8,16%
Netherlands	357.555.122	308.394.710	4,50%
Canada	265.618.127	280.077.248	4,08%
Poland	120.844.608	157.682.247	2,30%
Czech Republic	110.010.955	128.753.680	1,88%
Italy	112.356.662	112.891.076	1,65%
Other countries	2.275.490.869	1.133.956.918	16,54%
Total	6.431.378.698	6.857.091.847	100,00%

Source: COMTRADE.

Table 2. Tomato exports vs. Vegetable exports (percentage and thousand of euros)

WORLD	2005	2006	2007	2008	2009	2010	Annual growth.
Tomato exp.	4.090.042	4.536.856	4.909.763	5.013.257	4.920.738	5.926.697	3,7%
Vegetables exp.	26.846.290	30.748.850	32.789.068	33.388.731	34.613.431	41.557.936	4,7%
Share	0,13180423	0,12628729	0,13940693	0,14129004	0,15669943	0,15669943	3,8%

Source: COMTRADE.

Table 3. Main exporting countries (thousand of euros)

Importing countries	2005	2006	2007	2008	2009	2010
World	4.090.042	4.536.856	4.909.763	5.013.257	4.920.738	6.130.259
Netherlands	959.381	1.059.578	1.122.297	1.187.126	1.120.542	1.263.916
Mexico	707.788	891.037	779.595	819.249	868.156	1.201.393
Spain	849.489	803.924	860.220	851.292	816.544	817.684
Turkey	117.050	138.737	216.799	264.103	291.479	359.122
US	181.795	209.766	226.933	226.285	227.116	281.369
Canada	230.621	241.276	197.445	211.126	212.367	268.408
France	138.484	130.740	197.652	196.111	214.572	267.431
Morocco	113.295	96.137	148.768	180.414	182.836	226.656
Italy	137.046	155.504	181.110	164.353	146.938	216.270
Belgium	210.777	210.752	216.138	197.251	171.297	212.084
Other countries	444.316	599.405	762.806	715.947	668.891	1.015.926

Source: COMTRADE.

Table 4. Main exporting countries share (%)

% Exporting	2005	2006	2007	2008	2009	2010
World	100,00	100,00	100,00	100,00	100,00	100,00
Netherlands	23,46	23,35	22,86	23,68	22,77	20,62
Mexico	17,31	19,64	15,88	16,34	17,64	19,60
Spain	20,77	17,72	17,52	16,98	16,59	13,34
Turkey	2,86	3,06	4,42	5,27	5,92	5,86
US	4,44	4,62	4,62	4,51	4,62	4,59
Canada	5,64	5,32	4,02	4,21	4,32	4,38
France	3,39	2,88	4,03	3,91	4,36	4,36
Morocco	2,77	2,12	3,03	3,60	3,72	3,70
Italy	3,35	3,43	3,69	3,28	2,99	3,53
Belgium	5,15	4,65	4,40	3,93	3,48	3,46
Other countries	10,86	13,21	15,54	14,28	13,59	16,57

Source: COMTRADE.

Tomato represents a 15% share of global vegetable exports, and between 2010-2005 they have grown by 3,7%, a figure lower than vegetable exports (4,7%). The main tomato exporting countries of tomato are the Netherlands, Mexico, Spain, accounting for 5356% of global exports.

REVEALED COMPARATIVE ADVANTAGE INDEX FOR TOMATO TRADE

Methodology

Balassa´s proposed index is:

$$RCAa,i = (Xi,j / Xn,j) / (Xi,t / Xn,t)$$

where X represents exports, i is a country, a is a commodity (or industry), t is a set of commodities (or industries).

In this case:

Xi,j: Tomato exports of certain contry.
Xn,j: Vegetable global exports of a certain country.
Xi,t: Tomato global exports
Xn,t: Vegetable global exports

COUNTRIES

We find analytical studies in this vein in the papers of Briz [7] and De Pablo Valenciano [8], which analyzes global trade of fruits and vegetables. In theese studies it is seen that trade is growing faster than production, and on very specific areas with high purchasing power, such as the EU and the U.S. In our case study we will focus on the ten largest exporting countries: the Netherlands, Mexico, Spain, Turkey, the U.S., Canada, France, Morocco, Italy and Belgium, They account for 83% of global exports. For this analysis, we used Balassa's Revealed Comparative Advantage index (RCAI) and historical data from the International Trade of United Nations database (United Nations Commodity Trade Statistics Database.-COMTRADE).

THE NETHERLANDS

This is one of the main global food importing countries, giving great importance to the integration of the food chain. The objectives of this integration are to improve competitiveness by expanding the economies of scale, consolidating its position in the global and European market and improve the bargaining position with buyers and expand financial capacity. It should be added their geographic location (centrally located in Western Europe), excellent communication systems with all the large European cities and the port of Rotterdam (the largest in Western Europe), that is connected by excellent communication systems with all major European cities, are in the country's favour. Tomato exports have grown by 3,%, with vegetable exportsup 4.3%. RCAI is 1,38, similar to Spain, but, better than in this country. As we can see, the Netherlands has an increased advantage over Spain in the last few years, but we must not forget that this country sells Spanish tomato, especially to its main market Germany(40,80% of its exports), and its second market the U.K. (7,65%).

Table. 5. Tomato and vegetables exports. Netherlands

	2005	2006	2007	2008	2009	2010	ANNUAL GROWTH
Tomato exports	959381	1059578	1122297	1187126	1120542	1263916	3,4%
Vegetable exp.	3418741	4040771	4465897	4506138	4258668	5863309	4,3%
Share	0,28062407	0,26222174	0,25130383	0,26344644	0,2631203	0,2155636	-1,0%
RCAI	2,13	2,08	1,80	1,86	1,68	1,38	-4,6%

Source: Own elaboration from COMTRADE.

Table 6. Tomato and vegetable exports. Mexico

	2005	2006	2007	2008	2009	2010	ANNUAL GROWTH
Exp. Tomato	707788	891037	779595	819249	868156	1201393	2,5%
Exp. Vegetales	2506778	2769218	2595544	2629575	2649017	3256356	0,4%
Share	0,28234969	0,32176484	0,300359	0,31155187	0,32772761	0,36893786	2,1%
RCAI	2,14	2,55	2,15	2,21	2,09	2,35	-1,6%

Source: Own elaboration from COMTRADE.

MEXICO

Sinaloa is the main tomato-producing state, followed in importance by Baja California, San Luis Potosi, Michoacán and Areora. Its main client is the US. Exports have grown by 2.5% annually due to the FTA with the U.S. and the devaluation of the Mexican peso. Its RCAI is 2.35.

SPAIN

The tomato export market in 2010 represents 817.684 thousand euros. Tomato production is distributed as follows.

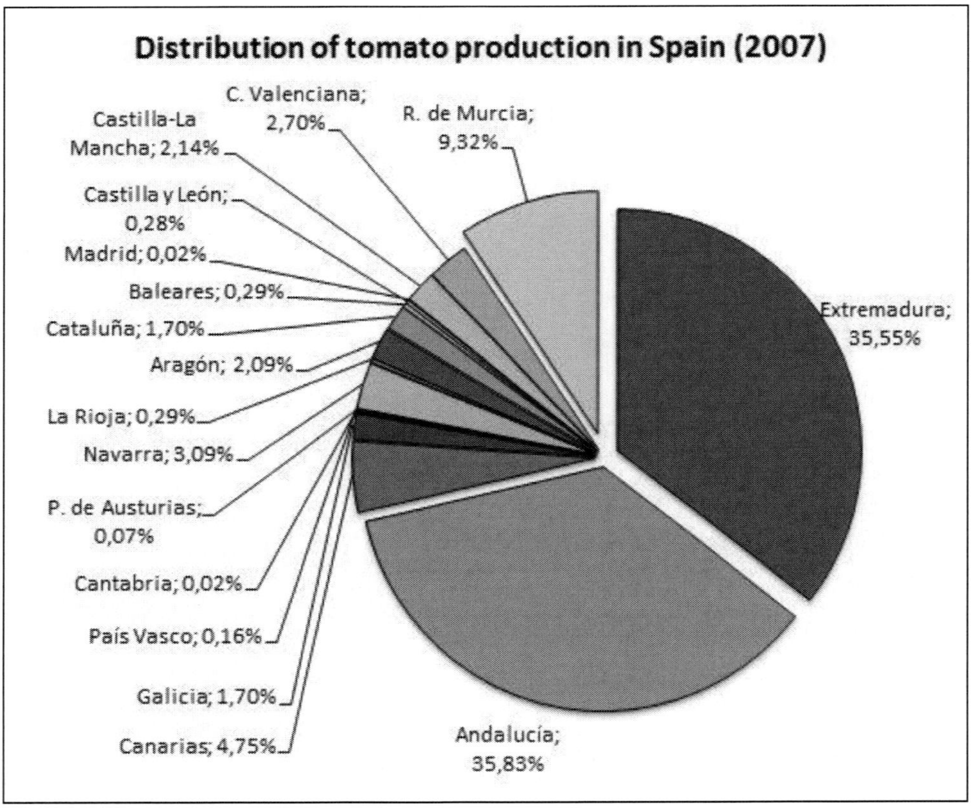

Source: MARM. Our own compilation of data.

We can easily see that Andalusia and Extremadura are the large tomato growing areas accounting for 35.54% and 33% respectively, equivalent to, in the case of Andalusia 18.943ha (being 10,100 Almería) and in Extremadura to 17.582ha (of which belong to 14.841ha Badajoz).These two regions are the largest producers of tomato, accounting for 71.40% of total production of this vegetable, which is equivalent to 2,914,067 tons of tomato (data from MARM 2007).

The comparative advantages of Spain are lower production costs than their European competitors and above all, the excellent weather, which make it a heavy producerin winter.

The main countries which Spain exports tomato to are reflected in Table 7.We can see that the majority are destined for Germany, the UK, France and the Netherlands. These account for 69.78% of total exports so it shows how important these marketsare for Spain.

Germany: it is, as we know, the great community importer of fruits and vegetables because it has low production compared to its needs and it is focused on summer.

Table 7. Main destinations for Spanish Tomato

IMPORTERS	2002	2009	2009%
Germany	236.452	160.875	19,39%
U.K.	191.592	159.082	19,18%
France	149.545	134.909	16,26%
Netherlands	185.604	123.966	14,94%
Poland	21.020	44.694	5,39%
Italy	21.918	37.532	4,52%
CzechRepublic	26.502	28.181	3,40%
Portugal	42.123	25.871	3,12%
Belgium	18.341	17.312	2,09%
Sweden	8.081	16.491	1,99%
Rest of world	73.221	80.627	9,72%
World	974.399	829.540	100,00%

Source: Own elaboration.

UK: One of the big consumers of vegetables and the second largest importer of tomato from Spain. Its focus is cultivation in summer and the dependence on foreign countries is absolute.

France: presents mixed conditions. It is a producer country in the period from March to October and an importing country throughout the year, although its import season is more pronounced during the winter months. Although a significant percentage of tomato is imported from Spain, this percentage has decreased due to the entry of other competitors, as in the case of Morocco.

The Netherlands: on the one hand, it is a producer and exporter-importer of vegetables destined for fresh consumption, and has a focus on advanced techniques of greenhouse crops. On the other hand, is the main receptor and re-export centre of European vegetables.

Table 8. Spanish tomato and vegetable exports

	2005	2006	2007	2008	2009	2010	ANNUAL GROWTH
Exp. Tomato	849489	803924	860220	851292	816544	817684	-0,2%
Exp. Vegetales	3459260	3510163	3674559	3756883	3971811	3991962	2,8%
Share	0,24556957	0,22902754	0,23410156	0,22659 529	0,20558 481	0,20483 261	-2,9%
RCAI	1,86	1,81	1,68	1,60	1,31	1,31	-6,2%

Source: Own elaboration from COMTRADE.

As we can see almost the total export in the four main countries: Germany, the UK, France and the Netherlands. There is no significant change in this period. In general, as can be seen, there has been a decline in total exports primarily because of new competitors. The exports of tomato account for 20% of total exports of vegetables, and their RCAI is 1.31, but the trend is for it to decrease (6.2%). An interesting issue to be considered by Spanish entrepreneurs, is the question of strategic positioning which is not easy since Spain is a country accustomed to selling in bulk, competing by volume and costs,. However, there are companies that have changed this approach and work from the standpoint of the focus or niche, differentiation, and do not neglect their costs linked with the incorporation of technology and the experience curve.

TURKEY

If we analyze its intrinsic characteristics, we see quickly that because of its privileged geographical position, climate and terrain, Turkey is a clear competitor to EU production. For this reason, Turkey is the third country after China and the U.S., in global tomato production, devoting about 20% of its tomato exports to EU countries. Currently, there is a customs agreement between the EU and Turkey, being exempt from supplying the same fruit and vegetables, so bilateral agreements are necessary for the importation of tomato and the product is regarded as sensitive by experts in intra-community trade of countries like the Netherlands and Spain.

According to the Statistical Institute of Turkey, the rate of opening of the Turkish economy at current prices was 42.09% of GDP in 2007. The trend should be an even greater increase of the degree of openness as they delve into the customs union. This competitiveness is reflected in both the increase in annual exports by 22.1%, and in the Balassa index that has showed continued growth in recent years, amounting in 2010 at 6.5%. There is widespread European protection of tomato against tomato from the Turkish community, so it can only access the market in the months in which there is no production from countries that traditionally produce it, thus the price of this product hasno variation that would occurif entry was free.

Tariffs on this product are 12% and have a subsidy of 5%, thus equating the price of Turkish tomato with Spanish or Dutch. The main customers of tomato exports from Turkey to the EU are countries recently incorporated to the community, as is the case of Romania (50.01%), Bulgaria (24.96%), Austria (7.60%), Hungary (4.48 %), Poland (3.90%) and Greece (2.48%).

Table 9. Turkish tomato and vegetables exports

	2005	2006	2007	2008	2009	2010	ANNUAL GROWTH
Exp. Tomato	117050	138737	216799	264103	291479	359122	22,1%
Exp. Vegetales	428142	562445	628407	646944	731939	834809	9,9%
Share	0,2733906	0,24666767	0,34499775	0,40823162	0,39822854	0,43018463	10,6%
RCAI	2,07	1,95	2,47	2,89	2,54	2,75	6,5%

Source: Own elaboration from COMTRADE.

All these six countries account for 93.43% of Turkish exports to the EU of tomato. This leads us to conclude that the Turkish tomato market is centred in neighbouring countries and new entrants, so transport costs are lower due to its proximity to target markets.

It is also noteworthy that the accession process of Turkey to the EU is showing the country adhering to Western standards and procedures. The reluctance of EU founding countries to the union with Turkey is slowing the process considerably, but it should be noted that, although with some misgivings, there are certain countries within the EU greatly interested in expanding trade with this country, and claim to be doing what the Dutch did with Spain; that is the Netherlands produced in its territory an innovative and efficient way in order to import products with lower costs.

U.S.

California and Florida agglutinate 67% of tomato production. The natural export market is Canada, but they also export to Mexico and European countries (Belgium and the U.K.). Exports in the last years analyzed have grown in 41%, (RCAI 0.63).

Table 10. U.S. tomato and vegetable exports

	2005	2006	2007	2008	2009	2010	ANNUAL GROWTH
Exp. Tomato	181795	209766	226933	226285	227116	281369	4,1%
Exp. Vegetables	1944149	2133847	2195849	2356872	2438449	2851664	4,5%
Share	0,09350878	0,09830414	0,10334636	0,09601073	0,09313953	0,09866836	-0,3%
RCAI	0,71	0,78	0,74	0,68	0,59	0,63	-3,7%

Source: Own elaboration from COMTRADE.

Canada

Ontario and Québec regions are the most important producers of tomato. From March to December they are cultivated in greenhouses, and from July to October they are cultivated in outdoor fields. Their main client is U.S.

Exports have decreased 2,4%, its RCAI is 0,68.

Table 11. Canadian tomato and vegetable exports

	2005	2006	2007	2008	2009	2010	ANNUAL GROWTH
Exp. Tomato	230621	241276	197445	211126	212367	268408	-2,4%
Exp. Vegetales	1376017	1520573	1735759	2065388	2167811	2534101	10,3%
Share	0,1676004	0,15867439	0,11375139	0,10222099	0,0979638	0,10591843	-11,9%
RCAI	1,27	1,26	0,82	0,72	0,63	0,68	-15,1%

Source: Own elaboration from COMTRADE.

FRANCE

This country is located twenty-third in the world ranking of tomato producers, 60% of its production is destined for fresh consumption, while 40% is processed. There is a trend of productivity growth for the coming years to increase production efficiency through increased investment in heated greenhouses, new varieties and clear specialization in production. It is important to frame this country as an importer where self-sufficiency in consumption moves around 55% and has a downward trend.

Table 12. French tomato and vegetable exports

	2005	2006	2007	2008	2009	2010	ANNUALG ROWTH
Exp. Tomato	138484	130740	197652	196111	214572	267431	10,7%
Exp. Vegetales	1455240	1584782	1773586	1666752	1558682	1795785	1,4%
Share	0,09516231	0,08249715	0,11144202	0,11766058	0,13766246	0,1489215	9,4%
RCAI	0,72	0,65	0,80	0,83	0,88	0,95	5,2%

Source: Own elaboration from COMTRADE.

For this last reason, France is the second relevant European buyer -after Germany-, and is increasing imports especially from Morocco. For Morocco, average exports to France in the period 2002-2010 has been 193,438.37 tons per year, equivalent to 80% of total tomato exports.

Their exports have increased in 10,7%, and tomato represents 15% of global vegetable exports. RCAI is 0.95, with an annual growth of 5,25%.

MOROCCO

Morocco has recently signed an FTA with the EU; in the medium term it will involve the liberalization of alaguita tomato in this commercial area, with consequent damage to European producing countries (Spain, Netherlands and France). Under the agreement, the import shares of this product will increase from the current 233,000 tonnes per year to 254,000 tons. This figure will increase to stand at a maximum of 285,000 tonnes in 2014.

Morocco has a preferential entry price for its tomato exports to the EU, set at 0.46 euros/kg, established by the Agreement and for imports from other countries, at 0.85 euros/kg. It also has other agreements: it is the first and second Arab-African country (after Jordan) to have this agreement with the U.S. This agreement is more political than economic and the U.S. is only the sixth largest trading partner of Morocco, with an export volume of 3%, compared with 69% of the EU. In terms of imports, 3.1% of Moroccan products go to the US, compared to 74% from the EU.

In the tomato market the situation is further accentuated by the percentage of Moroccan tomato exports to developed countries of the EU of more than 90%, especially to France.

Exports from this country have increased in the years analyzed in 13.75%. Tomato represents 50% of the exports of vegetables in this country; its RCAI is 3.17%, higher than other countries. Comparative advantages that can be noted are: a) Production costs: lower wages.

Table 13. Moroccan tomato and vegetable exports

	2005	2006	2007	2008	2009	2010	ANNUALGR OWTH
Exp. Tomato	113295	96137	148768	180414	182836	226656,23	13,7%
Exp. Vegetales	248334	233258	334828	377470	345121	456374	9,3%
Share	0,45622025	0,41214878	0,44431171	0,47795586	0,52977362	0,4966458	3,7%
RCAI	3,46	3,26	3,19	3,38	3,38	3,17	-0,1%

Source: Own elaboration from COMTRADE.

But we consider that unions are becoming more powerful. One example was the closure of two greenhouses of the Azura company (Maraissa group) b) Employees of some companies ask if they are required to sub-assembly. This does not mean working from sunrise to sunset, but if in seven hours they have completed the goal of ending the work, they don't have to work eight hours, c) Policy: preferential agreements with the EU, western governments seek political stability in North Africa (curb radical Islam), d) Communication and transport logistics: improved infrastructure, the biggest and best technical means and communication makes for easy trading.

We point out disadvantages such as: a) Cultural: lack of work methodology. In this sense there has been an important change. BRC (requires the job control in handling centres) and GLOBALGAP Nature's Choice and forces control in greenhouses, b) Political: problems with local governments, administrative problems, legal uncertainty and lack of transparency in economic legislation; c) Technical: lack of teams with the right skills, higher costs of agrimaterials d) Relatives: problems with transfers. e) High cost of transport services, exchanges take place in certain seasons which makes transportation work only for export periods, and returns are made without charge to ensure rotations. One aspect not taken into account, at least in appearance, is the huge exchange rate advantage of this country in relation to European producers.

ITALY

Italian market characteristics to be highlighted are: a high level of consumption but stable, and very heterogeneous and disorganized production. It positions itself as the European leader in the processed tomato segment. A third of its production is for fresh tomato, while the rest is industrialized, mostly for peeling and to a lesser extent, concentrate, crushed, pulp or sauce. This segment goes to the Spanish market; most of it is destined for fresh tomato exports

Table 14. Italian tomato and vegetable exports

	2005	2006	2007	2008	2009	2010	ANNUALGR OWTH
Exp. Tomato	137046	155504	181110	164353	146938	216270	1,5%
Exp. Vegetales	870248	964025	1032918	1063060	1030779	1317146	3,5%
Share	0,15747925	0,16130702	0,17533822	0,15460369	0,14255044	0,16419592	-1,8%
RCAI	1,19	1,28	1,26	1,09	0,91	1,05	-5,1%

Source: Own elaboration from COMTRADE.

Exports have grown 1,5% annually, with tomato making up 16,4% of global exports. Its RCAI is 1,05, decreasing in 2005-2010 by 5,1%.

Italian exports concentrates to Germany (31%) and Austria (17%) followed by the U.K. and the Netherlands with lower percentages.

BELGIUM

Belgium and Netherlands act as a forwarding market. Its exports have decreased in recent years. Its RCAI is 0.77.

Table 15. Belgian tomato and vegetable exports

	2005	2006	2007	2008	2009	2010	ANNUAL GROWTH
Exp. Tomato	210777	210752	216138	197251	171297	212084	-3,6%
Exp. Vegetales	1446241	1577841	1690293	1732354	1651864	1755943	2,8%
Share	0,14574127	0,13356986	0,12787014	0,11386299	0,10369921	0,12078069	-6,5%
RCAI	1,11	1,06	0,92	0,81	0,66	0,77	-9,7%

Source: Own elaboration from COMTRADE

GLOBAL VISION OF TOMATO CONSUMPTION METHODOLOGY

Available data from1998 and 2003 from the Food and Agriculture Organization of the United Nations (FAOSTAT, Italy), allows us to analyze food consumption through the Analysis Model (ACAM) methodology, fresh tomato and process tomato consumption with respect to vegetables. This methodology was presented at the International Food and Agribusiness Management Association (IFAMA) congress which was held in Frankfurt (Germany) in June, 2011. The sustainable market section presents an increase in consumption of the food analyzed; in this case tomato and the rest like onion, lettuce, green beans, cabbage and other vegetables. The tough market is a section where only consumption of tomato increases and no other vegetables. In the vulnerable market, tomato consumption decreases and the intake of other vegetables increases; while in a declining market the trend of tomato and other vegetables declines. The greatest challenge for a producer region that exports to countries which are positioned in different quadrants of the market is that each requires a different strategy: cost, distribution and logistics, value-added, differentiation, etc.

RESULTS

The countries analyzed in this chapter represent 80.17% of global consumption as reported by FAO and COMTRADE, and Figure 1 represents the results of ACAM between

2000 and 2007. In 2007, 76.13% of the volume consumed was termed 'sustainable'-the high volume of consumption of some countries explains the high proportion, this is the case of China, U.S., United Kingdom, Russia, Pakistan, Canada,, Brazil, Sweden Egypt, Iran and India. In 'vulnerable'4.39% is consumed, and includes countries like Spain, Netherlands, Korea, Portugal an Hungary.

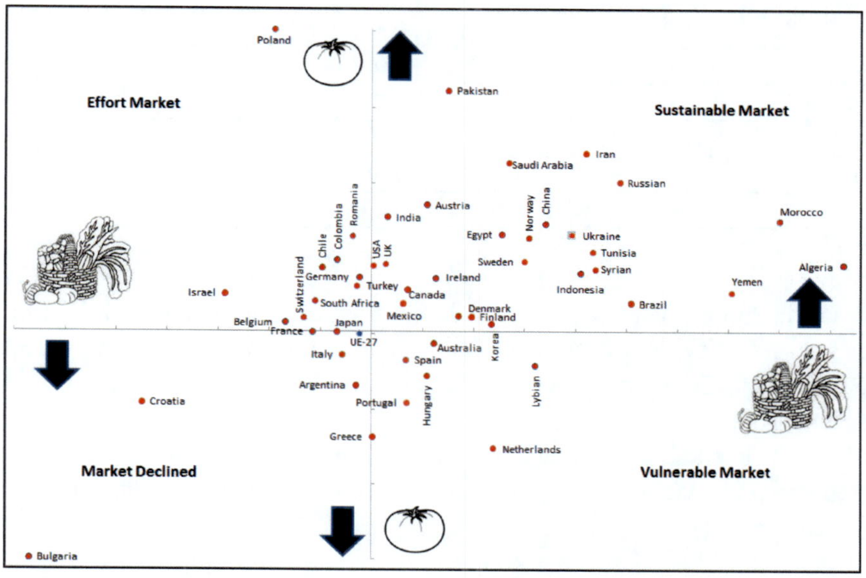

Chart 1. Trends in global tomato consumption (2000-2007).

In 'effort', 13.16% is consumed in such a manner in countries like Japan, Colombia, Germany, Turkey and Belgium; and finally in 'declining' consumption is6.32% and includes important countries as Italy and Argentina.

CONCLUSION

The data analyzed in this study allow us to conclude that there is a hard and competitive tomato trade inside the European Union and between Mexico and the U.S. The U. S., Germany and the U.K. account for 52% of world imports. Note that virtually all import markets are located in the northern hemisphere, which could be an advantage for off season production. The Netherlands, Mexico and Spain account for 54% of global exports. At a European level, stand the Netherlands and Spain and with products from the Mediterranean Basin countries, especially Morocco. In the case of North America, Mexico has its natural market in the U.S.

Considering RCAI of studied countries, we can see that Morocco, even being out of the EU and therefore, having no commercial facilities that exist in the EU, has quite a significant advantage when it is compared with other countries and, in 2010, 91,74% of Moroccan tomato exports were destined for EU countries (France 80%). It is based on weather, workforce…etc. Another interesting fact is that the Netherlands and Spain, direct competitors between them, have a similar RCAI.

Table 16. RCAI of main countries (2010)

Morocco	3,17
Turkey	2,75
Mexico	2,35
Netherlands	1,38
Spain	1,31
Italy	1,05
France	0,95
Belgium	0,77
Canada	0,68
U.S.	0,63

Source: Owndata

As for consumption trends, 72,4% of volume is sustainable, like China, Russia, France, Germany, Brazil, Turkey, Egypt, Iran and India. In an effort market, there are countries like Japan, Colombia and Belgium (2,5%). In the latter two cases there has been an increase in tomato consumption.

This does not happen in vulnerable markets (17,9% of consumption) that include such important countries as Mexico, the U.S. and Morocco, neither in declining markets where Spain, Italy, etc are located.

REFERENCES

[1] B. Balassa, "*Trade liberalization and revealed comparative advantage*" The Manchester School of Economic and Societal Studies 33:99-123 . (1965).

[2] T. Vollrath,"A theoretical evaluation of alternative trade intensity measures of revealed comparative advantage" *Weltwirtschaftliches Archiv*. pp. 264-280, (1991).

[3] L. Scott, T. Vollrath,. "Global competitive advantages and overall bilateral complementary in agriculture". *USDA/IRS Statistical Bulletin°* 850,(1992).

[4] I. Pérez Infante . "Costes laborales y competitividad de la economía española" *Revista de Economía y Sociología del Trabajo* nº 25 (1994).

[5] M. A. Sastre, E. Aguilar. "Un modelo de competitividad empresarial basado en los recursos humanos" *Economía Industrial* nº 332 (2000).

[6] A.V. Ayala-Garay, G. Almaguer-Vargas, N. K.de la Trinidad-Pérez, I. Caamal-Cauich, R. Rendón, "Competitividad de la producción de mango (Mangifera indica L.) en Michoacán" Revista Chapingo. Serie horticultura, (2009) Vol. 15, Núm. 2, mayo-agosto, 2009, pp. 133-140. at http://redalyc.uaemex.mx/src/inicio/ArtPdfRed.jsp?iCve =60912457004.

[7] A. Chebil. y J.Briz "*Escenario competitivo del sector hortícola español*" Distribución y Consumon, 52 pp. 25-46, (1999).

[8] A. Sarris. "European community enlargement and world trade in fruits and vegetables" *American Journal of Agricultural Economics*. Nº 65 (1983).

[9] B. Avendano, R. Schwentesius. "Factores de competitividad en la producción y exportación de hortalizas: El caso del Valle de Mexicali, B. C., México", *Problemas del Desarrollo*, vol.36 nº 140, (2004).

[10] J.C. Sánchez "Aplicación del Índice de Ventajas Comparativas Reveladas (IVCR) al Comercio entre El Salvador y Estados Unidos" Banco Central de Reserva de El Salvador. *Boletín Económico*(2006).

[11] J. Arias, O Segura "Índice de ventaja comparativa revelada: un indicador del desempeño y de la competitividad productivo-comercial de un país",*InterCambio*, (2004).

[12] J. Briz "Principales competidores en el mercado internacional: Fortalezas y debilidades de los más importantes concurrentes" Curso de post-cosecha y comercialización de productos hortícolas: Técnicas y canales de distribución. *FIAPA-Junta de Andalucía*, Almería, (1995).

[13] J. de Pablo, J. C. Pérez, R. García. "*Caracterización de la comercialización hortofrutícola española: una aplicación a la producción intensiva*". Ed. Thomson*Civitas (2004).

II. Tomatoes Composition and its Effects

In: Tomatoes: Cultivation, Varieties and Nutrition
Editor: Tadahisa Higashide

ISBN: 978-1-62417-915-0

Chapter 11

THE CHEMICAL COMPOSITION AND NUTRITIONAL VALUE OF TOMATOES

M. Hernández Suárez, E. M. Rodríguez Rodríguez and C. Díaz Romero*

[1] Agriculture Technology National Centre “Extremadura” CTAEX, Ctra. Villafranco-Balboa km 1.2, Villafranco del Guadiana, Badajoz, Spain

[2] Analytical Chemistry, Nutrition and Food Science Department. La Laguna University, Av. Astrofísico Francisco Sánchez s/n, La laguna, Santa Cruz de Tenerife, Spain

[3] Analytical Chemistry, Nutrition and Food Science Department. La Laguna University, Av. Astrofísico Francisco Sánchez s/n, La laguna, Santa Cruz de Tenerife, Spain

ABSTRACT

The tomato has traditionally been classified as a *dessert vegetable*. However, at present, this view has shifted because of a new concept which claims that the tomato is a source of interesting nutrients and bioactive components, especially antioxidants, which have positive effects on human health. Several parameters such as soluble solids, sugars, acidity and pH are important to establish the quality of both fresh market and processing tomatoes. The main components responsible for tomato taste and aroma associated to the quality are sugars, organic acids, free amino acids, salts and antioxidants. Oxalic acid, ascorbic acid and lycopene have interesting effects on human health. As with other vegetable foods, tomatoes have an adequate mineral and trace element composition to benefit and maintain human health.

The scientific community is focused on indentifying the cause and the action mechanism which increase the concentrations of these biologically active compounds. The tomato, during ripening, undergoes a wide range of biosynthetic as well as hydrolysis reactions that dramatically affect the final fruit composition. Ascorbic acid and some phenolics tend to be accumulated from the green to mid-ripe stage, whereas the

* E-Mail address: cdiaz@ull.es.

total carotenes constantly increase their concentrations during the ripening process. Furthermore, under water and salt stress, tomatoes mature earlier and accumulate more lycopene in the period before harvest. The chemical composition depends on the simultaneous interaction of genetic aspects, environmental parameters and cultural practices. Multivariate analysis is a suitable tool to show the complex interactions between chemical composition and factors of production.

INTRODUCTION

Wild tomatoes are native from western South America. The generic status of wild tomatoes within the family of *Solanaceae* has been a matter of controversy since the eighteen century. Linnaeus in 1753 classified tomatoes in *Solanum* genus while Miller, a contemporary of Linnaeus, classified tomatoes in a new genus *Lycopersicon*. The *Solanum* genus includes many other species and cultivars, with a wide biodiversity. At present, there are a large number of tomato cultivars with a wide range of morphological, chemical, nutritional and sensorial characteristics [1]. Different tomato cultivars are available on the market, in response to fresh consumption and the demands of industry [2]. However, most traditional cultivars are disappearing as they are being replaced by modern cultivars or hybrids, which has resulted in a decline in the broad diversity of tomato cultivars and their quality characteristics.

In this sense, the International Year of Biodiversity (IYB 2010) provided an opportunity to explore and recognize the following points: (i) the existing biodiversity in horticultural crop cultivars and their nutritional composition; (ii) the use of the nutrient content among the criteria used in cultivar selection and promotion; and, (iii) the importance of nutritional differences within the species and the micronutrient superiority of some lesser-known cultivars and wild cultivars over some more widely-used cultivars of horticulture- based crops.

According to chemical composition, tomatoes have often been labeled as *"dessert vegetable"* for their sugar content. However, the tomato has recently been considered as a functional food due to its special composition of bioactive compounds, as it is a good source of minerals, fibre, vitamins and antioxidants such as lycopene [3, 4].

Tomatoes have many and varied antioxidant compounds, such as carotenoids, ascorbic acid, phenolic compounds and α-tocopherol. Therefore, the tomato is recognized as a major source of dietary antioxidants with additional benefits for human health above their nutritional value [5, 6]. Many population studies have established a link between dietary intake of tomatoes, a major source of the antioxidant lycopene, and a reduced risk of suffering from cancer and cardio vascular diseases [7]. The consumption of tomatoes and their derived products increase lycopene levels in blood and tissue, which is an effective singlet oxygen quencher that reduces oxidative stress. This reduction plays a role in the prevention of chronic disease but there are also other action mechanisms which can explain the benefits to human health: hormone and immune modulation, gene function regulation and carcinogen metabolism [8].

Many factors are known to affect the nutrient content, including cultivar, climate, geography, geochemistry and agricultural practices. This explains the rather large number of studies aimed a evaluating and improving the quality of the tomato fruit. Some of those studies will be discussed in the following sections.

Chemical Composition

The usefulness of the chemical composition data of cultivars is becoming increasingly acknowledged. These data are important for the sectors of health, agriculture, trade and the environment [9]. For example, it is recognized that different cultivars of horticultural crops can be distinguished by their differing levels of nutrients and other bioactive compounds which is important in the authenticity test.

In the past, generic food composition data were considered sufficient for most purposes. Davies and Hobson [10] defined the overall chemical composition of the tomato as 5–7.5% dry matter. About 50% of tomato dry matter (DM) is sugars and organic acids. The main sugars are glucose (22% DM) and fructose (25% DM), plus some sucrose (1% DM) [11, 12]. However, other sugars, such as xylose, mannose, maltose and raffinose have been found using modern analysis methods [13, 14]. Fernández-Ruíz et al. [15] analyzed 28 accessions of tomatoes from Ecuador, Peru and Spain. Fructose content ranged from 3.51 to 40.42 $g \cdot kg^{-1}$, glucose ranged from 3.19 to 43.56 $g \cdot kg^{-1}$ and sucrose was detected in all samples and varied between 0.45-6.11 $g \cdot kg^{-1}$. These authors concluded that this variation is useful for selecting interesting donor parents according to their usefulness for internal quality breeding programs. Besides, the glucose/fructose ratio is important data in the authenticity test of fruits and, in the case of tomatoes, this ratio must be about 0.8–1[16].

Growers often use the value of Brix degree to indicate sugar contents in tomato fruit. However, the Brix degree is not a reflection only of sugars, as there are also other soluble solids in the fruit, including organic acids, soluble nitrogen compounds or minerals that influence this value. Therefore, the use of the Brix degree as a representation of total sugar content can be misleading. On the other hand, sweetness not only reflects the sugar concentration because the perception of sweetness is different for each sugar; fructose is the sweetest natural sugar and glucose has only 60% of the sweetness of the fructose [17].

Sugars and organic acids are metabolically linked. The sugar/acid ratio has been widely used an important indicator of the acceptability of the fruit as reducing sugars and organic acids are components involved with the sweet and sour tastes of the tomato, respectively [18]. Organic acids are important components in tomatoes representing ≅ 13% DM, and they strongly influence the fruit taste, and as a consequence the overall quality. Citric acid is the major fruit acid and, therefore, it is the main responsible agent of the acidity of tomato; it also forms complex molecules with oxidant metal ions, and therefore, the catalytic activity of these ions is inhibited.

Even though acidity mainly depends on citric acid, there are other organic acids that contribute, to a certain extent, to the acid taste of the vegetable, as well as have interesting effects on human health [14]. Thus, malic acid is an important indicator of the freshness of fruits or tubers, and its concentration increases throughout the storage period to the detriment of the citric acid concentration [19]. According to Gardner [20], malic acid is approximately 14% more sour than citric acid; however, the influence on the tomato taste is limited because its concentration is lower than the citric acid concentration. A similar conclusion was drawn by Cebolla-Cornejo et al. [21].

Oxalic acid is important from a nutritional point of view. This dicarboxylic acid forms insoluble salts with calcium and other essential divalent cations producing a decrease in the bioavailability of these mineral nutrients. Ascorbic acid is another important organic acid that may play a key role as an antioxidant, as it delays the pathogenesis of a variety of degenerative diseases such as cardiovascular disease, certain cancers, cataracts and it also prevents DNA mutation induced by oxidative stress [22-24]. Adalid et al. [25] analyzed the contents of ascorbic acid in many accessions of the *Solanum* genus cultivated in different regions. In the case of *S. esculentum,* ascorbic acid ranged from 51 to 183 $mg{\cdot}L^{-1}$ for BGV011359 from Taiwan and BGV007022 from Ecuador, respectively. Ascorbic acid, in *S. esculentum var. cerasiforme,* varied between 38 $mg{\cdot}L^{-1}$ in BGV008169 from China and 311 $mg{\cdot}L^{-1}$ in BGV012627 from Colombia and 30-144 $mg{\cdot}L^{-1}$ for BGV008068 (Peru) and BGV008116 (Ghana), respectively for *S. pimpinellifolium*. Although the ascorbic acid content may vary according to several factors, Gould [26] and Roselló et al. [27] showed that the commonly accepted average level in the commercial tomato is about 200 $mg{\cdot}kg^{-1}$. Therefore, a consumption of 100 g of fresh tomatoes contributes ≅3% of the recommended dietary intake of vitamin C (60 mg) [28].

High sugar concentrations, together with relatively high organic acids, are required for the best flavour; low sugar content and high organic acid content produce an acidic tasting tomato; high sugars and low organic acids produce a bland taste; and both low sugar and organic acids contents result in a tasteless fruit [29]. However, the tomato is not only a *dessert vegetable* as cited above, some authors have suggested that the main components responsible for the tomato taste and aroma associated with quality are sugars, organic acids, free amino acids and salts [30].Moreover, Lindemann [31] found that sugars and organic acids, certain free amino acids such as L-Glutamate would considerably affect the tomato taste. Pratta et al. [32] reported that glutamine and glutamate were the main amino acids in tomatoes (Table 1).

Table 1. Relative molar content of free amino acids, and total amino acids (AAs) in the pericarp tissue of the Caimanta tomato cultivar at two ripening stages: mature green (MG) and red ripe (RR)

Amino acid (%)	MG RR	
Aspartate	0.38	9.18
Threonine	7.58	1.25
Serine	8.02	1.74
Asparagine	10.40	3.53
Glutamate	13.35	58.26
Glutamine	25.58	13.66
Glycine	2.02	0.79
Alanine	2.48	2.75
Valine	4.16	1.24
Isoleucine	2.62	0.38
Leucine	1.54	0.53
Tyrosine	1.88	0.41
Phenylalanine	3.88	1.05

Amino acid (%)	MG RR	
GABA	12.38	3.83
Others	3.73	1.40
Total AAs (μmol·mg^{-1})	**25.03**	**24.66**

Hernández et al. [33] defined the tomato taste as the combination of sugar, acidity, sodium and ascorbic acid. Sodium is responsible for the salty taste and this aspect has not been considered by other authors. Perhaps, this is due to the fact that the tomatoes analyzed by these authors came from areas near the coast and the marine spray affected the sodium content.

As regards other minerals, only a few works have examined the mineral content regardless of agricultural techniques; which is why some examples are discussed in the Cultural practices section. Fanasca et al. [6] showed that when tomato plants take up high amounts of K from the soil or nutrient solution, the quality attributes of tomato fruit increased, including the antioxidant contents (especially lycopene), while the high proportion of Mg in the solution improved the total antioxidant activity; whereas a high proportion of Ca improved tomato fruit yield and reduced a physiological disorder called Blossom-end rot (BER). BER first appears as one or more tiny lesions slightly below the surface near the blossom-end of the fruit, close to the base of the style, the affected tissue later appears watery and danker green [34].

Fruits and vegetables, from a nutritional point of view, usually contribute 35, 24, and 11% of the total K, Mg, and P to the dietary intake of humans, respectively [35]. The consumption of one serving (≈ 200 g) of tomato is for ≅11% of the adequate intake for K in individuals older than 14 years of age. The contribution to the P and Mg intakes from a serving of tomato accounts for 5–10% of the Recommended Dietary Allowances (RDA) of these elements for adults, while the contributions to the Na and Ca intake were low, representing only 1–2% of the adequate intake and RDA for adults [28]. In general, the contribution to the trace element intake from the consumption of one serving (≈200 g) of the tomato is very low. A serving of tomato accounts for only 5–7% of the RDA of Cu and Mn for adults [28]. The contributions to the daily intake of the rest of the trace elements are lower.

Tomato is a good source of fibre. Claye et al. [35] observed that the tomato fibre was composed of 87% insoluble fibre and 13% soluble fibre. Hernández et al. [37] found that the mean content of total fibre was 1.82% in salad tomatoes grown in the Canary Island (Spain), which was slightly higher than other data found in the literature [38, 39].

Scientific research, during the last few decades, has been focused on the tomato as an antioxidant source and on its role as a functional food. Dumas et al. [40] showed that the nutritional quality of tomatoes depends on their antioxidant contents. The antioxidant profile has been recognized as being an important parameter for predicting the impact of food on human health. This antioxidant profile plays an important role by scavenging reactive oxygen species, so the research focus has shifted more to the identification of the change of those biologically active components [41].

The main antioxidant of tomato is lycopene which plays an important role in the prevention of cardiovascular diseases and several types of cancer and this has been widely reported [42]. In particular, lycopene has been positively correlated with the prevention of prostate cancer; and probably also inhibits disease progression in patients with benign prostate hyperplasia [43]. Moreover, it is possible that phenolics and other bioactive compounds in the tomato fruit can act synergistically with lycopene in preventing cell damage.

Lycopene is the most abundant carotenoid in the ripened tomato, accounting for approximately 80–90% of the total pigments with the rest being β-carotene and other carotenoids [7]. Lycopene is a straight chain hydrocarbon containing 11 conjugated and 2 unconjugated double bonds and is an effective singlet oxygen quencher, two and ten times greater than β-carotene and vitamin E, respectively [44, 45]. Lycopene is present in its natural form in tomatoes like all trans isomeric compounds. However, processing tomatoes into tomato products induces transformation to several cis isomers of lycopene, increasing its bioavailability [46-48].

Adalid et al. [25] analyzed the contents of lycopene in many accessions of the *Solanum* genus cultivated in different regions. In the case of *S. esculentum,* the lycopene content ranged from 0.4 to 52 mg·kg $^{-1}$ for BGV004209 from the Czech Republic and BGV012620 from Taiwan, respectively. Lycopene content for *S. esculentum var. cerasiforme* varied between 0.5 mg·kg^{-1} in BGV008008 from Bolivia and 98.4 mg·kg^{-1} in BGV006872 from Ecuador and 2-120 mg·kg^{-1} for BGV007825 (Mexico) and BGV008068 (Peru), respectively for *S. pimpinellifolium*. Hernández et al. [49] analyzed the lycopene concentration of five tomato cultivars from Tenerife (Spain), ranging between 18.9 and 25.6 mg·kg^{-1}, which agrees with most of the data reported in the literature.

Another important carotenoid is β-carotene which represents the precursor of vitamin A and is essential for animals. It has been shown that β-carotene protects the unsaturated lipids from free radical oxidation and is an effective quencher of singlet oxygen. It is generally believed that the simultaneous presence of β-carotene and lycopene has a synergic antioxidant action [50]. In the case of *S. esculentum,* the β-carotene content ranged from 0.79 to 13.14 mg·kg $^{-1}$ for BGV004209 from the Czech Republic and BGV011512 from Portugal, respectively. β-carotene content for *S. esculentum var. cerasiforme* varied between 0.94 mg·kg^{-1} in BGV008008 from Bolivia and 12.3 mg kg^{-1} in BGV008354 from Costa Rica and 3.9-14.6 mg·kg^{-1} for BGV012638 (Ecuador) and BGV008230 (Honduras), respectively for *S. pimpinellifolium* [25]. The carotenoid (lycopene and β-carotene) content varies according to ripening stages and this will be discussed in the next section.

The contribution to the vitamin E intake of one serving of fruit ranged from 0.4% to 12% of the recommended daily intake [51], while the contribution of tomato to the α-tocopherol intake and the antioxidant properties is of minor relevance. Moreover, in tomatoes, vitamin E is present mainly within seeds, which are not readily digested by humans. So, the contribution of the tomato to the intake of vitamin E for is lower than the expected contribution according to its concentration, which is due to its low bioavailability.

Other antioxidant compounds should be considered, such as hydroxycinnamic acids (caffeic, ferulic, *p*-coumaric and chlorogenic acids). Chlorogenic acid is the major hydroxycinnamic acid commonly found in tomato, with concentrations in the range of 2.5-32.7 mg·kg^{-1}; caffeic and ferulic acids ranged between 0.3-14 mg·kg^{-1} and between 0.4-2.8 mg·kg^{-1}, respectively; while *p*-coumaric had concentrations below 0.8 mg·kg^{-1} [49]. Hydroxycinnamic acids are found most frequently in plants such as simple esters [52]. The amount of total phenolics is much higher than the sum of hydroxycinnamic acids detected [49]. This is due to the fact that there are other phenolic compounds in tomatoes such as flavonoids. Tomato consumption constitutes a significant proportion of the total phenolics ingested in a normal diet, since they are readily absorbed by the digestive tract.

Causes of Change in Chemical Composition

The chemical composition of fruits and vegetables depends on the simultaneous interaction of maturity degree, environmental parameters and cultural practices, but not all chemical compounds respond in the same way. In general, compounds from secondary metabolism, especially antioxidants, vary more than those from primary metabolism. Therefore, the content of antioxidants varies according to several factors such as ripening stage, tomato cultivar, fertilization, cultivation method and environmental effects (Table 2).

Maturity Degree

The tomato fruit, during ripening, undergoes a wide range of biosynthetic as well as hydrolysis reactions that dramatically affect the final fruit composition. These changes are highly coordinated and they take place in every sub-cellular compartment.

After weight, fruit color is considered as a marker for the ripeness of tomato fruits. According to Srivastava and Handa [58], fruit development is divided into the following stages: stage 0 (immature green); stage I (mature green), in which fruits are nearly their final fresh weight; stage II (breaker), in which fruits show a yellowish and orange color; stage III (turning); and stage IV (red and ripe), in which fruits are edible and tasty.

Synthesis of lycopene, β-carotene, ascorbate, and flavonoids is stimulated during ripening by their antioxidant or photoprotective activity. Tomato fruits possess different photoreceptors (phytochromes and a blue light photoreceptor, also called cryptochrome), which are involved in the regulation of carotene biosynthesis. Bramley [59] observed that the carotene content increased between 14 and 28 fold during ripening, mainly due to the accumulation of lycopene and, to a lesser extent, phytoene, phytofluene, and β-carotene.

Oms-Oliu et al. [14] studied the metabolic characterization of the tomato and observed the following changes: 1) The contents of galactose, galacturonic acid and mannose increase throughout maturation and ripening in parallel to the decreasing firmness, indicating a breakdown of the cell wall complex; 2) the levels of galactinol, myo-inositol and raffinose during the first stages of development are higher than during fruit maturation and the concentrations of amino acids vary with different stages of maturity and ripening; 3) throughout ripening there is an increase in glutamic and aspartic acids, methionine, phenylalanine and threonine; whereas, at the same time, there is a reduction in the levels of γ-aminobutirate acid (GABA), β-alanine, glycine and valine; 4) other amino acids, such as asparagine, glutamine, leucine, isoleucine and serine increase at the breaker stages and decrease at subsequent postharvest stages; and 5) amongst the free amino acids, GABA, glutamic acid, glutamine and aspartate show the highest relative increase in tomato fruit throughout ripening.

Table 2. Antioxidant contents in Solanum esculentum (mg·100g^{-1})

Sample	Lycopene	β-carotene	Lutein	Chlorogenic acid	Caffeic acid	Ferulic acid	p-Coumaric acid	Ascorbic acid	Reference
Ripening stage:									
Immature				9.57					[53]
Breaker	2.23-7.4	0.17-0.83		7.9				5.1-18.3	[53]
Orange	0.05	1.2		6.71				10.4	[53]
Deep red	2.9-5.17	0.33-0.67		5.73				8.5	[53]
Kind of tomato:									
Conventional	1.08	1.28	0.18						[54]
Genetically modified	1.01	1.61	0.15						[54]
Fertilization with high proportion of:									
NO_3^-	20.8							97.5	[55]
NH_4^+	12.7-18.6						120.2-128.8		[55]
K	2.80-3.75	0.56-0.57	0.13-0.18		0.25-0.35	1.40-1.51		43.1-44.4	[6]
Ca	1.86-2.16	0.39-0.41	0.14-0.15		0.24-0.39	1.38-1.50		37.9-54.8	[6]
Mg	1.96-3.04	0.48-0.51	0.11-0.14		0.30-0.52	0.90-1.05	44.1-48.26		[6]
Cultivation method:									
Conventional	1.64-2.45			0.86-1.39	0.09-0.12	0.14-0.26	0.15-0.19	13.1-16.8	[49, 56]
Organic	2.01-2.76			0.91-1.36	0.11-0.18	0.12-0.27	0.17-0.21	13.1-17	[49, 56]
Hydroponic	1.87-2.63			0.97-1.07	0.07-0.13	0.17	0.16-0.17	13-16.3	[49, 56]
Effects of environmental (biotic and abiotic factors):									
Water stress	5.16-7.50	0.53-0.62		1.35-2.40					[57]
Nematode attack	9.04-9.27	0.75-0.84		1.17-2.25					[57]
Temperature:									
20°C	4.05	0.75							[40]
30°C	0.11	0.66							[40]
25.2/17.8 (d/n)	6.48	0.22							[40]
13.9/2.8 (d/n)	2,42	0.37							[40]

Another symptom is the change in sugar content. Starch accumulates at the beginning of fruit development. After this, the reducing sugars (glucose and fructose) increase due to the starch degradation [60]. Titratable acidity strongly decreased, probably due to reduced malic and citric acid content from the breaker to the deep red stage [10]. Chlorophyll content, during ripening, decreased and carotene content increased because chloroplasts are transformed into chromoplasts [61]. Both glucose and fructose occur in equal quantities in the tomato fruit during ripening, which is generally attributed to the invertase activity responsible for breaking down sucrose [62]. As a result, ripe tomatoes have increased levels of fructose and glucose, whereas the sucrose decreases progressively during ripening of fruit. The sucrose concentration decreases by more than 90% during the final postharvest stages of the fruit.

Important changes in the organic acid contents occur during the ripening of tomatoes. It has been reported that the concentrations of organic acids falls with over-ripening, which could be due to the use of organic acids as respiratory substrates or their conversion to sugars during ripening [63]. Citric acid content increases with maturation up to a maximum during the postharvest stage, whereas malic acid concentration decreases during ripening. When the tomato reaches maturity and begins to overripe, the catabolic activity of the malic enzyme increases, which decarboxylates malic acid to pyruvate. On the other hand, the activities of malate dehydrogenase and citrate synthase continue. Both processes result in a decline in malic levels and the preferential accumulation of citric acid [14]. Fumaric acid, a precursor of malic acid in the Krebs cycle, declines during maturation and disappears during postharvest. However, ascorbic acid, dehydroascorbic acid, isocitric acid and succinic acid contents seem to increase during ripening but decrease again during postharvest storage down to pre-ripening concentrations [14]. DeBolt et al. [64] observed that tartaric acid is accumulated during ripening and ascorbic acid decreases at late maturity stages, which could be related to the synthesis of tartaric acid from ascorbic acid.

According to Pratta et al. [32] the highest relative amino acid content corresponded to glutamate, glutamine, GABA and asparagine in the mature green stage though glutamate was the most abundant in the red ripe stage (Table 1).

Most of the total amino acid content found in the pericarp, in developing fruit, belongs to the glutamic acid family [65]. GABA is the predominant N-form in the earlier stages of the tomato fruit, showing a decrease from the breaker to the postharvest stages and glutamic acid represents the most abundant free amino acid during ripening and postharvest. The content of the latter is very low in small green, immature and mature green fruit but its levels progressively increase with fruit ripening [66, 67].

Influence of Temperature and Light during Development and Ripening of the Fruit

Little is known about the environmental regulation of tomato metabolites that are responsible for variations in fruit nutritional quality [68]. Compounds from secondary metabolism, especially antioxidants, vary more than those from primary metabolism because antioxidants play an important role by scavenging oxygen-active species generated during ripening [69], under excessive radiation [70] and cold or heat stress [71].

Fruit ripening occurs at relatively mild temperatures and with low light intensity levels. High temperatures, during fruit development, can affect photosynthesis, respiration, aqueous relations and membrane stability as well as levels of plant hormones, and primary and secondary metabolites. Most of the temperature effects on plants are mediated by their effects on plant biochemistry. The immediate effects of heat treatments have generally been to reduce protein synthesis, and increase protein breakdown [72].

Gautier et al. [61] observed that light and temperature had no significant effects on final sugar or organic acid content and that sucrose content does not vary with temperature, irradiation, or ripening duration. Besides, these authors also observed that increasing fruit temperature from 21 to 26 ºC was effective for improving fruit taste quality because it increases the fruit sugar/organic acid ratio. This is due to the fact that the sugar content was increased and titratable acidity was reduced by up to 25%. Lobit et al. [73] reported a 50% decrease in malate accumulation when temperatures increased from 15 to 25 ºC. Indeed, Rosales et al. [74] observed a lower sucrose content and higher hexose content linked to the higher sucrolytic activity in fruits produced under high irradiative and temperature conditions.

Some authors found that rutin content was considerably increased (+40%) at 32 °C, but lycopene and ascorbate contents were reduced and, consequently, fruit nutritional quality could be modified [40]. Rivero et al. [71] found a 2-fold increase in total phenols for plants grown at temperatures under 35 °C instead of 25 °C. Total ascorbate, lycopene, β-carotene, rutin, and caffeic acid derivates reached maximum levels in the deep red tomatoes. Ascorbic acid and some phenolics tend to accumulate from the green to midripe stage, whereas the total carotenes increase constantly during the ripening process. In contrast, chlorogenic acid decreases during fruit ripening [75]. Its role as a potent antioxidant during abiotic stress such as UV-B exposure, which may explain its induction in fruit from plants under water stress [76]. Chlorogenic acid is a precursor of other phenolics such as caffeic acid derivates or caffeic acid glucoside. These phenolic compounds accumulate during ripening and the decrease of chlorogenic acid content is consistent with a possible decrease in phenylalanine ammonialyase activity and the use of chlorogenic acid as a precursor of other phenolics [77]. Hernández et al. [49] observed a clear tendency to the differentiation of the tomato according to period of sampling although the tomato samples had the same maturity degree, when the content of lycopene and hydroxycinnamic acids was studied (Figure 1). Climatic conditions during ripening such as temperature, light or relative humidity had an influence on the lycopene, chlorogenic and ferulic acid contents of tomatoes [49]. The soluble phenol content of plants grown under high light is approximately double the content under low-light [78].

Hunt and Baker [79] also reported that light stimulated the production of phenolic acids and flavonoids in the cuticles of the tomato fruit. The phenolic constituents of the cuticles of the fruit of three tomato cultivars (Ailsa Craig, Alicante and Grower's Pride) have been identified as chalconaringenin, naringenin, naringenin-7-glucoside, and *m*- and *p*-coumaric acids. Light increases the rate of biosynthesis of phenolics in plants by increasing the activities of enzymes, especially phenylalanine ammonialyase which plays an important role in converting phenylalanine (formed from shikimic acid pathway) into coumaric acid. This acid is the initial precursor of molecules involved in the synthesis of phenolic components in plants (flavonoids and hydroxycinnamates). Therefore, the field grown tomato fruits, which receive a higher amount of light and UV radiation, have been reported to contain a higher amount of flavonoids (quercetin and kaempferol) than greenhouse grown tomatoes [10].

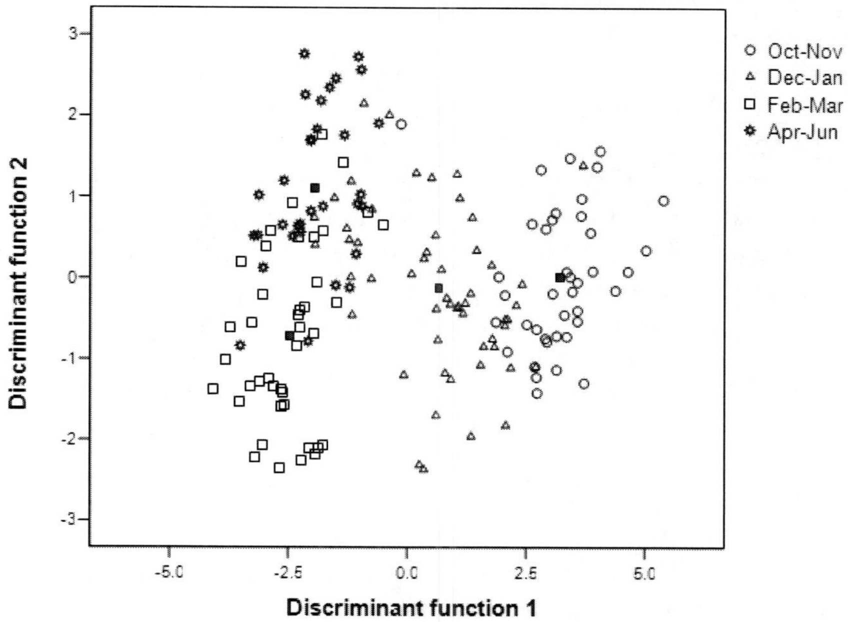

Figure 1. The antioxidant content can be used to differentiate harvest periods with the same maturity degree.

It is well established that phytochrome regulates light-induced lycopene accumulation [80] because red light enhances the first step of carotenogenesis by modulating the phytoene synthase activity, which is an important control step of carotene biosynthesis [81]. Gautier et al. [61] observed that the carotene content decreased when the temperature increased from 27 to 32 °C. The highest carotene content was found for fruits kept at 21 °C under light (88 $mg \cdot kg^{-1}$) and the lowest for fruits kept at 32 °C under darkness (47 $mg \cdot kg^{-1}$). Regardless of the experiment, phytoene, phytofluene, lycopene, γ-carotene, and β-carotene contents significantly increased with irradiation. However, increasing fruit irradiation during ripening significantly increased total carotene content by 45% and 42% at 21°C and 32 °C, respectively.

Rosales et al. [82] reported increased oxidation of reduced ascorbate as a result of temperature and irradiance increase during the summer. In that study, the discrimination between light and temperature effects shows that increasing irradiation promoted ascorbate accumulation, whereas increasing fruit temperature (from 27°C to 32 °C) limited its accumulation. Similarly, transferring plants from shade to sunshine leads to an increase of ≅ 66% in the ascorbic acid content of ripe fruits. However, ascorbate may also decrease with increased fruit solar exposure, probably due to increased ascorbic acid degradation with higher temperatures [83].

The greenhouse grown tomatoes are usually found to have lower levels of ascorbic acid than those grown outdoors because of the lower light intensity in the greenhouses [40]. Fruit exposed to direct sunlight during its development had higher carotene levels than shaded fruit. The carotenoid contents of fresh tomato fruit produced under glass, plastic tunnels and in the open field were compared. Fruit grown under a glass or plastic tunnel yielded a β-carotene content lower than in the open field. The lycopene content increased according to the following sequence: glass tunnel < plastic tunnel < field.

Ascorbate content decreases with fruit shading during ripening, in agreement with the data reported by El-Gizawy et al. [84]. This confirms that light reaching the fruit itself, during ripening, significantly increases fruit total ascorbate content. This could be due to lower oxidation or to an enhanced synthesis of ascorbate under increased irradiation.

As regards total antioxidant capacity, Kubota et al. [85] observed that tomatoes, at a red ripe stage, increased trolox equivalent antioxidant capacity (TEAC) under low temperature storage (12ºC for the 6 days and 5ºC for the following 7 days). The tomatoes, during this time, presented almost double the TEAC when compared with the level before storage. Tomatoes stored at room temperature did not show a significant change in TEAC, although lycopene concentration increased over time.

Cultural Practices

Crop Management with organic or conventional techniques is the main cultural practice with the greatest effect on the chemical composition of tomato. In conventional agriculture, the decision to use a pesticide is generally based on its effectiveness against pests, application costs and the economic value of the crop, as well as the relative risks to the crop of using the pesticide (phytotoxicity, resistance, etc.) [86].

It has been reported that plants cannot simultaneously allocate resources to growth and defense, as there is competition between protein and phenolic synthesis in plants for the common precursor involved in their biosynthesis, L-phenylalanine [87]. Modulation of the level of these compounds by pesticides has been observed by Barret et al. [88]. Plant secondary metabolism can be affected by pesticides, which may either increase or decrease concentrations of phenolic compounds in plants, depending on the action mechanism of the pesticide. Phenolic compounds are probably involved in the plant defense mechanism by acting as a chemical barrier against phytopathogens [89, 90].

Another aspect to consider is the mineral nutrition as a preharvest factor key that determines the yield and fruit quality of tomato plants. According to the 'C/N balance theory', when N is readily available, plants will primarily make compounds with high N content (e.g., proteins for growth). When N availability is limited, metabolism changes more towards carbon-containing compounds such as starch, cellulose, and non-N containing secondary metabolites such as phenolics and terpenoids [91]. Gao et al. [92] observed that the application of high NH_4^+ and low NO_3^- levels resulted in improved tomato quality. There was no significant difference in the soluble antioxidant activities of tomatoes grown using NO_3^- and chicken manure and grass-clover mulch. However, the antioxidant activity was significantly lower in tomatoes grown using NH_4^+ fertilisers as compared to other treatments. Heeb et al. [93] observed that the tomatoes grown using grass-clover mulch and NH_4^+ fertiliser had higher scores for acidity, flavour intensity and flavour persistence. The higher titratable acidity of these tomatoes is most likely to be responsible for their higher taste scores as there was no significant difference observed in the total soluble solids of tomatoes grown using different forms of fertilizers. Besides, a supplementary N, especially at high rates, generally tends to decrease the tomato fruit vitamin C content, possibly due to the increased shading caused by the greater development of plant foliage due to high N availability [40].

Dixon and Paiva [94] observed that phenolic compounds may be accumulated due to a deficiency in N, P, and Fe. A similar conclusion was drawn by Bourn and Prescott [95] when they observed that increasing nitrogen fertilization seems to decrease the level of phenolic compounds. Supplementation with P improves the colour of the fruit and increases the lycopene content. Increasing the P supply in hydroponics does not affect the vitamin C content of the fruit [40]. López et al. [96] reported that limitation of S supply in tomato plants can affect the colour of the tomatoes and also cause reduced a photosynthetic rate and chlorophyll content. These results suggest that low levels of S can inhibit the biosynthesis of lycopene in tomatoes. However, the exact cause for this observation is not known and requires further investigation.

Trudel and Ozbun [97] reported an increase in the lycopene concentration when the K concentration in the nutrient solution was increased from 0 to 8 mM, whereas a lower β-carotene concentration was observed. Fanasca et al. [6] suggest that it is of great interest to understand the effect of cationic proportions (K, Ca, and Mg) and their interaction to improve fruit quality and antioxidant content in tomato, especially with the hydroponic system. The high proportion of Mg in the solution improved the total antioxidant activity. Paiva et al. [98] showed that an increase in Ca concentration of the nutrient hydroponic solution results in a decrease in lycopene content (29%) due to the antagonism between Ca and K or the possible Ca influence on ethylene biosynthesis.

However, long term use of fertilizers produces an increase in soil salinity. It is widely believed that fruits from tomato plants grown under saline conditions have a higher quality. The organoleptic quality of marketable fruit improved significantly with increasing salinity [84, 99]. However, while salinity generally enhances fruit quality, the reduction in fruit size can affect the price received by the growers although soluble solids content and titratable acidity increase whereas pH decreases. Increases in total soluble solids content and titratable acidity enhance perceived fruit flavour. The most striking difference between saline and non-saline conditions is the increased starch accumulation [90]. The accumulation of proline in response to several environmental stresses is very common in plants [100]. Besides, Zushi and Matsuzoe [101] observed that salt stress induced changes in the antioxidant content and antioxidant systems. However, salt stress cannot improve the content of antioxidants such as ascorbic acid and glutathione reducted (GSH), which do not always increase under salt stress conditions. Tomatoes, under water and salt stress, mature earlier and accumulate more lycopene during the time before harvest. The altered water flow to the fruit under high electrical conductivity (≈ salinity) appears to contribute even more to the deposition of lycopene on a fresh weight basis [85].

Hernández et al. [102] analyzed the concentrations of minerals (P, Na, K, Ca and Mg) and trace elements (Fe, Cu, Zn and Mn) in the Dorothy tomato cultivar. These authors also considered the cultivation method and period of sampling, and they observed that the period of sampling had an important influence on the mineral and trace element content (Table 2 and 3). As regards intensive cultivation, there was a tendency to lower mineral mean concentrations in the October 2004–January 2005 period than those tomato samples from February to June 2005. The results were opposite with respect to organic tomatoes and no clear tendency was observed in the hydroponic cultivation, (Table 3).

Table 3. Concentration (mg·kg^{-1}) of the minerals analyzed in the Dorothy cultivar according to the cultivation method and the sampling period

Method	Period	P	Na	K	Ca	Mg
Intensive	Oct04-Jan05	210±58	84.1±39.6	2239±251	61.8±14.9	109±24
	Feb05-Jun05	228±42	171±71	2496±620	76.2±14.5	110±17
Organic	Oct04-Jan05	239±38	50.7±22.2	2280±441	64.1±18.6	108±30
	Feb05-Jun05	265±37	65.7±37.3	2846±548	78.7±18.4	118±10
Hydroponic	Oct04-Jan05	254±48	87.7±75.5	2481±261	70.7±9.2	113±7
	Feb05-Jun05	294±39	58.2±36.7	2894±604	82.5±18.9	103±21

One can observe, in Table 4, that Cu, Mn and Fe concentrations were higher during the February–June period than those in the October–January period, except for Fe in the intensive method. Zn behavior was the opposite. The organic system had the highest mean values.

Table 4. Concentration (mg·kg^{-1}) of the trace elements analyzed in the Dorothy cultivar according to the cultivation method and the sampling period

Method	Period	Fe	Cu	Zn	Mn
Intensive	Oct04-Jan05	1.58±0.25	0.303±0.105	0.673±0.164	0.625±0.133
	Feb05-Jun05	2.18±0.68	0.291±0.126	0.841±0.299	0.543±0.137
Organic	Oct04-Jan05	2.01±0.49	0.403±0.127	0.846±0.090	0.663±0.169
	Feb05-Jun05	1.65±0.68	0.350±0.119	0.860±0.092	0.607±0.109
Hydroponic	Oct04-Jan05	1.70±0.24	0.308±0.071	0.699±0.165	0.614±0.081
	Feb05-Jun05	1.62±0.17	0.247±0.065	0.770±0.192	0.551±0.078

Use of Multivariate Techniques to Establish Tomato Quality

Although sensory evaluation is the best method to characterize fruit quality, these tests are expensive, time-consuming, and require a panel with a large number of experts, and panelists are often the first source of variation. Thus, when a high number of samples are to be analyzed, this type of evaluation is substituted by chemical determinations [103].

The influence various factors exert on the chemical composition of the tomato have been described in the previous sections. In some cases there has been some disagreement, perhaps because not all factors were considered altogether: genetic, agricultural practices and environmental. In this regard, recent studies have applied multivariate analysis which revealed complex relationships between different components which helps in the interpretation of the results.

Many studies have attempted to clarify how pre-harvest factors are related to the characteristics of the final harvested product. However, the interactions of the genetic, environmental and agricultural practices involved are so intricate that a complete understanding would require not just an exhaustive experimental design, but input of multidisciplinary scientific knowledge and a suitable statistical method to search for behavioural patterns [104]. Some examples of the application of statistical analysis are given by Arvanitoyannis and Vaitsi [105]. These authors reviewed several papers for tomato authenticity where the results were analyzed by means of multivariate analysis (principal component analysis, discriminant analysis, cluster analysis, canonical analysis and factor analysis).

The two most frequently used techniques of multivariate statistics to process this kind of data matrix are principal component analysis (PCA) and factor analysis (FA). The fundamental utility of PCA lies in the fact that it simplifies the analysis of a multidimensional phenomenon when some or many of the variables in the study are correlated. Basically, its objective is to construct uncorrelated linear combinations that exhibit the greatest variance from the representative variables of a multidimensional phenomenon.

However, in some cases it is not useful to replace the original variables with these linear combinations. In multi-environment trials (MET), bi-plot analysis is being increasingly used in the analysis of this kind of data. The "Bi-plot" methods proposed by Gabriel [106] are a graphical display of multivariate data in two dimensions. This is done by representing the variables as vectors in the same plane and the correlations between them as the angles between those vectors. GGE-Bi-plot and HJ-Bi-plot may be two interesting alternatives. A GGE-Bi-plot simultaneously displays the genotype main effect (G) and the genotype x environment (GE) of a two-way table, and can visually address many questions relative to cultivar and test environmental evaluation [107]. In spite of their many useful features, GE bi-plots have been criticized for not being able to incorporate information on genetic co-variables that may explain the GE patterns. In order to amend this, factorial regression using genetic and/or environmental covariables has been suggested to complement bi-plot analysis [108].

Cebolla-Cornejo et al. [103] applied GGE-Bi-plot on tomato samples and observed that in the case of the main organic acids, the citric contents were not environment dependent. On the contrary, malic acid contents were significantly affected by the growing system showing higher accumulation under protection. In general, the protected cultivation in greenhouse, with lower light intensity and higher temperature, tended to offer lower sugar concentrations with similar acid contents and, thus, probably a lower organoleptic quality.

Regarding the HJ-Bi-plot, it is a symmetric simultaneous representation technique that allows an exploratory data analysis to look for hidden patterns [109]. Hernández [110] applied it to visualize the effect of the interaction of genetics, agricultural technique and environment on the taste of tomatoes from Tenerife (Spain). This author observed that, during the warmer months and using conventional farming practices, the sugar/acid ratio is higher,

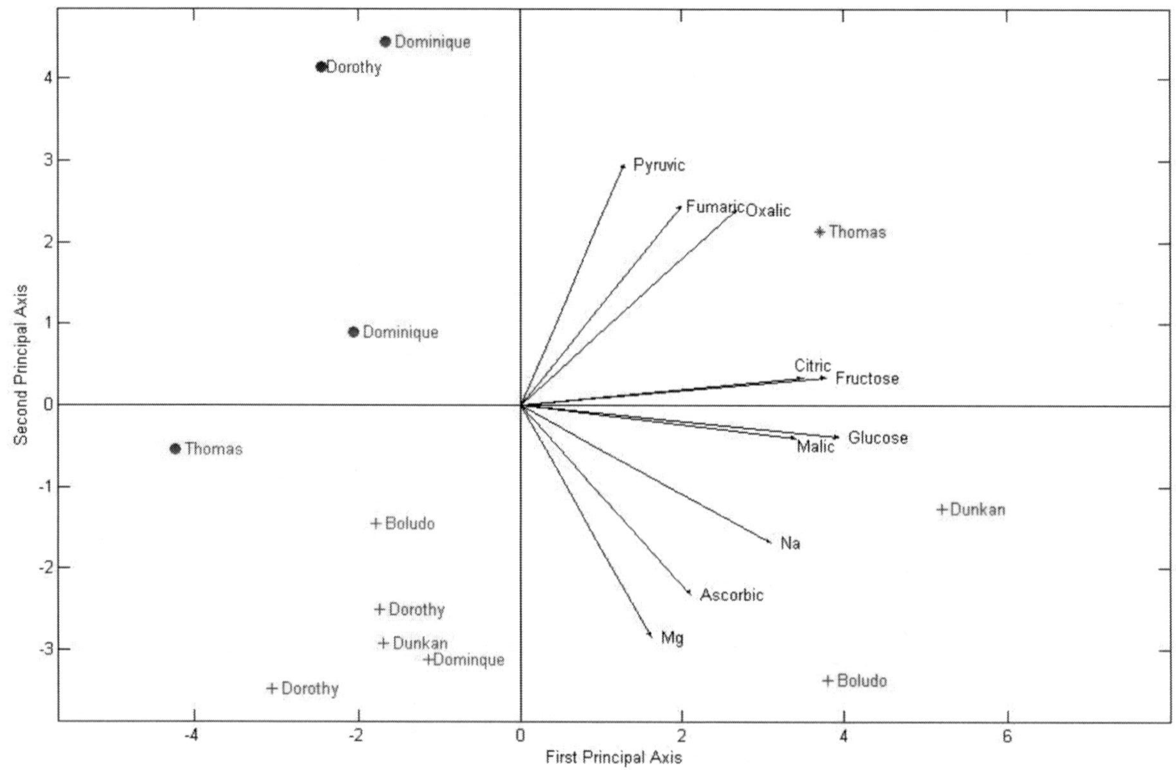

The conventional agriculture is shown in red, organic in blue, and hydroponic in blue while samples from October are shown with the symbol •, February with + and April with *

Figure 2. Results of HJ-Bi-plot applied to salad tomatoes from Tenerife (Spain) [110].

while the ascorbic acid content increases during the cooler months and with the use of organic farming practices. The best option to improve the sugar/acid ratio and ascorbic acid content is through conventional farming techniques with the Dunkan cultivar, especially if the climatic conditions are similar to those of February, or can be made so by cropping under cover (Figure 2). This author also applied STATIS-dual analysis (Structuration des Tableaux A Trois Indices de la Statistique, which could be translated in English as structuring three-way data sets in statistics) (Figure 3) to evaluate the effect of the harvest period on the taste of the tomato and confirmed that the sugar/acid ratio was high during the warmer months and ascorbic acid content was high during the cooler months for tomatoes from Tenerife (Spain).

Several studies have developed mathematical relationships between sensory descriptors and instrumental measures of the fresh tomato flavour to provide reliable analytical tools [111- 114]. Network analysis has proved to be a powerful tool to distil data into meaningful information. In network analysis, the traits are represented by nodes that are connected by links, with each link representing the interactions between two components. A biological network using sensory profile analysis and agronomic features was constructed by Carli et al. [115] to identify relationships among tomato metabolites. Various types of interactions, including amino acid networks, phenotypic and metabolic associations with sensory attributes, and links among metabolites of organoleptic importance, were revealed and essential relationships visualized. Carli et al. [115] observed, in their results, that the main feature of the network of tomato characteristics was the presence of three interconnected nodes (dry matter, pH, and °Brix) which are interconnected with other traits and nodes with widely different degrees (scales) mainly inside the same module. Dry matter, pH, and °Brix were identified as important drivers of fruit tomato quality components. Moreover, few metabolic traits seem to have a direct influence on important sensory traits, such as citric acid on tomato smell, glutamic acid on juiciness and dry matter on granulosity.

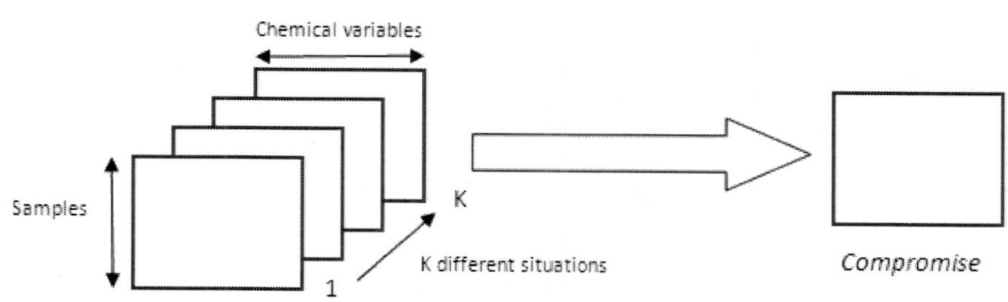

The goal of STATIS is 1) to compare and analyze the relationship between the different data sets (with the same row and column in each one, for example, same samples and same chemical variables), 2) to combine them into a common structure called a *compromise* (it is not a data set with mean values) which is then analyzed via PCA to reveal the common structure between the observations, and finally 3) to project each of the original data sets onto the compromise to analyze communalities and discrepancies [117].

Figure 3. Cube structure of data matrix for a STATIS-Dual analysis.

Nordén et al. [116] performed a study to characterize biochemical changes during the development, ripening and postharvest shelf-life of the tomato fruit, covering both the pre- and postharvest phases. A Partial Least Squares Discriminant Analysis (PLS-DA) was conducted to visualize the data in relation to the different developmental stages sampled. The different metabolites, for the PLS-DA, were considered as X-variables while the categorical variables maturity stage: 1–15 and pre-harvest versus postharvest fruit: 0–1were considered as Y-variables.

CONCLUSION

The tomato is one of the most important crops worldwide, and there is a vast amount of biodiversity within this specie. Therefore, in recent decades there has been a growing interest in knowing the chemical composition of tomatoes, for its role in the human diet, as well as the causes of change in its composition. During ripening, the tomato fruit undergoes a wide range of biosynthetic as well as hydrolysis reactions that markedly affect the final fruit composition. Crop management using organic or conventional techniques is the main cultural practice with the greatest effect on the chemical composition of the tomato. Little is known about the environmental regulation of tomato metabolites that are responsible for variations in fruit nutritional quality. Tomatoes exposed to direct sunlight in the field often develop a poor colour, mainly because the fruits exposed to high temperatures have low lycopene content. However, ascorbate may also decrease with increased fruit solar exposure, probably due to increased ascorbic acid degradation with elevated temperatures. Tomatoes, under water and salt stress, mature earlier and accumulate more lycopene during the time before harvest.

Studies on chemical composition have improved due to the application of multivariate statistical analysis. The different types of multivariate analysis are making it possible to recognize patterns of behavior, which was not possible using by classical techniques. There are observable controversies between the results obtained by several authors depending on the statistical method applied for the data analysis. Knowledge of the physiology of the tomato plant is also required to interpret the data correctly.

REFERENCES

[1] F. Nuez, J. Prohens and J. M. Blanca, *Am. J. Bot.* 9, 86 (2004).
[2] B. Burlingame, R. Charrondiere and B. Mouille, *J. Food Compos. Anal.* 22, 361 (2009).
[3] J. T. Esquinas-Alcázar and F. Nuez, *El cultivo del tomate*, Nuez, Madrid (1995).
[4] S. Sato, S. Sakaguchi, H. Furukawa and H. Ikeda, *Sci. Hortic.-Amsterdam* 109, 248 (2006).
[5] G. Giovanelli, V. Lavelli, C. Peri and S. Nobili, *J. Sci. Food Agr.* 79, 1583 (1999).
[6] S. Fanasca, G. Colla, G. Maiani, E. Venneria, Y. Rouphael, E. Azzini and F. Saccardo, *J. Agric. Food Chem.* 54, 4319 (2006).
[7] S. Agarwal and A. V. Rao, *Can. Med. Assoc. J.* 163, 739 (2000).
[8] A.V. Rao, *Tomato, Lycopene and Human Health. Preventing Chronic Diseases*. Rao, 2006.
[9] B. Burlingame, *J. Food Compos. Anal.* 17, 251 (2004).
[10] J. N. Davies and G. E. Hobson, *Crit. Rev. Food Sci.* 15, 205 (1981).
[11] T. C. Garvey and J. D. Hewitt, *J. Am. Soc. Hortic. Sci.* 116, 77 (1991).
[12] D. Miron and A. A. Schaffer, *Plant Physiol.* 95, 623 (1991).

[13] J. Osvald, N. Petrovič and J. Demšar, *Acta Alimentaria* 30, 53 (2001).
[14] G. Oims-Oliu, M. L. A. T. M. Hertog, B. Van de Poel, J. Ampofo-Asima, A. H. Geeraerd and B. M. Nicolaï, *Postharvest Biol. Tec* 62, 7 (2011).
[15] V. Fernández-Ruíz, M. C. Sánchez-Mata, M. Cámara and M. E. Torija, *HortScience*, 39, 339 (2004)
[16] AIJN – European Fruit Juice Association, Rue de la Loi 221 box 5, B-1040 Brussels, BE, B-1040, http://aijn.org/
[17] L. M. Hanover and J. S. White, *Am. J. Clin. Nutr.* 58, 724S (1993).
[18] M. A. Stevens, A. A. Kader, M. Albright-Holton and M. Algazi, *J. Am. Soc. Hortic. Sci.* 102, 680 (1977).
[19] W. G. Burton, a. van Es and K. J. Hartmans, *The physics and physology of storage*, Harris, London (1992).
[20] W. H. Gardner, *Food acidulants*, New York (1966).
[21] J. Cebolla-Cornejo, S. Roselló, L. Galaina-Balaguer and F. Nuez, *Actas Portuguesas de Horticultura*, 8, 302 (2005)
[22] T. Byers and N. Guerrero, *Am. J. Clin. Nutr.* 62, 1385S (1995).
[23] E. A. Lutsenko, J. M. Carcamo and D. W. Golde, *J. Biol. Chem.* 277, 16895 (2002).
[24] R. Marchioli, C. Schweiger, G. Levantesi, L. Tavazzi and F. Valagussa, *Lipids*, 36, S53 (2001).
[25] A.M. Adalid, S. Roselló and F. Nuez, *J. Food Compos. Anal.* 23, 613 (2010).
[26] W. A. Gould, *Tomato Production, Processing and Technology*, Baltimore (1992).
[27] S. Roselló, L. Galiana-Balaguer, A. M. Adalid and F. Nuez, *Actas de Horticultura* 45, 73 (2006).
[28] The National Academies Press, 500 Fifth St. N. W., Washington DC 2001, http://nap.edu/
[29] D. Grierson and A. A Kader, *The Tomato Crop, a Scient Basis for Improvement*; Atherton and Rudich, London (1986).
[30] C. Salles, S. Nicklaus and C. Septier, *Food Chem.* 8, 395 (2003).
[31] B. Lindemann, *Nature*, 413, 219 (2001)
[32] G. R. Pratta, G. R. Rodríguez, R. Zorzoli, L. A. Picardi and E. M. Valle, *Am. J. Plant S.* 2, 255 (2011).
[33] M. Hernández, E. Peña Méndez, B. Rodríguez Galdón, E. Rodríguez Rodríguez and C. Díaz Romero, *Agricultural Sciences*, 2, 424 (2011)
[34] M. C. Saure, *Sci. Hortic.-Amsterdam* 90, 193 (2001).
[35] O. A. Levander, *HortScience*, 25, 1486 (1990)
[36] S. S. Claye and A. C. W. Weber, *Plant Foods Hum. Nutr.* 49, 257 (1996).
[37] M. Hernández, E. M. Rodríguez and C. Díaz, *Food Chem.* 106, 1046 (2008).
[38] R. Ortega Anta, A. López Sobaler, A. Requejo Marcos and P. Carvajales, *La composición de los alimentos*, Madrid (2004).
[39] O. Moreiras, A. Carvajal, L. Cabrera and C. Cuadrado, *Tablas de composición de alimentos*, Madrid (2005).
[40] Y. Dumas, M. Dadomo, G. D. Lucca and P. Grolier, *J. Sci. Food Agr.* 83, 369 (2003).
[41] K. Canene-Adams, J. K. Campbell, S. Zaripheh, E. H Jeffery and J. W. Erdman, *J. Nutr.* 135, 1226 (2005).
[42] A.O. Omoni and R. E. Aluko, *Trends Food Sci. Technol.* 16, 344 (2005).

[43] S. Schwarz, U. C. Obermuller-Jevic, E. Hellmis, W. Koch, G. Jacobi and H. K. Biesalki, *J. Nutr.* 138, 49 (2008).
[44] P. Di Mascio, S. Kaiser and H. Sies, *Arch. Biochem. Biophys.* 274, 532 (1989).
[45] M. Díaz, B. Frei, J. A. Vita, JA and K. JF, *NEJM.* 337, 408 (1997).
[46] S. Clinton, C. Emenhoser and S. Schwartz, *Cancer Epidemiol. Biomar. Prev.* 5, 823 (1996).
[47] C. Gärtner, W. Stahl and H. Sies, *Am. J. Clin. Nutr.* 66,116 (1997)
[48] W. Stahl and H. Sies, *J. Nutr.* 122, 1220 (1992).
[49] M. Hernández, E. Rodríguez and C. Díaz, *J. Agric. Food Chem.* 55, 8604 (2007).
[50] D. Giuntini, G. Graziani, B. Lercari, V. Fogliano, G. F. Soldatini and A. Ranieri, *J. Agric. Food Chem.* 53, 3174 (2005).
[51] SINU (Societa Italiana di Nutrizione Umana), c/o Promo Leader Service Congressi Via della Mattonaia 17, Firenze, It 50121, http://sinu.it/
[52] K. Herrmann, *Crit. Rev. Food Sci. Nutr.* 28, 315 (1989).
[53] H. Gautier, V. Diakou-Verdin, C. Bénard, M. Reich, M. Buret, F. Bourgaud, J. L. Poëssel, C. Caris-Veyrat and M. Génard, *J. Agric. Food Chem.* 56, 1241 (2008).
[54] E. Venneria, S. Fanasca, G. Monastra, E. Finotti, R. Ambra, E. Azzini, A. Durazzo, M. S. Foddai and G. Maiani, *J. Agric. Food Chem.* 56, 9206 (2008).
[55] R. K. Toor, G. P. Savage and A. Heeb, *J. Food. Comps. Anal.* 19, 20 (2006).
[56] M. Hernández Suárez, E. Rodríguez Rodríguez and C. Díaz Romero, *Eur. Food Res Technol.* 226, 423 (2008).
[57] N. J. Atkinson, t. P. Dew, c. Orfila and P. E. Urwin, *J. Agric. Food Chem.* 59, 9273 (2011).
[58] A. Srivasta and A. K. Handa, *J. Plant Growth Regul.* 24, 67 (2005).
[59] P. M. Bramley, *J. Exp. Bot. 53*, 2107 (2002).
[60] G. A. Tucker, *Biochemistry of fruit ripening*, Seymour, Taylor and Tucker, London (1993).
[61] H. Gautier, T. Guillemette, P. Simoneau and B. Lapied, *Plant and fungal secondary compounds as potential insecticide agents*, Goudey-Perrière, Benoit, Goyffon and Marchot, Paris (2006).
[62] D. L. Richardson, H. V. Davies, H. A. Ross and G. R Mackay, *J. Exp. Bot.* 41, 95 (1990).
[63] S. Islam, T. Matsui and Y. Yoshida, *Sci. Hortic.- Amsterdam* 65, 137 (1996).
[64] S. DeBolt, D. C. Cook, C. M. Ford and CM. *PNAS.* 103, 5608 (2006).
[65] E. M. Valle, S. B. Boggio and H. W. Heldt, *Plant Cell Physiol.* 39, 458 (1998).
[66] S. B. Boggio, J. F. Palatnik, H. W. Heldt and E. M Valle, *Plant Science*, 159, 125 (2000).
[67] G. Le Gall, I. J. Colquhoun, A. L. Davis, G. J. Collins and M. E. Verhoeyen, *J. Agric. Food Chem.* 51, 2447 (2003).
[68] E. Giovannucci, *Exp. Biol. Med.* 227, 860 (2002).
[69] A. Jimenez, G. Creissen, B. Kular, J. Firmin, S. Robinson, M. Verhoeyen and P. Mullineaux, *Planta*, 214, 751 (2002)
[70] N. Smirnoff, Philos. *Trans. R. Soc. London* B355, 1455 (2000).
[71] R. M Rivero, J. M. Rui, P. C. García, L. R. López-Lefebre, E. Sánchez and L. Romero, *Plant Sci.* 160, 315 (2001).
[72] B. Ferguson, S. Lurie and J. H. Bowen, *Plant Physiol.* 104, 1429 (1994).

[73] P. Lobit, M. Génard, P. Soing and R. Habib, *J. Exp. Bot.* 57, 1471 (2006).
[74] M. A. Rosales, M. M. Rubio-Wilhelmi, R. Castellano, N. Castilla, J. M. Ruiz and L. Romero, *Sci. Hortic.-Amsterdam* 11, 244 (2007).
[75] M. Winter and K. Herrmann, *J. Agric. Food Chem.* 34,616 (1986).
[76] C. Cle, L. M Hill, R. Niggeweg, C. R. Martin, Y. Guisez, E. Prinsen and M. A. K. Jansen, *Phytochemistry*, 69, 2149 (2008).
[77] A. Fleuriet and J. J. Macheix, *Phytochemistry*, 24, 929 (1985)
[78] R. T. Wilkens, J. M. Spoerke and N. E. Stamp, *Ecology*, 77, 247 (1996)
[79] G. M. Hunt and E. A. Baker, *Phytochemistry*, 19, 1415 (1980)
[80] R. Alba, M. M. Cordonnier-Pratt and L. H. Pratt, *Plant Physiol. 123*, 363 (2000).
[81] A. Schofield and G. Paliyath, *Plant Physiol. Biochem.* 43, 105 (2005).
[82] M. A. Rosales, J. M. Rui, J. Hernández, T. Soriano, N. Castilla and L. Romero, *J. Sci. Food Agr.* 86, 1545 (2006).
[83] C. A. Torres, P. K. Andrews and N. M. Davies, *J. Exp. Bot.* 57, 1933 (2006).
[84] A.M. El-Gizawy, M. M. F. Abdallah, H. M. Gomaa and S. S. Mahamed, *Acta Horticulturae*, *323*, 349 (1993).
[85] C. Kubota, C. A. Thomson and M. Wu, *HortScience*, 41, 522 (2006)
[86] Van der Werf, *Photoassimilate Distribution in Plants and Crops*, Zamski and Schaffer, New York (1996).
[87] M. Riipi, V. Ossipov, K. Lempa, J. Haukioja, S. Ossipova and K. Pihlaja, Oecologia, 130, 380 (2002)
[88] D. M. Barrett, C. Weakley, J. V. Dia and M. and Watnik, *J. Food Sci.* 72, C441 (2007).
[89] Brandt and J. P. Molgaard, *J. Sci. Food Agr.* 81, 924 (2001).
[90] A.E. Mitchell, Y. J. Hong, E. Koh, D. M. Barrett, D. E. Bryant, R. F. Denison and S. Kaffka, *J. Agric. Food Chem.* 55, 6154 (2007).
[91] E. Haukioja, V. Ossipov, J. Koricheva, T. Honkanen, S. Larsson and K. Lempa, *Chemoecology*, 8, 133 (1998)
[92] Z. Gao, M. Sagi and H. Lips, *Plant Physiol. Biochem.* 34, 807 (1996).
[93] A. Heeb and B. Lundegardh, T. Ericsson, G. P. Savage, *J. Sci. Food Agr.* 85, 1405 (2005).
[94] R. A. Dixon and N. L. Paiva, *Plant Cell*, 7, 1085 (1995)
[95] D. Bourn and J. Prescott, *Crit. Rev. Food Sci.* 42, 1 (2002).
[96] Lopez, N. Tremblay, W. Voogt, S. Dube and A. Gosselin, *Sci. Hortic.-Amsterdam* 67, 207 (1996).
[97] J. Trudel and J. L. Ozbun, *J. Am. Soc. Hortic. Sci.* 96, 763 (1971).
[98] E. A. S. Paiva, R. A. Sampaio and H. E. P. Martinez, *J. Plant Nutr.* 21, 2653 (1998).
[99] J. Cuartero and R. Fernández-Muñoz, *Sci. Hortic.-Amsterdam* 78, 83 (1999).
[100] D. P. S. Verma, *Osmotic stress tolerance in plants: role of proline and sulfur metabolism*, Shinozaki and Yanaguchi-Shinozaki, Austin TX (1999).
[101] K. Zushi and N. Matsuzoe, *Sci. Hortic.-Amsterdam* 120, 181 (2009).
[102] Hernández, E. M. Rodríguez and C. Díaz, C. *Food Chem.* 104, 489 (2007).
[103] J. Cebolla-Cornejo, S. Roselló, M. Valcárcel, E. Serrano, J. Beltrán and F. Nuez, *J. Agric. Food Chem.* 59, 2440 (2011).
[104] S. R. Grattan and C. M. Grieve, *Sci. Hortic.-Amsterdam* , 78, 127 (1999).
[105] Arvanitoyannis and O. B. Vaitsi, *Crit. Rev. Food Sci.* 47, 675 (2007).
[106] R. Gabriel, *Biometrika*, 58, 453 (1971)

[107] W. Yan and N. A. Tinker, *Crop Sci.* 45, 1004 (2005).

[108] Brancourt-Humel, G. Doussinault, C. Lecomte, P. Berard, B. Le buance and M. Trottet, *Crop Sci.* 43, 37 (2003).

[109] Galindo, *Questíio*, 1, 13 (1986)

[110] M. Hernández Suárez. Preprint, 2011 at http://gredos.usal.es/ jspui/bitstream/10366 /108780/1/TFM-Marcos Hdez.pdf

[111] A. Krumbein and H. Auerswald, *Food/Nahrung*, 42, 395 (1998)

[112] Krumbein, P. Peters, S. Widell, H. Auerswald, B. Brückner and M. Linke, *Fruit and vegetable quality: an integrated view*, Shewfelt and Brürkner, Florida (2000).

[113] F. Maul, S. A. Sargent, C. A. Sims, E. A. Baldwin, M. O. Balaban and D. J. Huber, *J. Food Sci.* 65, 1228 (2000).

[114] S. Tandon, E. A. Baldwin, J. W. Scott and R. L. Shewfelt, *J. Food Sci.* 68, 2366 (2003).

[115] Carli, S. Arima, V. Fogliano, L. Tardella, L. Frusciante and M. R. Ercolano, *J. Exp. Bot.* 60, 3379 (2009).

[116] Nordén, P. Broberg, C. Lindberg and A. Plymoth, *Chem. Biodivers.* 2, 1487 (2005).

[117] H. Abdi and D. Valentin, D. *Encyclopedia of Measurement and Statistics*, Neil Salkind, California (2007).

In: Tomatoes: Cultivation, Varieties and Nutrition
Editor: Tadahisa Higashide

ISBN: 978-1-62417-915-0

Chapter 12

FACTORS AFFECTING THE ANTIOXIDANT CONTENT OF FRESH TOMATOES AND THEIR PROCESSED PRODUCTS

Norma Francenia Santos-Sánchez, Rogelio Valadez-Blanco and Raúl Salas-Coronado*

Instituto de Agroindustrias, Universidad Tecnológica de la Mixteca, Huajuapan de León, Oaxaca, México

ABSTRACT

Worldwide, *Solanum lycopersicum* (tomato) is the most important vegetable, mostly due to its sensorial properties and also because it is a rich source of carbohydrates, fiber and antioxidants. In recent years there has been a growing interest in the study of antioxidants in vegetables due to the effects of these compounds on human health. The antioxidant activity of tomato fruits is attributed to vitamin C, phenolic compounds and carotenoids, particularly lycopene. This latter compound is of essential importance in tomato´s definition as a functional food, given its proven properties to fight prostate cancer. The antioxidants in tomato fruits have different degrees of liability in terms of exposure to air, storage and processing temperatures and enzymatic activity in fresh tomato. The degradation of these compounds in tomato fruits and its products is synergistic. Also, it has been observed that the concentration of antioxidants in fresh tomato depends on the variety and growing conditions. In processed tomato fruits, the antioxidant content is additionally influenced by the sample preparation (juice, puree, slices, etc.), processing temperature and time, and extent of water removal. Different methods have been developed to reduce the degradation of antioxidants due to processing. These methods include the use of coadjuvants, such as maltodextrins in juice spray drying, sodium metabisulfite in tray drying of tomato slices, etc. Therefore, the study of the antioxidant content in the fruit from fresh tomato to its final products is a topic of paramount interest to the farming and food processing industries; however, it has not been comprehensively addressed in the literature. Thus, this chapter describes and

* E-mail address: rsalas@mixteco.utm.mx. Phone: +52(953)5320214 ext.400.

analyzes the factors affecting the antioxidant content in fresh tomatoes and its processed products.

INTRODUCTION

Tomato is the most important vegetable worldwide due to its large consumption and it is the second most widely cultivated fruit. This vegetable is a key component of the "Mediterranean diet", which has been directly associated with the reduction in the risk of chronic degenerative diseases [1].

Tomato fruit is an important natural source of antioxidants that contribute to human diet. This fruit is eaten raw in salads and sauces, and also as a processed food in juices, purees and soups. Diverse studies have shown that consumption of raw tomatoes and its processed derivatives is associated with a reduction in the risk of cancer incidence and cardiovascular diseases. Monitoring of *in vivo* antioxidant activity has demonstrated that consumption of tomato products improves the antioxidant activity of blood plasma and reduces low density lipoproteins (LDL) oxidation in healthy men [2]. These properties were attributed to the antioxidant compounds present in tomato. Among these compounds are carotenoids such as β-carotene (a vitamin A precursor), lutein and lycopene; the latter being responsible of the tomato fruit red pigmentation. Tomato fruits also contain vitamins such as ascorbic acid and tocopherols, and phenolic compounds such as flavonoids and hydroxycinnamic acid derivatives.

The antioxidant compounds in tomato fruits, and tomato nutrients in general, can be affected by many pre- and post-harvest factors. Pre-harvest factors include: light intensity, temperature, ripening stage at harvest and irrigation system. Tomato´s lycopene content strongly depends on the plantation (open field and greenhouse crops) temperature range. Plantation temperatures between 12°C and 32°C have been recommended to enhance tomato lycopene content since temperatures out of this range inhibit lycopene accumulation. Light exposure favors the production of ascorbic acid and growing season regulates the production of phenolic compounds [3].

Since tomato is a climacteric fruit, it continues to ripen after its harvest. Thus, the storage conditions of tomato fruit can also affect the antioxidants content. A wide number of treatments have been used to process tomato, such as drying in halves, slices or quarters, or smashing to produce puree which is an unstructured food. Puree is used as a base ingredient to prepare other foods such as pizzas, sauces, soups and juices. Both temperature and processing time have a significant effect on the antioxidant content of tomato products. Thermal processing induces oxidative degradation of antioxidants including lycopene, and also promotes isomerization of this compound. In raw tomato, lycopene is present in its *trans* form which is poorly absorbed by the human body. It appears that *trans* isomers need to be transformed to their *cis* isomers before lycopene can be readily absorbed. High concentrations of the *cis* isomer are present in human serum and prostate tissue, suggesting that isomerases present in the tissues might catalyze *in vivo* lycopene isomerization from all-*trans* to *cis* forms, Figure 1 [4].

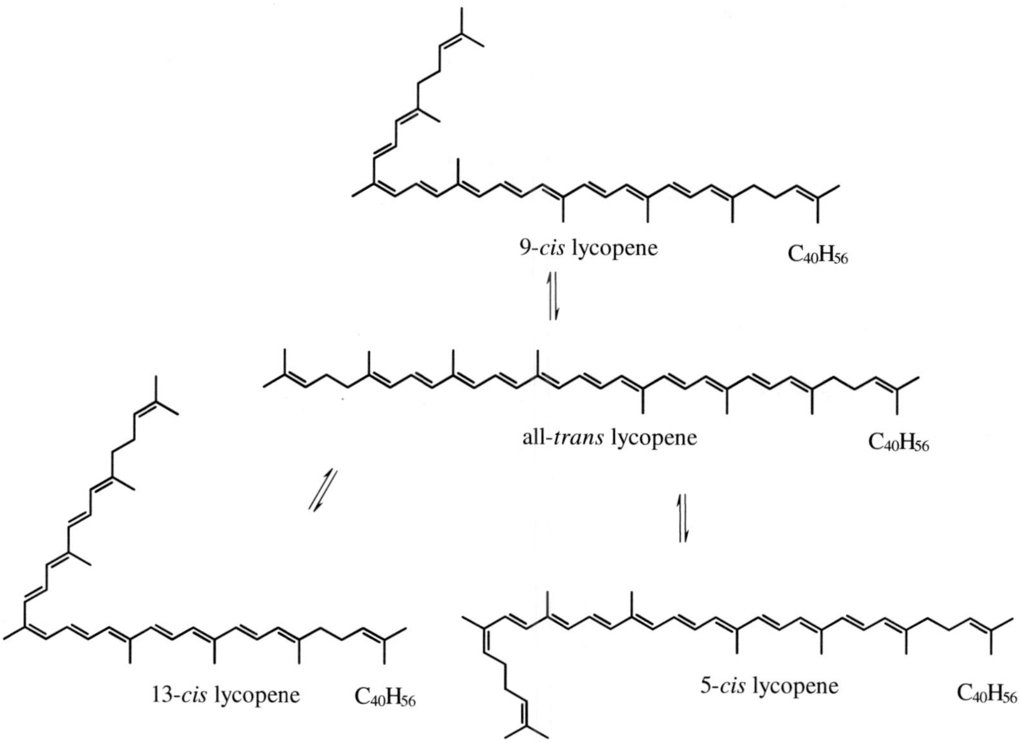

Figure 1. Chemical structures of lycopene isomers.

Stahl et al. [5] observed that *cis*-lycopene is more bio-accessible than the all-*trans* isomer. Furthermore, they concluded that thermally-processed tomato products had higher activity against prostate cancer than fresh tomato due to the fact that thermal processing induces the formation of *cis*-lycopene isomers. Clinton et al. [6] and Boileau et al. [7] found that *cis*-lycopene isomers are better absorbed in the body compared to the all-*trans* compounds. Lovrić et al. [8] studied lycopene isomerization as a function of time and temperature. These authors observed that no *trans-cis* isomerization occurred during storage; *trans* isomeric regeneration and lycopene oxidation reactions took place instead, Figure 2.

Summarizing, the content of antioxidants in tomato fruits is determined by a wide range of physicochemical and biochemical factors. Such factors can be roughly classified into pre- and post-harvest treatments which include: plantation temperature, light intensity, fruit ripening stage, storage time and processing methods. This chapter critically reviews these factors and treatments as well as their effect on the antioxidant content of tomato products.

PRE-HARVEST EFFECTS

The intensity of the light irradiated to tomato plants has a significant effect on the antioxidant content of tomato fruits. It is well known that an increase in the exposure time of a tomato plant to light results in an increment of the fruit carbohydrate content, as it occurs in the biosynthesis of ascorbic acid from sugars supplied by the photosynthesis (Figure 3). Gautier et al. [9] reported a direct correlation between sugar content and ascorbic acid.

Synthetized carbohydrates also serve as substrates for flavonoid biosynthesis through the shikimic acid (Figure 4) and phenylpropanoids (Figure 5) pathways.

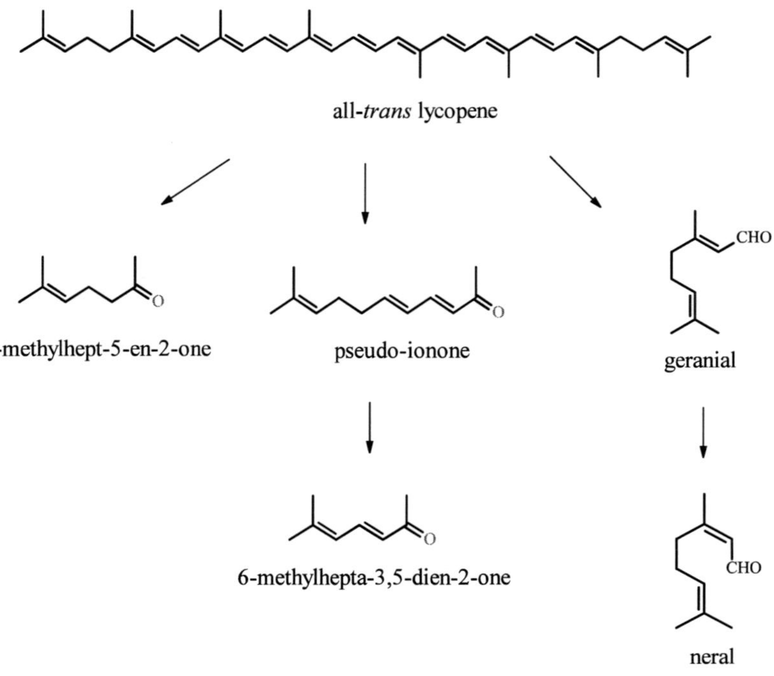

Figure 2. Lycopene oxidation products.

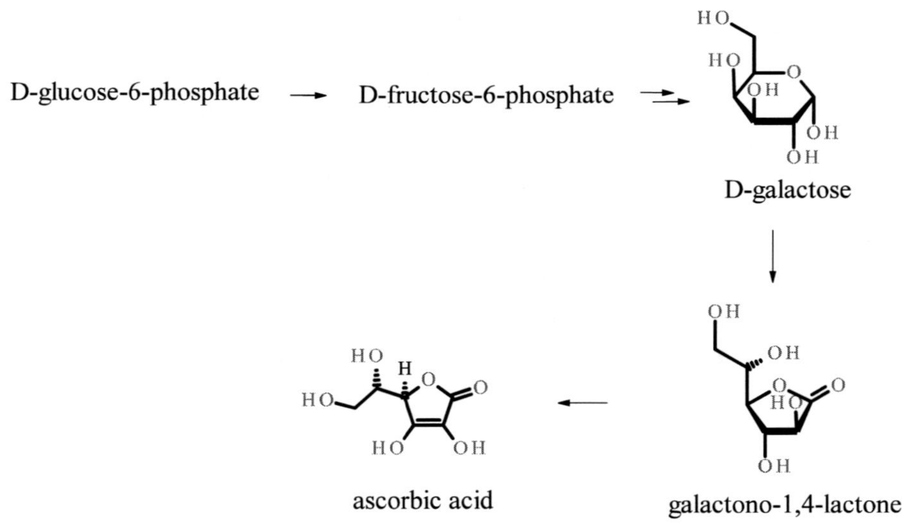

Figure 3. Ascorbic acid biosynthesis.

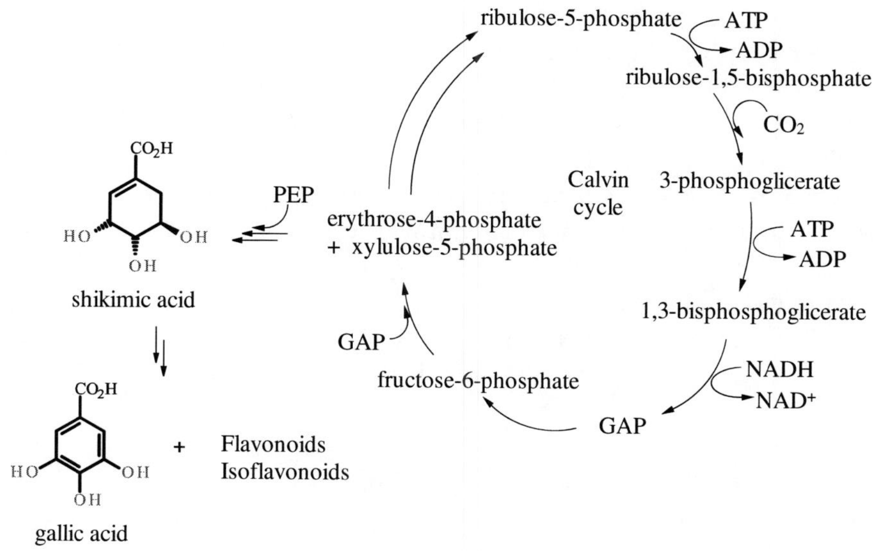

Figure 4. Shikimic acid biosynthesis from erythrose-4-phosphate (Calvin cycle).

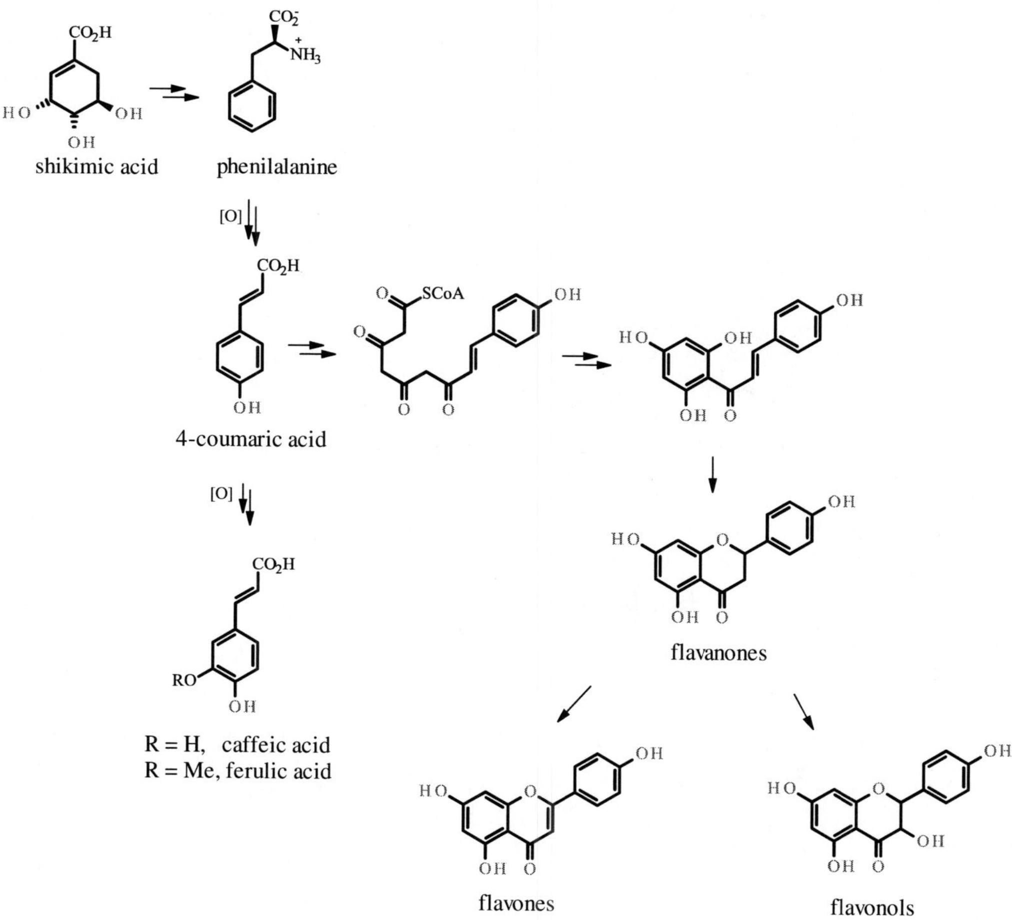

Figure 5. Phenylpropanoids biosynthesis.

The biosynthesis of tomato carotenoids (Figure 6) is affected by light. According to Thomas and Jen [10] both light quality and quantity determine the amount of carotenoids produced during the ripening stage. Jen [11] found that blue light was the most effective to enhance carotenoid biosynthesis which is regulated by the phytochrome photoreceptor during tomato ripening. Aiming at controlling the rate of ripening, Lee et al. [12] reported the quality of light required as well as the optimum levels of irradiance and irradiation time. These authors found that an irradiation treatment with red light alone or in combination with blue light during 3 to 5 min intervals was effective to increase the rate of lycopene production (indicated by a red color in the fruit). On the other hand, a far-red irradiation treatment of 38.9 W•m^{-2} during 3 to 5 min caused a retarding effect in lycopene production as the reaction was diverted to the formation of phytoene (indicated by a yellow color in the fruit).

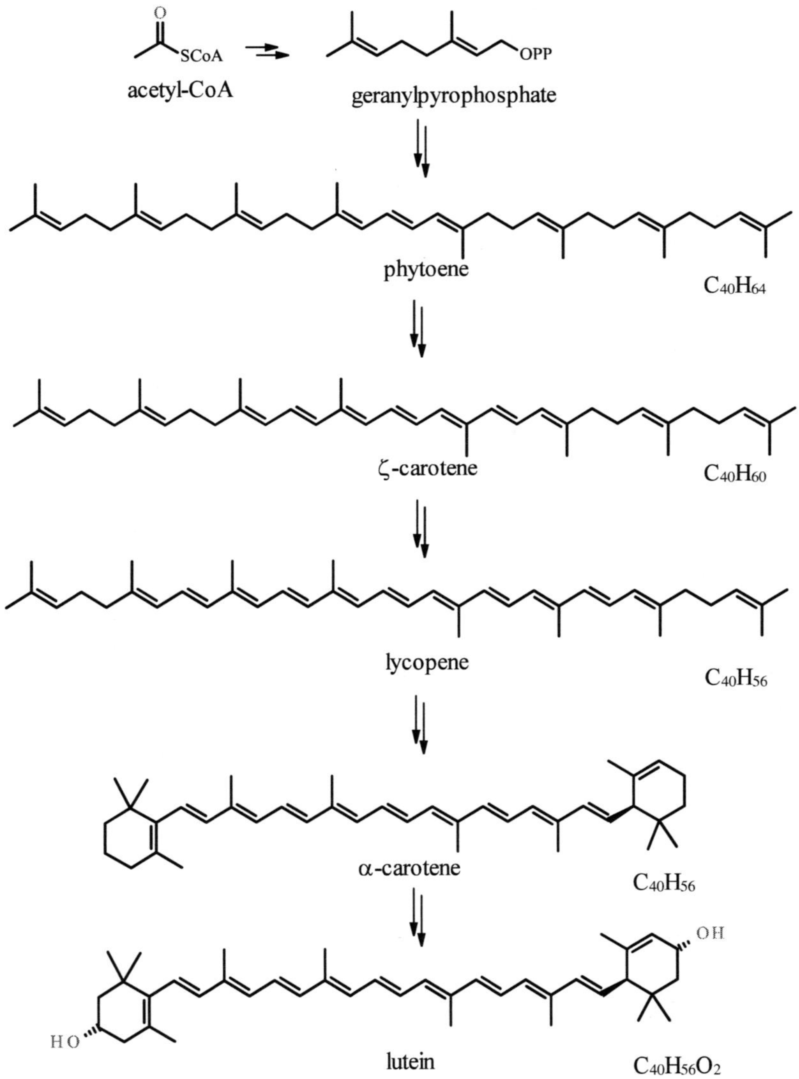

Figure 6. Carotenoids biosynthesis.

These results can be explained considering that lycopene biosynthesis demands a larger amount of energy than that of phytoene. This energy is provided by blue light, which first causes the production of phytoene as an intermediate followed by a series of subsequent steps which yield the formation of lycopene. On the other hand, the formation of phytoene as an end-product only needs the energy provided by the far-red light.

In tomato cultivars grown in open fields, plant leaves significantly affect the temperature of the fruit. Tomato fruits directly exposed to an intense solar radiation could reach 10ºC more than the fruits that are covered by leaves. Strong direct solar radiation on the fruits (2990 $\mu mol \bullet m^{-2} \bullet s^{-2}$ for 1.5-4.0 h) is harmful. Lycopene production is inhibited when the temperature of the fruit exceeds 30°C. Biosynthesis of lycopene is favored at temperatures between 12ºC and 21°C [13]. On the other hand, low-intensity light results in a reduction of the lycopene content thus producing fruits with a non desirable heterogeneous color [14]. Cox et al. [15] studied the effects of the duration of light exposure on the lycopene content of tomato fruits. Their result showed that in the breakers stage, ripening of tomatoes exposed to an 8 h photoperiod presented lower lycopene content than those exposed to a 24 h photoperiod.

Pernice et al. [16] investigated the effect of irrigation on the antioxidant content profile in tomato cultivars. They applied three irrigation regimes (none, reduced and normal irrigation) to the cultivation of two Corbarini small tomato varieties (biotypes PC01 and PC05). The results showed differences in carotenoid and flavonoid contents between the two varieties. The carotenoid content increased 50.6% and 34.9% for the biotypes PC01 and PC05, respectively, when tomato plantations were not irrigated in comparison with normally irrigated plants. Carotenoids were quantified as the sum of the individual determinations of phytoene, phytofluene, lycopene, β-carotene and ζ-carotene. On the other hand, the reduced irrigation regime produced an increment of 70.4% and 57.0% in the flavonoid content of biotype PC01 and PC05 fruits, respectively, with respect to the normal irrigation treatment. In this study, total flavonoids content were expressed as the sum of the amounts of quercetin and naringenin in the samples [16]. The results show that it is possible to modulate the antioxidant concentration in the tomato fruit using a controlled irrigation system. Particularly, the contents of carotenoids and flavonoids in tomato fruit are greatly influenced by both the irrigation regime and the cultivar´s genotype.

García-Valverde et al. [17] quantified the effects of the ripening stage and cultivar variety on the fruit´s antioxidant content. The study considered four ripening stages (green, breaker, pink and red), and four tomato varieties: Ronaldo, Pear, Zoco and Cherry Pear. The authors found significant differences in the antioxidant content in terms of both ripening stage and cultivar variety. The average lycopene contents of the four varieties were 0.63, 12.20, 26.76 and 116.66 $mg \bullet kg^{-1}$ of fresh weight tomato (FW), respectively. Ascorbic acid contents were not dependent on the ripening stage and ranged from 2.79 (breaker Cherry Pear) to 297.62 (red Cherry Pear) $mg \bullet kg^{-1}$ FW. In agreement with this, Jiménez et al. [18] and Cano et al. [19] observed that ascorbic acid content remained relatively constant during the first three ripening stages, with a slight ascorbic acid increase in the last stage when the fruit was completely ripened. However, Abushita et al. [20] and Giovanelli et al. [21] found an ascorbic acid increase during the first ripening stages which was followed by a decrease in concentration in the last ripening periods. Similarly, Periago et al. [22] reported significant increments during all the ripening stages. Opara et al. (2011) studied the effect of the ripening stage in the antioxidant content of three tomato varieties (Cherry, Monika and Isabella) [23]. They found no correlation between the changes on ascorbic acid content and the ripening stage (green, breaker, turning, pink, red and deep red). The authors suggested that the onset and amount of ascorbic acid biosynthesis in tomato depend primarily on its variety. In conclusion, ascorbic acid content does not depend on the ripening stage but on the cultivar variety.

Total phenolic compounds content varied between 186.92 and 558.63 mg gallic acid equivalents (GAE)•kg^{-1} FW, corresponding to green Ronaldo and red Cherry Pear tomato, respectively. In the same study, the authors found high content of the flavonoid rutin in the pink Cherry Pear variety (43.3 mg•kg^{-1} FW), which accounts for more than twice the rutin content present in the other varieties. Chlorogenic acid was also present in the four tomato varieties. Similar results were found by Kotíková et al. [24] in the analysis of eight varieties of tomato collected at four ripening stages of the fruit (green, breaker, pink and red). These authors reported that hydrophilic antioxidants, particularly ascorbic acid, had a predominant contribution to the fruit antioxidant activity. Lipophilic antioxidants represented only 17% of the total antioxidant activity. In addition, synergic effects between lycopene-lutein, lycopene-β-carotene and α-tocopherol-β-carotene were observed. In the same accord, Fuhrman [25] observed synergic effects of α-tocopherol and lycopene on the inhibition of the oxidation of LDL.

Liu et al. [26] found that ascorbic acid assists in lycopene regeneration. Ilahy et al. [27] studied the changes in the antioxidant compounds in an ordinary cultivar (Rio Grande) and three high-lycopene tomato cultivars (HLY 13, HLY 18 and Lyco 2) at four different fruit ripening stages (green, breaker, pink and red). They quantified ascorbic acid, dehydroascorbic acid, total vitamin C (ascorbic acid + dehydroascorbic acid), phenolics, flavonoids, and total carotenoids. Ascorbic acid and dehydroascorbic acid had antioxidant properties, but the hydrolysis product of dehydroascorbic acid (2,3-diketogulonic acid) was inactive (Figure 7). The stage of ripening significantly influenced both total carotenoid and lycopene content. During ripening, the fruits of cultivars HLY 13 and HLY 18 exhibited more than 100% higher total carotenoid and lycopene contents as compared to those found in Rio Grande fruits. Cultivar HLY 18 fruits accumulated the highest amounts of phenolics and flavonoids. On the other hand, Lyco 2 tomato showed the highest levels of dehydroascorbic acid and total vitamin C.

The content of phenolic acid compounds is generally expressed in terms of equivalents of gallic acid since there is not a simple analytic method to quantify total phenols in an exact and universal manner. This is due to the fact that there is a great structural diversity among phenols depending on the plant´s nature. In addition, total flavonoids can also be determined in terms of a specific group of phenolic compounds. However, current antioxidant science aims for the identification and quantification of individual phenolic compounds.

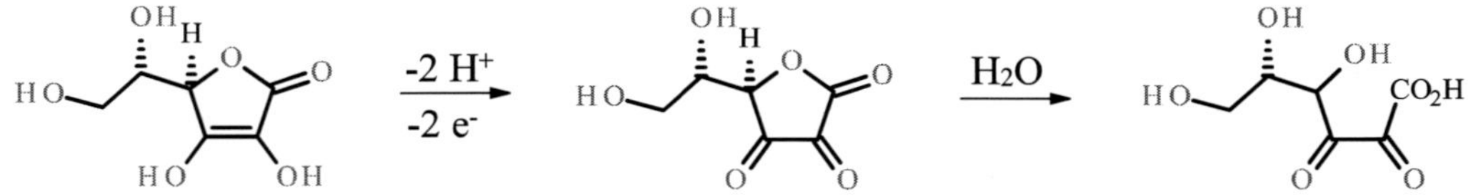

ascorbic acid dehydroascorbic acid 2,3-diketogulonic acid

Figure 7. Reaction of degradation of ascorbic acid.

García-Valverde et al. [17] quantified flavonoids and hydroxycinnamic acids in tomato samples. Rutin and naringenin were the main flavonoids present in the fruits; additionally, the following flavonoids were identified: quercetin, quercitrin, kaempferol, kaempferol-3-*O*-rutoside, naringin and some chalcones (Figure 8). Chlorogenic acid was the phenolic acid present in larger amounts. These results are in accord with the studies of Slimestad et al. [28, 29] who quantified this acid as the most abundant phenolic compound in a wide range of tomato varieties. The authors also reported the presence of the following phenolic compounds: caffeic, ferulic and 4-coumaric acids (Figure 9). Also in agreement to Valverde-García et al. [17] results, the authors identified rutin as the most abundant flavonoid in the fruits, with concentrations varying according to the variety and ripening stage of the fruit from 6 to 24 mg•kg^{-1} FW in green and red tomato, respectively. In general, individual phenolic compounds increased gradually during the green and the pink stages and decreased in the red stage. Accordingly, Salunkhe et al. [30] concluded that the amount of phenolic compounds decreases with tomato ripening.

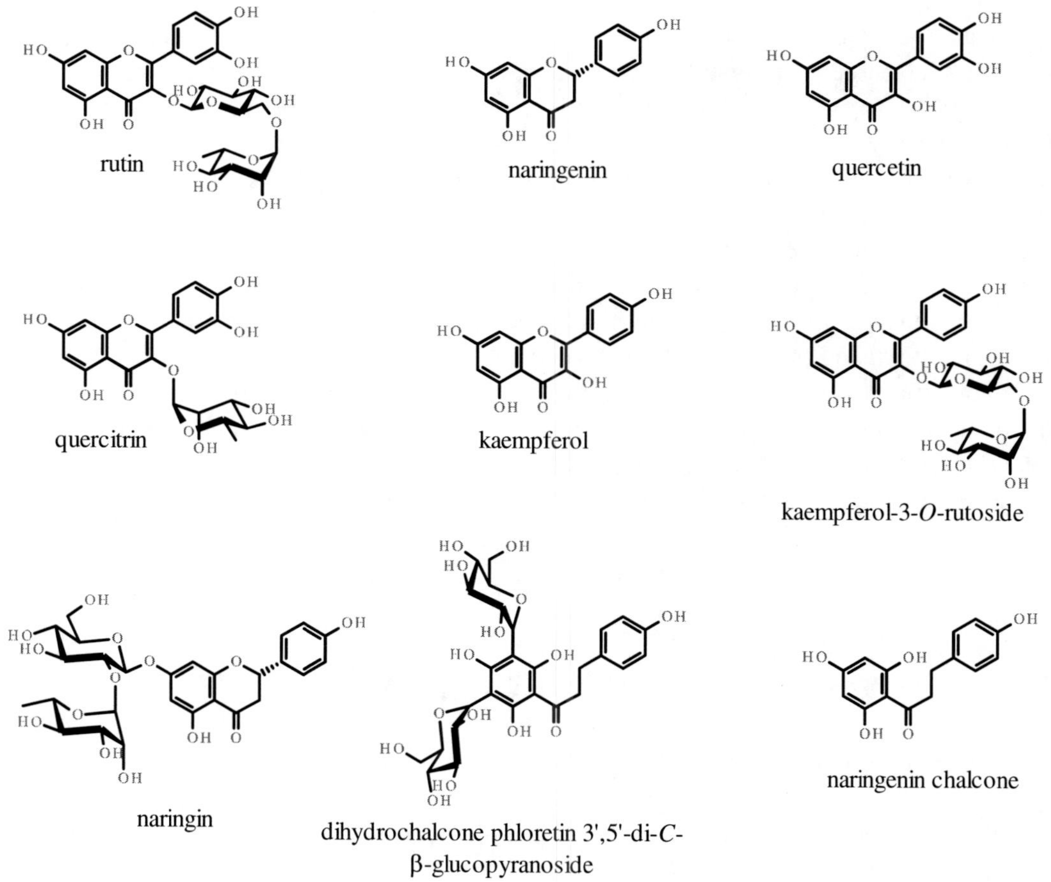

Figure 8. Structures of flavonoids and chalcones isolated from tomato fruits.

Antioxidant structural analysis at the different ripening stages revealed the biochemical processes involved in tomato ripening. Tomato fruit ripening can be described as an oxidative phenomenon which requires the turnover of reactive oxygen species such as H_2O_2 and superoxide anion [31]. A balance must necessarily exist between the production of reactive oxygen species and antioxidants compounds. Thus, it is likely that the antioxidant reactive system plays a crucial role in the ripening process.

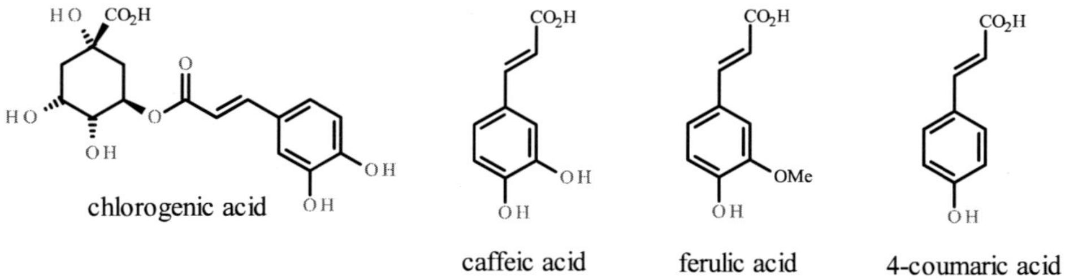

Figure 9. Structures of simple phenolic acids isolated from tomato fruits.

Jimenez et al. [18] investigated the oxidative process of a tomato saladette variety (*Solanum lycopersicum* Mill. Cv. Alisa Craig) during ripening. Significant changes in the antioxidative and oxidative activities were detected using H_2O_2 concentration, lipid peroxidation and protein oxidation as the indicators of the oxidative process. These parameters increased during the breaker stage while the activity of the antioxidant enzymes, catalase and superoxide dismutase, gradually decreased from 17.5 to 2.2 and from 25.6 to 16.0 µmol•(g dry weight)$^{-1}$•min^{-1}, respectively, during the green-to-breaker stage transition. The activity of the antioxidant enzymes increased slightly during the pink and red ripening stages. However, the decline of these regulatory enzymes in the early ripening stages is offset by a significant increase of the phenolic, carotenoid and ascorbic acid contents during the pink and red stages, as it was mentioned above.

Aherne et al. [32] reported that the plantation climate has a significant effect on the content of antioxidant bio-accessible compounds. The bio-accessibility study consisted on an in vitro digestion followed by the formation of an aqueous micellar fraction simulating human digestion conditions. Carotenoids were quantified in this fraction and were related to the initial carotenoid content. The authors reported that tomatoes grown in two different climates have a two-fold difference in the amount of bio-accessible carotenoids. For example, bio-accesible β-carotene in Cherry tomato fruits varied from 2.2 to 5.0% of the initial carotenoid content in two different plantation climates.

POST-HARVEST TREATMENTS

Treatments used to extend the shelf life of tomato fruit products are diverse and involve a number of physicochemical methods; they can be divided into thermal and non-thermal. Non-thermal treatments include selective radiation and dilution-immersion; both methods are used to eliminate the presence of pathogen agents.

Storage and Minimal Processing

Recently, modified atmospheres combined with low temperature conditions, have been employed during storage and ripening to extend the consistency, shelf life and phytochemical quality of tomato fresh products. UV radiation has also been used at this stage. Different treatments have been used to preserve the phytochemical and organoleptic properties of fresh tomato during storage and ripening: low temperature treatment, modified atmospheres and UV irradiation combined with low temperatures. Odriozola-Serrano et al. [33] studied the effect of a modified atmosphere packaging (5% O_2 + 5% CO_2) combined with storage at 4°C for different tomato fruit varieties. Lycopene concentrations in the fresh fruits of the different cultivars ranged from 20.0 to 80.5 mg•kg^{-1} FW for the Durinta and Bodar varieties, respectively. The authors reported no difference in the concentrations of the main health-related compounds (lycopene, ascorbic acid and phenolic compounds) between the whole untreated and modified-atmosphere fresh-cut tomatoes. Color and ascorbic acid content was preserved during 3 weeks storage. The authors reported that no apparent change occurred in the antioxidant compounds (lycopene, total phenolics, ascorbic acid) of the modified-atmosphere samples during 21 days of storage at 4°C.

Low temperatures have been used to preserve the physicochemical and organoleptic properties of fresh tomato by reducing enzymatic activity. However, processes conducted at temperatures below 11°C are prone to produce a chilling injury (CI) to the fruit [34]. This damage consists in an irregular ripening, pitting and the presence of water-soaked areas, which is evidenced after the fruit is transferred to a non chilling condition [35].

A few researchers have determined the extent of CI by measuring the changes in bioactive compounds of tomato fruits during and after cold storage. Yahia et al. [36] found that chilling temperatures (4°C) had a drastic effect on the antioxidant components of tomato fruits with respect to samples stored at 20°C. They reported a reduction on the content of β-carotene, lycopene, ascorbic acid, isoascorbic acid, dehydroascorbic acid and an increment in α-tocopherol. This is consistent with the study performed by Javanmardi and Kubota [37] who reported the effect of chilling storage temperatures on tomato antioxidant components. In this study, room temperature storage was compared to a two-step chilling process consisting of an initial 7 days period at 12°C followed by a 7 days period at 5°C. In the samples stored at room temperature, lycopene concentration increased 62% to a final concentration of 68 mg•kg^{-1} FW during the 7 d storage period. Lycopene content at 12°C remained constant, while a slight lycopene reduction was observed during the 5°C storage period.

Apart from modified atmospheres, two methods have been tested to reduce the extent of CI: hot treatment (34-40°C) previous to chilling [36, 38, 39] and wax application [40]. Yahia et al. [36] studied hot air treatments (34 and 38°C for 24 h) previous to chilling storage (4°C) as a method to reduce CI. The authors observed that heating of the fruit at 38°C for 24 h resulted in product heat injury with a significant reduction in β-carotene and lycopene. The 34°C treatment did not produce a negative effect in the antioxidant content of the product but was similar as the exposed to air at 20°C. In a follow-on study, Soto-Zamora et al. [38] tested the effect of a thermal treatment (38°C) in an oxygen-reduced atmosphere on the antioxidant components of tomato fruits. The combined treatment was not effective to alleviate the effect of CI on the content of antioxidant components of the fruit. In general it can be concluded that heat treatments (28-40°C) prior to chilling (2-5°C) have not been effective to reduce the CI effects on the reduction of the antioxidant main components. Mejía-Torres et al. [40] studied the effect of spreading wax on fresh tomato previous to chilling to prevent CI. The treatment had a positive impact in the chilling injury index (a qualitative parameter for CI) of tomato stored during 20 days at 5°C (chilling stage), followed by 7 days at 22°C (ripening stage). The lycopene content was monitored during the ripening stage. As expected, there was an increase in lycopene content with time (from 2.9 to 35 $\mu g \bullet g^{-1}$ FW). However, there was not a significant effect of the wax treatment on the final lycopene content.

Tzortzakis et al. [41] studied the effect of ozone-enriched atmospheres in the antioxidant profile of the product. Two ozone levels were used (0.05 and 1.0 $\mu mol \bullet mol^{-1}$) for preventing microbial spoilage of the fruit in cold storage (13°C, 95% RH) as an alternative to chemical treatments. Ozone treatments resulted in no significant changes in the behavior of the ascorbic acid and phenolic compounds with respect to the exposure to charcoal/particulate-filtered air, CFA (<0.005 $\mu mol \bullet mol^{-1}$ ozone). A significant reduction in the amount of phenols in the organic phase was observed for both the ozone and CFA treatments. In addition, an increase in lycopene and β-carotene concentrations was observed for both the ozone and the CFA samples. The content of lutein remained constant through the 12 days storage.

Jagadeesh et al. [42] investigated the effect of UV-C Hormesis irradiation of tomato fruits (3.7 $kJ \bullet m^{-2}$) on the antioxidant components during different storage periods at 13°C, followed by a ripening stage at 23°C. The UV treatment caused a significant reduction in lycopene content as compared with the untreated fruit, except for the case of the 10 days storage samples where no significant changes occurred. The behavior of ascorbic acid concentrations was similar for the ozone samples and the untreated fruit, with a slight ascorbic acid concentration increase with storage time. The authors attributed this effect to the inhibition of the ascorbate oxidase enzyme caused by UV irradiation. No significant differences were observed in the total phenolic content of tomato fruits for the UV treatments and the untreated samples. Nevertheless, a significant increase in the phenolics concentration was observed when the storage time was prolonged from 10 to 20 days. In conclusion the UV-C treatment seems an efficient sanitizing technique which according to the authors causes no harm to the antioxidant properties of the fruit and mildly enhances the content of ascorbic acid. However, this latter effect did not reflect an increase in the antioxidant activity of the treatments, probably because ascorbic acid is not the most important compound determining the hydrophilic antioxidant activity.

Thermal Processing

A large number of thermal treatments are employed such as homogenization followed by pasteurization to make puree. This product is used as a ubiquitous food ingredient to make pastas, soups, juices, sauces, powders and novel ready-to-use frozen products. Hot water immersion at temperatures of up to 100ºC is used to inactivate enzymes and eliminate pathogenic microorganisms in tomato products. Different tomato dehydration methods have been developed among which drying of tomato slices or cubes is a very important and convenient treatment. A wide range of commercial drying equipment can be used such as tray dryers, tunnel dryers, fluidized bed, etc. Spray-drying is an important dehydration treatment to obtain tomato powders. Conventional drying temperatures range from 50 to 150ºC whereas processing times may vary from seconds to hours depending on the drying process. Tomato´s antioxidant compounds are affected in different degrees as a function of the variables involved in each process.

Hot Water Immersion

Thompson et al. [43] studied the treatment of tomato by water immersion at high temperatures. These authors found that immersion time determines the final lycopene content of the product. This study showed that lycopene content in the cooked tomato was higher than in the raw fruit. The highest lycopene content was obtained for the samples subjected to an 8 min immersion period, whereas a 16 min treatment resulted in a considerable lycopene reduction, probably due to oxidative phenomena.

Dewanto et al. [44] determined the antioxidant compounds during cooking of tomato slurry at temperatures of 88ºC for 2, 15 and 30 min. Ascorbic acid was reduced in the thermally treated samples. In addition, an increasing degradation pattern of ascorbic acid with heating time was observed: 10.5, 15.8 and 28.9% for the corresponding heating periods of 2, 15 and 30 min. No significant differences were observed in the amount of total phenolics and total flavonoids (expressed as mg GAE and catechin equivalents, respectively) between the raw and cooked samples. Additionally, the phenolic and flavonoid compounds were stable at high temperatures without dependence on the processing time. The content of *trans*-lycopene increased with heating time by 54.4, 171.1 and 164.3% for the 2, 15 and 30 min treatments, respectively; similarly, 5.8, 17.4 and 34.9% increments in *cis*-lycopene were observed.

Tomato Drying

Tomato fruit drying generally uses hot air, and in some cases sunlight, to remove water from the solid matrix. Drying processes can cause severe damage to the product mainly by the depletion of ascorbic acid. On the other hand, lycopene remains highly stable during tomato drying. The extent of antioxidants degradation can be controlled by a careful selection of the method and the conditions of the drying process. Currently, there is a wide range of drying methods commercially available for whole and sliced tomato, such as sun-drying, tray drying and tunnel drying. Dry tomato powder can be obtained by spray drying of juices or pastes.

Sun Drying

In developing countries sun drying is generally used to preserve fresh foods which are not readily consumed or sold; whereas most of the people in developed countries regard sun dried tomatoes as gourmet products [45]. Due to the current consumer preferences for natural and healthy foods, sun dried tomato have gained wide acceptance in the food industry as a "natural" ingredient [46]. This fact combined with the consumer awareness of tomato´s antioxidant properties, will promote the popularity and acceptance of sun-dried tomato products among consumers and food companies.

However, during sun drying, tomato fruits are exposed to large amounts of visible and UV lights in the presence of air. The latter radiation can result in an alteration of the content of antioxidants. Chemical pretreatments have been used to reduce antioxidant degradation during sun drying; these include sulfur and SO_2 as well as sodium metabisulfite salts. Latapi et al. [47] studied the effect of the addition of sulfur, sodium metabisulfite and sodium chloride on the content of ascorbic acid and lycopene in Halley 3155 tomato halves. This tomato variety has been used in United States for the production of Campbell´s® soups. The sodium chloride treatment did not prevent the total depletion of ascorbic acid during an 8 h sun drying, while lycopene was reduced only in 10.7%. The use of SO_2 caused a reduction in ascorbic acid degradation and preserved 14.7% of lycopene. Despite sun drying provides an inexpensive and sustainable drying method, the prolonged exposure times cause significant ascorbic acid degradation. Therefore, in order to reduce such drawbacks Rajkumar et al. [48] attempted to reduce drying times by using a vacuum-assisted sun dryer. The authors compared the effect of vacuum and conventional sun drying on the drying time of tomato slices of 4, 6 and 8 mm thickness. Vacuum sun drying reduced the drying time by 20%. Product quality studies focused on color changes and rehydration ratio. No quantitative studies were performed about the effect of the two drying treatments on the antioxidant content of the product. Considering the reduced-O_2 atmosphere in vacuum drying, the vacuum process is likely to importantly enhance the preservation of the product antioxidant components.

Tray Drying

Tray drying of tomato halves, slices or cubes is a simple and popular dehydration method for tomato processing. This method consists of placing slices, halves or whole tomatoes on a set of removable trays which are located inside of a drying chamber. Typical drying conditions are processing times between 3 and 8 h, hot air supply with velocities of 0.5-2.0 $m•s^{-1}$ and temperatures of 40-110°C. Zanoni et al. [49] investigated the drying process of 6.5 cm-diameter tomato halves in a tray dryer with a 1.5 $m•s^{-1}$ air velocity and temperatures of 80 and 110°C. The treated tomato halves at 80°C suffered a 92.4% decrease in the ascorbic acid content; whereas no presence of this antioxidant was detected in the 110°C treated samples. On the other hand, there was not a significant lycopene reduction in the 80°C samples, while a 12% lycopene reduction was observed for the 110°C samples. Kerkhofs et al. [50] studied tray drying of tomato quarters using a temperature of 42°C and an air velocity of 2.0 $m•s^{-1}$. The authors quantified the contents of ascorbic acid, total phenols and lycopene contents in fresh and dried tomatoes of three New Zealand varieties (Aranka, Encore and Flavourine). They found that the drying process did not cause a marked decrease in the antioxidants of the Aranka and Encore varieties. However, in the case of the Flavourine variety, the dehydration caused a reduction in the content of ascorbic acid (74.4%) and total phenols (33.4%), while lycopene content increased up to 55.0%.

Employing a custom-made tray-rotating dryer, Santos-Sánchez et al. [51] evaluated the effect of the operation variables (temperature, air velocity and tray rotation) on the contents of lycopene, ascorbic acid and total phenols in the dried product. Drying runs with and without tray rotation were performed using temperatures of 45, 50 and 60°C and air velocities of 0.6 to 1.2 $m•s^{-1}$. In the experiment, 6 mm-thick tomato slices were pretreated with 1% sodium metabisulfite. The drying run performed at 60°C, 0.6 $m•s^{-1}$ and tray rotation, reflected the lowest antioxidant degradation compared with the rest of the experimental conditions: lycopene (2.9%), ascorbic acid (17.3%) and total phenols (2.1%).

The preservation of antioxidants during tomato tray drying depends on a number of factors among which the more important are: temperature, air velocity, product thickness, tomato pretreatment and tray rotation. Similarly to other drying processes, ascorbic acid is the antioxidant most affected by tray drying. Aiming at preserving the functional food quality of tomato dry products, processing temperatures should be in the 40-60°C range with air velocities of 1.0 to 2.0 m•s^{-1} and the use of an automatic tray rotation system is desirable.

Spray Drying

Tomato spray drying consists of atomizing tomato juice or puree into a hot gas stream to obtain finely dispersed drops, from which the water is rapidly evaporated. The dried solid particles are separated from the hot gas stream by gravity using two separation stages: a drier chamber and a cyclone. Dried powders are collected at the bottom of the separation stages. The processing conditions commonly employed for tomato spray drying range from 120 to 220°C [52] with drying times between 1 and 120 s, which depend on the feed rate and drying chamber dimensions. Because of the high sugar concentrations of the dried product, spray-dried tomato products become highly hygroscopic and thermoplastic. These characteristics promote that significant amounts of product stick on the dryer walls, thus reducing the process final yields [53]. Maltodextrins and cyclodextrins are drying aids used to reduce product sticking and also to prevent the degradation of antioxidant compounds during tomato spray drying. On this regard, Candelas-Cadillo et al. [52] studied the effect of the use of maltodextrins on the preservation of lycopene during tomato spray drying at 180°C. Two tomato feed formulations were prepared with the addition of 80 or 100% (weight/weight) of maltodextrins, the percentage was expressed with respect to the total solid content in tomato juice. These authors found that when 80% of maltodextrins was used, the product kept a 26.1% of lycopene, while the use of 100% of maltodextrins preserved the 85.5% of lycopene. Accordingly, Quaglia et al. [54] reported that an increment in the amount of β-cyclodextrin in the tomato juice feed considerably reduced lycopene degradation in the product. Therefore, it can be concluded that maltodextrins and cyclodextrins added to the feed can ameliorate lycopene degradation during spray drying.

High Hydrostatic Pressure

High hydrostatic pressure treatment (300 to 800 MPa) reduces the microbial load of raw tomato, thus extending the product shelf life. Since no heating is involved in this method, most of the antioxidant compounds are preserved. Patras et al. [55] compared hydrostatic pressure treatments at 400, 500 and 600 MPa with a thermal process at 70ºC for 2 min. The latter resulted in a decrease in the total phenol content (expressed as GAE), total carotenoids (expressed as β-carotene equivalents), and ascorbic acid by 5.4, 9.7 and 38.9%, respectively. The thermally treated product had a similar antiradical activity (measured by the DPPH assay) compared to that of the unprocessed samples. In a similar study, Hsu [56] reported that a hot-break treatment (92ºC, 2 min) caused a 42.9% reduction in the antiradical activity of the hydrophilic extracts (measured by the DPPH assay) and a 14% increase in the activity of the hydrophobic extracts. The activity of the hydrophilic extracts is caused by the polyphenols and phenols in the sample, whereas that of the hydrophobic extracts indicates primarily the presence of lycopene. High hydrostatic pressure treatments caused an increase in the antiradical activity at pressures of 500 MPa [56] and 600 MPa [55] with respect to the untreated sample, while no significant differences were found in the pressure ranges of 100-400 MPa [56] and 400-500 MPa [55]. Significant increments in the contents of total phenols and total carotenoids were determined at 600 MPa. Ascorbic acid concentration was reduced with the pressure treatments at 400 and 500 MPa, while it was maintained at 600 MPa as compared with the untreated samples [57].

Combined Treatments

Pérez-Conesa et al. [57] studied the conditions used in the production of industrial tomato puree. In an experiment, they established the following sequential processing steps: raw/thermal treatment, pressurized homogenization and pasteurization. The homogenization step (0.1-0.2 MPa) did not affect the ascorbic acid and all-*trans* lycopene content in comparison with the samples without pressure treatment. The use of a hot-break treatment (82°C, 2 min) combined with pasteurization (98°C, 40 s) caused a two-fold increase in the all-*trans* lycopene content and a drastic reduction in the ascorbic acid content (up to 93%), while the total phenols remained practically unchanged with respect to the samples with no initial thermal treatment.

CONCLUSION

Tomato varieties can be resistant to pests, water stress, high temperature cultivation, high levels of solar radiation, among others. All these factors combined with the diverse tomato genotypes greatly affect the content of antioxidant compounds in tomato fruits and thus they also determine tomato´s beneficial properties for human health. Moreover, the high demand of tomato as a fresh product as well as in the food industry has prompted the development of innovative processes that can greatly prolong the product shelf life, while minimizing antioxidants degradation. Studies on the antioxidant content in different tomato genotypes have revealed that specific varieties can produce higher amounts of lycopene, β-carotene or ascorbic acid. Carotenoids, including lycopene and β-carotene, are very susceptible to the plantation conditions, while phenolic compounds depend not only on these conditions but also on the fruit variety. Minimally processed treatments such as modified atmospheres (low O_2 and enriched CO_2 and ozone atmospheres), low temperatures, and UV irradiation, have been successfully employed to maintain the level of the main antioxidant components of fresh fruits during long storage periods. On the other hand, thermal processing at temperatures below 60°C does not significantly degrade phenolic compounds. According to several reports, lycopene increases with thermal processing; however, a clear explanation to this has not been suggested. Therefore, fundamental studies on this phenomenon may help to elucidate the mechanisms involved. Due to its high reactivity in solution, ascorbic acid is the antioxidant which is more prone to thermal degradation. Few researchers have studied the changes in tocopherols, the antioxidant group present in lower concentration in tomato fruits, in terms of plantation conditions, tomato genotype and processing methods; thus, this subject may constitute an area of opportunity for future studies. Tomato drying has been extensively used for the production of long-shelf-life processed products. The use of chemical pre-treatments provides a relatively simple and inexpensive method to inhibit antioxidant degradation. Tomato slices or halves are commonly pretreated with sulfur dioxide and sodium metabisulfite prior to drying; while for the production of tomato powder, maltodextrins and more recently cyclodextrins have been used. Finally, there are emerging methods that may be useful for the preservation of antioxidants in processed tomato products, including high hydrostatic pressure processing (200-600 MPa).

ACKNOWLEDGMENTS

We thank L.C.P. Juliana Angélica de los Santos Vera for her English assistance in preparing this manuscript.

REFERENCES

[1] J. Pinela, L. Barros, A. M. Carvalho, and I. C. Ferreira, *Food Chem. Toxicol. 50*, 829 (2012).

[2] A. Lee, D. I. Thurnham, and M. Chopra, *Free Radical Bio. Med. 29*, 1051 (2000).

[3] H. M. Chandra, and S. Ramalingam, *Food Sci. Biotechnol. 20*, 15 (2011).

[4] F. Khachik, L. Carvalho, P. S. Bernstein, G. J. Muir, D.-Y. Zhao, and N. B. Katz, *Exp. Biol. Med. 227*, 845 (2002).

[5] W. Stahl, and H. Sies, *J. Nutr. 122*, 2161 (1992).

[6] S. K. Clinton, C. Emenhiser, S. J. Schwartz, D. G. Bostwick, A. W. Williams, B. J. Moore, and J. W. Jr. Erdman, *Cancer Epidem. Biomar. 5*, 823 (1996).

[7] T. W. M. Boileau, A. C. Boileau, and J. W. Jr. Erdman, *Exp. Biol. Med. 227*, 914 (2002).

[8] T. Lovrić, Z. Sablek, and M. Bošković, *J. Sci. Food Agric. 21*, 641 (1970).

[9] H. Gautier, A. Rocci, M. Buret, D. Graselly, Y. Dumas, and M. Causse, *Can. J. Plant Sci.* 439 (2005).

[10] R. L. Thomas, and J. J. Jen, *J. Food Sci. 40*, 566 (1975).

[11] J. J. Jen, *J. Food Sci. 39*, 907 (1974).

[12] G. Lee, J. M. H. Y. J. Bunn, and D. R. Decoreau, *T. ASAE 39*, 169 (1996).

[13] M. L. Tomes, *Bot. Gaz. 124*, 180 (1963).

[14] L. C. Raymundo, D. C. Chichester, and K. L. Simpson, *J. Agric. Food Chem. 24*, 59 (1976).

[15] S. E. Cox, C. Stushnoff, and D. A. Sampson, *Can. J. Plant Sci. 83*, 913 (2003).

[16] R. Pernice, M. Parisi, I. Giordano, A. Pentangelo, G. Graziani, M. Gallo, V. Fogliano, and A. Ritieni, *Sci. Hortic-Amsterdam 126*, 156 (2010).

[17] V. García-Valverde, I. Navarro-González, J. García-Alonso, and M. J. Periago, *Food Bioprocess Technol.* (2011). doi:10.1007/s11947-011-0687-3.

[18] A. Jimenez, G. Creissen, B. Kular, J. Firmin, S. Robinson, and M. Verhoeyen, *Planta 214*, 751 (2002).

[19] A. Cano, M. Acosta, and M. B. Arnao, *Postharvest Biol. Tec. 28*, 59 (2003).

[20] A. A. Abushita, E. A. Hebshi, H. G. Daood, and P. A. Biacs, *Food Chem. 60*, 207 (1997).
[21] G. Giovanelli, V. Lavelli, C. Peri, and S. Nobili, *J. Sci. Food Agric. 79*, 1583 (1999).
[22] M. J. Periago, J. García-Alonso, K. Jacob, A. B. Olivares, M. J. Bernal, D. Iniesta, C. Martinez, and G. Ros, *Int. J. Food Sci. Nutr. 60*, 694 (2009).
[23] U. L. Opara, M. R. Al-Ani, and N. M. Al-Rahbi, *Food Bioprocess Technol.* (2011). doi:10.1007/s11947-011-0693-5
[24] Z. Kotíková, J. Lachman, A. Hejtmánková, and K. Hejtmánková, *Food Sci. Technol-Leb. 44*, 1703 (2011).
[25] B. Fuhrman, N. Volkova, M. Rosenblat, and M. Aviram, *Antioxid. Redox Sign. 2*, 491 (2000).
[26] D. Liu, J. Shi, A. C. Ibarra, Y. Kakuda, and S. J. Xue, *Food Sci. Technol-Leb. 41*, 1344 (2008).
[27] R. Ilahy, C. Hdider, M. S. Lenucci, I. Tlili, and G. Dalessandro, *J. Food Compos. Anal. 24*, 588 (2011).
[28] R. Slimestad, T. Fossen, and M. J. Verheul, *J. Agric. Food Chem. 56*, 2436 (2008).
[29] R. Slimestad, and M. Verheul, *J. Sci. Food Agric. 89*, 1255 (2009).
[30] D. K. Salunkhe, S. J. Jadhav, and M. H. Yu, *Plant Foods Hum. Nutr. 24*, 85 (1974).
[31] T. Brennan, and C. Frenkel, *Plant Physiol. 59*, 411 (1977).
[32] S. A. Aherne, M. A. Jiwan, T. Daly, and N. M. O´Brien, *Plant Foods Hum. Nutr. 64*, 250 (2009).
[33] I. Odriozola-Serrano, R. Soliva-Fortuny, and O. Martín-Belloso, *Food Sci. Technol-Leb. 41*, 217 (2008).
[34] T. S. Cheng, and R. L. Shewfelt, *J. Food Sci. 53*, 1160 (1988).
[35] M. Sharom, C. Willemot, and J. E. Thompson, *Plant Physiol. 105*, 305 (1994).
[36] E. M. Yahia, G. Soto-Zamora, J. K. Brecht, and A. Gardea, *Postharvest Biol. Tec. 44*, 107 (2007).
[37] J. Javanmardi, and C. Kubota, *Postharvest Biol. Tec. 41*, 151 (2006).
[38] G. Soto-Zamora, E. M. Yahia, J. K. Brecht, and A. Gardea, *Food Sci. Technol-Leb. 38*, 657 (2005).
[39] B. D. Whitaker, *Postharvest Biol. Tec. 4*, 75 (1994).
[40] S. Mejía-Torres, M. Vega-García, J. Valverde-Juárez, J. López-Valenzuela, and J. Caro-Corrales, *J. Food Quality 32*, 735 (2009).
[41] N. Tzortzakis, A. Borland, I. Singleton, and J. Barnes, *Postharvest Biol. Tec. 45*, 317 (2007).
[42] S. L. Jagadeesh, M. T. Charles, Y. Gariepy, B. Goyette, G. S. V. Raghavan, and C. Vigneault, *Food Bioprocess Technol. 4*, 1463 (2011).
[43] K. A. Thompson, M. R. Marshall, C. A. Sims, C. I. Wei, S. A. Sargent, and J. W. Scott, *J. Food Sci. 65*, 791 (2000).
[44] V. Dewanto, X. Wu, K. K. Adom, R. H. Liu, *J. Agric. Food Chem. 50*, 3010 (2002).
[45] A. O. Olorunda, O. C. Aworth, and C. N. Onuoha, *J. Sci. Food Agric. 52*, 447 (1990).
[46] E. A. Sloan, *Food Technol. 53*, 24 (1999).
[47] G. Latapi, and D. M. Barrett, *J. Food Sci. 71*, S32-S37 (2006).
[48] P. Rajkumar, S. Kulanthaisami, G. S. V. Raghavan, Y. Gariepy, and V. Orsat, *Dry. Technol. 25*, 1349 (2007).
[49] B. Zanoni, E. Pagliarini, and R. Foschino, *J. Sci. Food Agric. 80*, 2203 (2000).

[50] N. S. Kerkhofs, C. E. Lister, and G. P. Savage, *Plant Foods Hum. Nutr. 60*, 117 (2005).
[51] N. F. Santos-Sánchez, R. Valadez-Blanco, M. S. Gómez-Gómez, A. Pérez-Herrera, and R. Salas Coronado, *Food Sci. Technol-Leb. 46*, 298 (2012).
[52] M. G. Candelas-Cadillo, M. G. J. Alanís-Guzmán, M. Bautista-Justo, F. Del Río-Olague, and C. García-Díaz, *Rev. Mex. Ing. Quím. 4*, 299 (2005).
[53] A. M. Goula, and K. G. Adamopoulos, *J. Food Eng. 66*, 35 (2005).
[54] F. Quaglia, G. Fusco, G. De Rosa, F. Ungaro, A. Miro, and M. I. La Rotonda, *J. Incl. Phenom. Macrocycl. Chem. 57*, 669 (2007).
[55] A. Patras, N. Brunton, S. Da Pieve, F. Butler, and G. Downey, *Innov. Food Sci. Emerg. 10*, 16 (2009).
[56] K.-C. Hsu, *Food Sci. Technol-Leb. 41*, 450 (2008).
[57] D. Pérez-Conesa, J. García-Alonso, V. García-Valverde, M.-D. Iniesta, K. Jacob, L. M. Sánchez-Siles, G. Ros, and M. J. Periago, *Innov. Food Sci. Emerg. 10*, 179 (2009).

In: Tomatoes: Cultivation, Varieties and Nutrition
Editor: Tadahisa Higashide
ISBN: 978-1-62417-915-0

Chapter 13

LYCOPENE IN TOMATO FRUITS AND ITS CONTENT VARIATION DURING THE RIPENING PROCESS AFTER HARVEST

Francelina Costa[1], Maria de Lurdes Baeta[2], David Saraiva[2], Maria da Conceição Castilho[2], Manuel Teixeira Verissimo[3] and Fernando Ramos[2*]

[1]Centro de Saúde de Torres Novas, R. Dr. Pedro GM Salazar, Torres Novas, Portugal

[2]CEF – Centro de Estudos Farmacêuticos da Universidade de Coimbra Health Sciences Campus, Pharmacy Faculty, Coimbra University, Azinhaga de Santa Comba, Coimbra, Portugal

[3]Serviço de Medicina Interna dos Hospitais da Universidade de Coimbra, Faculdade de Medicina Universidade de Coimbra, Coimbra, Portugal

ABSTRACT

Lycopene is a carotenoid with no pro-vitamin A activity. However, lycopene is a very powerful singlet oxygen scavenger with an important protective effect against oxidative damage that also acts as an immune system stimulator. Among other benefits, lycopene intake appears to reduce the incidence of several cancers, including prostate, stomach and lung cancers and also plays a role in the prevention of certain cardiovascular diseases.

Thus, in this chapter, a review of the properties of lycopene with respect to human health was described, and a discussion of the lycopene content in tomato fruits during ripening after harvest was conducted on the basis of a comparative study.

The lycopene contents were determined in three types of tomato fruits (round, elongated and cherry). The extraction of the lycopene was achieved with a mixture of hexane, acetone and ethanol. An aliquot of the organic layer was evaporated to dryness.

* Corresponding author: Phone number: +(351) 239 488492, Fax number: +(351) 239 488503, Email address: fjramos@ci.uc.pt

The dry residue was recovered in the mobile phase and injected into an HPLC-UV system with detection at λ = 450 nm. The results revealed that the lycopene content varies after harvest throughout the ripening process. A significant increase in the lycopene content occurs at the maturation stage of turning tomatoes. Although lycopene synthesis occurs after harvest, the concentrations of lycopene in green tomatoes were still consistently below those in red tomatoes. However, an increase in the lycopene content was not observed after the harvest of red tomatoes. A maturation period greater than eleven days after the harvest consistently led to a decrease in the total lycopene content.

INTRODUCTION

The word "tomato" refers to the fruit of the tomato plant (*Solanum lycopersicon L), a member of the* Solanaceae family. This plant is native to the Andean region, in South America's western coast, between Southern Colombia and Northern Chile, including Bolivia, Peru and Ecuador. It was, however, domesticated by the Mexican Aztecs and introduced in Europe by Spaniards [1-3].

Currently, the production of tomato fruits is widespread, as the fruits have been incorporated into a multitude of dishes typical of the diets of different cultures from around the world.

The rise of tomato fruit production in the last few decades has been a result of its increased consumption in both fresh and processed forms (in juices, jams, ketchup, sauces, soups and other culinary preparations). This trend was due to the growth of the restaurant business (particularly fast-food and self-service). Additionally, the consumer now has the perception that the tomato fruit is on the list of healthier foods [3, 4]. Due to its highly perishable nature, the tomato fruit was the first vegetable to be industrialized.

The present chapter aims to contribute to a better understanding of the potential benefits of carotenoids, particularly lycopene. However, because tomatoes are usually harvested before reaching maturity, it is important to understand the variation in lycopene content during the maturation period after the harvest. Thus, a comparative study was conducted during maturation after harvest in three of the most widely consumed types of tomato fruits in Portugal (round, elongated and cherry).

TOMATO FRUIT TYPES

There are several types of tomato fruits, distinct in both shape and color which are more or less adapted to local tastes and certain geographical characteristics.

In Portugal, there are several types of tomato fruits being produced, but the round, elongated and cherry types are the most popular (Figure 1) [5].

Regarding growing tomato fruits in Portugal, there is a predominance of greenhouse rather than outdoor cultivation. For tomato fruits produced in the open air, the production campaign runs between June and September on the west coast and through October on the Algarve coast. Greenhouse tomatoes are grown throughout the year [5].

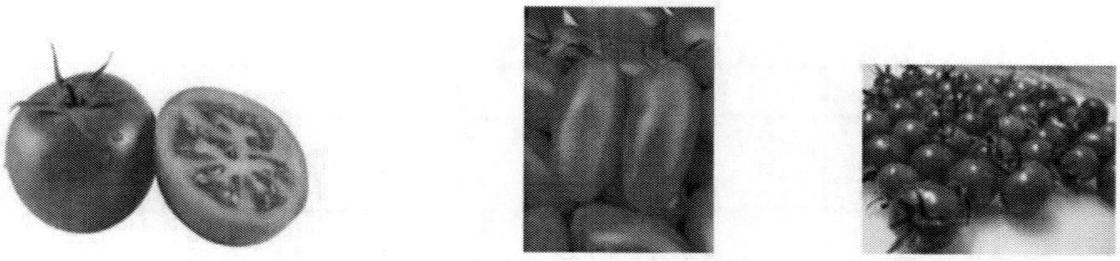

Round **Elongated** **Cherry**

Figure 1. Most popular types of tomato fruits in Portugal.

TOMATO FRUIT PRODUCTION

FAO (Food and Agriculture Organization) data show that the worldwide production of tomato fruits and its derivatives involves more than 180 countries. It is clear that the largest portion of the production is in Asian countries, such as China and India (with 35.6% of the production and 34.6% of the planted area). The American continent follows, with 20% of the production share (of which 10% are grown in the USA). In Europe, Turkey, Italy and Russia are the greatest producers, which, combined, account for 12.9% of the production and 12.8% of the planted area. The largest African producers are Egypt and Nigeria [6].

TOMATO FRUIT COMPOSITION

The composition of a tomato fruit depends on the type, ripeness and production conditions (temperature, fertilization and irrigation). The fruit has a low calorie value that is associated with low levels of protein (0.8%), fat (0.3%) and carbohydrates (3.5%). The water content is approximately 94%. In Table 1, a detailed tomato fruit composition can be observed.

Table 1. Composition of raw tomato fruit [7]

Macro components	%
Water	93.5
Protein	0.8
Lipids	0.3
Carbohydrates	3.5
Dietary fiber	1,3
Ash	0.54
Minerals	**mg/100g**
Potassium	253
Calcium	11
Phosphorus,	17
Magnesium	11
sodium	13

Table 1. (Continued)

Minerals	**mg/100g**
Zinc	0.1
Iron	0.7
Vitamins	**mg/100g**
Vitamin C	20
Thiamine	0.050
Riboflavin	0.030
Niacin	1.30
Vitamin B6	0.14
Vitamin E	1.2
Vitamin A	85 (ER) µg/100g

CAROTENOIDS

Physico-Chemical Properties

Carotenoids are natural pigments that comprise a large family of lipophilic chemical compounds that are orange, red and yellow. These compounds exist in many fruits and vegetables, such as mango, papaya, apricot, carrot, squash, spinach, orange, peach, watermelon, cherry, chili, tomato fruit and others. Carotenoids can be divided into two groups: the carotenes (hydrocarbons), containing only carbon and hydrogen, are the low polarity carotenoids (e.g., alpha- and beta-carotene, lycopene) and the xanthophylls, the oxygenated derivatives, which have in their structures at least one oxygen atom (e.g., lutein and zeaxanthin) [8,9]. The molecular structure for certain carotenoids is shown in Figure 2 [10].

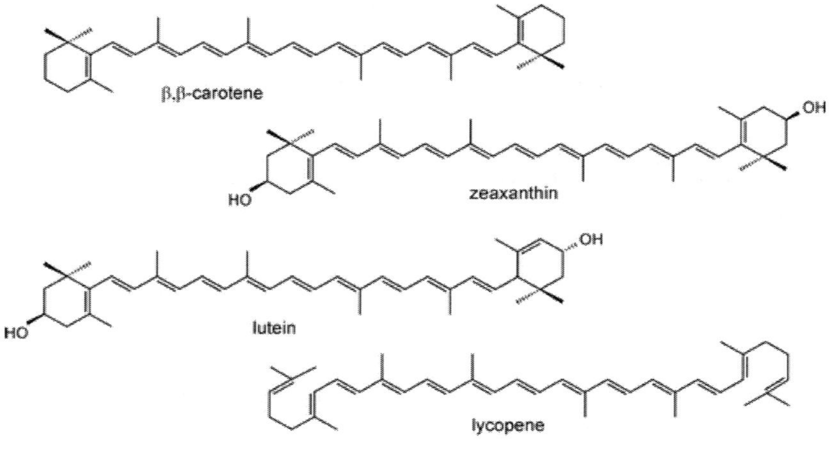

AR von Lintig J. 2010.
Annu. Rev. Nutr. 30:35–56

Figure 2. Molecular Structures of some carotenoid.

Carotenoids may or may not be selectively absorbed upon consumption. Those carotenoids that are absorbed may be converted into vitamin A. In either case, the absorbed carotenoids are deposited in the tissues, either in their original forms or modified to form carotenoids typical of animals (e.g., astaxanthin).

Properties

The antioxidant power of carotenoids can act directly in the neutralization of the action of free radicals. These compounds can participate indirectly in enzyme systems that have this function. Thus, carotenoids appear to act as protectors of cells and organisms against photo-oxidation by fighting free radicals and excited species that can cause irreversible damage to the DNA and lipid phases [11].

As expected, these properties of carotenoids, in conjunction with those of vitamins, have attracted great interest, not only due to their role as pro-vitamin A. Their antioxidant capacity is associated with a reduced risk of certain diseases that are assumed to be triggered by free radicals, such as cancer, atherosclerosis, cataracts, age-related macular degeneration (AMD), multiple sclerosis and others [12]. As an example, the inhibition of carcinogenesis may result from different mechanisms of action of carotenoids, such as [13]:

- capturing singlet oxygen,
- removing peroxy radicals,
- protecting lipid structures,
- inhibiting cell proliferation,
- stimulating communication between cells,
- increasing the immune response.

Lycopene

Lycopene ($C_{40}H_{56}$) is one of 650 carotenoid pigments in nature; being one of the most abundant of the 25 that appear in human plasma and tissues [14-16]. Lycopene is synthesized from mevalonic acid, which, after several steps, leads to phytoene. This compound undergoes a series of processes and ultimately is transformed into lycopene.

Lycopene is characterized by a symmetrical, acyclic structure, and its molecule contains only carbon and hydrogen atoms. It may, however, undergo cyclization, induced by the enzyme lycopene-cyclase, to produce alpha and beta-carotene, followed by oxidation that will lead to the production of xanthophylls, as can be observed in Figure 3.

The synthesis of lycopene is influenced by various environmental factors, such as induction by light, temperature and agronomic factors, including variety, maturity, type of fertilizer and irrigation. When tomato fruits are harvested, for example, there is a higher content of lycopene when the fruits are picked ripe than when they are ripened in camera [1, 18].

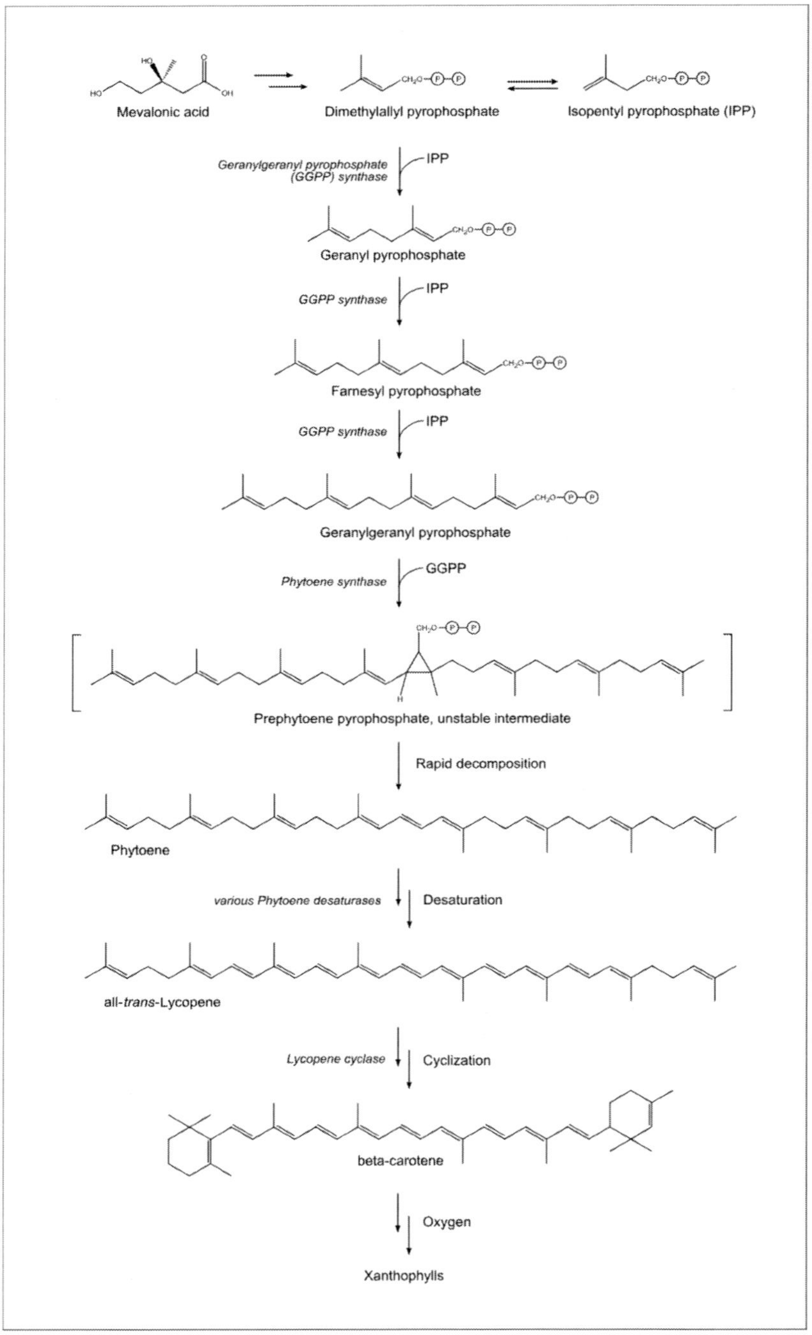

Figure 3. The role of lycopene in the biosynthesis of other carotenoids [17].

Lycopene has 11 conjugated double bonds and 2 unconjugated bonds (Figure 4) [19]. The structure of lycopene is responsible for the red-orange coloration of the fruits and vegetables in which it is present.

(All-E)-lycopene is the predominant geometric isomer found in plants. (Z)-isomers of lycopene are also found in nature, including the (5Z)-, (9Z)-, (13Z)- and (15Z)- isomers. The lycopene found in human plasma is a mixture of approximately 50% (Z)-lycopene and 50%

(all-E)-lycopene. The lycopene in processed foods exists mainly in the form of (all-E)-lycopene.

Lycopene has no pro-vitamin A activity, due to the absence of ß-ionone structure.

Among the various evaluated carotenoids, lycopene, due to its chemical structure (the presence of two non-conjugated double bonds in the molecule), is one of the best biological suppressors of free radicals. This compound can donate electrons to neutralize singlet oxygen molecules and other oxidizing molecules before they can damage cells. Therefore, it has twice the antioxidant capacity of other carotenoids and 100 times that of vitamin E [20, 21].

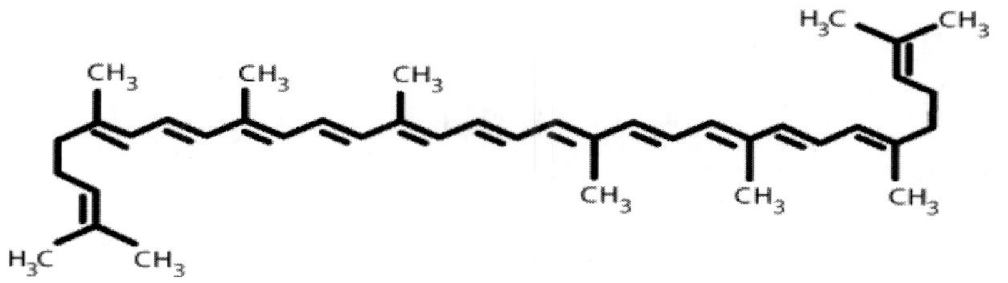

Figure 4. Chemical structure of lycopene.

Lycopene exists in a limited number of foods. The tomato fruit and its derivatives are the best dietary sources, but watermelon, papaya, red guava and cherry also contain this compound. Today's richest sources of lycopene in the diet, however, are the derivatives of tomato fruit, rather than the fresh produce. The lycopene content of these products depends on the composition of the food source and the processing conditions. A fresh tomato fruit has, on average, 3.0 mg lycopene/100 g, tomato sauce 6.2 mg / 100 g, tomato juice between 5.0 and 11.6 mg / mL, and ketchup contains, on average, 10.0 mg/100 g [1, 22].

BIOAVAILABILITY OF LYCOPENE

The bioavailability of the constituents of any food is a complex process that involves digestion, intestinal uptake, absorption and tissue distribution.

There are several factors that affect the bioavailability of carotenoids, such as the food matrix, food processing, the isomeric form of the lycopene, the amount and type of dietary fat consumed in addition, the absorption process, the interactions between different carotenoids and the presence of dietary fiber.

In the case of lycopene, its bioavailability appears to be related to the isomer it presents, a modification that is caused by heat. During intestinal uptake, certain carotenoids, such as lutein and beta-carotene, compete for the same absorption sites and may thus limit the bioavailability of lycopene [23].

Clinton [24] has shown that 79% to 91% of the lycopene present in tomato fruits and its derivative products are found in the trans form, in contrast with the lycopene levels present in serum and tissue, which are more than 50% in the cis form. Lycopene that is ingested in its natural form (trans) is poorly absorbed.

Lycopene absorption is affected by the amount of fat in the meal, and it appears to be higher when the tomato fruit is subjected to thermal processing, as this procedure bursts the cellular wall, enabling lycopene to be extracted from the chromoplasts [25]. During digestion and absorption, lycopene is separated from the remainder of the nutrients and incorporated into micelles. It is possible that the isomerization of lycopene occurs during this separation, changing its configuration from trans to cis. The cis isomers are better absorbed, as they do not aggregate due to their higher solubility in the micelles [26]. Several types of fiber found in foods, such as pectin, increase the viscosity of the medium and, therefore, may reduce the bioavailability of lycopene, with a consequent reduction in absorption.

The bioavailability of lycopene can be influenced by several factors, including the food matrix, food processing, dietary fat, the presence of fiber in meals, lycopene absorption, and different forms of lycopene, as discussed below.

The matrix in which lycopene is found in foods can affect its availability; thus, the first step in the absorption of lycopene must be its release from the food matrix. The process of the absorption of lycopene occurs passively, i.e., without energy depletion. Lycopene reaches the small intestine, where it is incorporated into micelles formed by bile acids and lipids. Because bile production is stimulated by the intake of dietary fat, consuming a fattier meal increases the efficiency of lycopene absorption. The fat content required for ideal lycopene absorption appears to be between 3 and 5 g per meal. However, the type of dietary fat is also important. Meals rich in medium-chain triglycerides (MCTs) reduce the bioavailability of lycopene because they are absorbed into the portal vein, which will reduce the formation of chylomicrons after the meal. It also appears that the consumption of fat substitutes reduces the absorption of carotenoids by anywhere from 20 to 120%. This rate of absorption is dependent on the type and amount of carotenoid and of product. Lycopene appears to have a "preference" for substitutes rather than for micelles, which create a hydrophobic barrier in the lumen of the intestine, making it unavailable for absorption. Once absorbed by the intestinal mucosal cells, lycopene is incorporated into lipoproteins, specifically in chylomicrons and VLDLs and, subsequently, in the LDLs, which are the largest carriers of beta-carotene and lycopene [15]. Drugs and phytosterols that are responsible for lowering cholesterol may also interfere with the incorporation of lycopene into micelles, potentially reducing the efficiency of its absorption.

The study of food processing has shown that there is an increase in the bioavailability of lycopene due to its release from the food matrix. This release is why tomato sauce and mashed tomatoes are better bioavailable sources of lycopene compared with other, uncooked food sources, such as fresh tomato fruits [15]. The physical disruption of the cell structure in processed tomato products compared with fresh tomato fruits partially explains the difference in the bioavailability of lycopene.

It appears that the thermal treatment and the grinding of tomato fruits increase the bioavailability of lycopene to the body tissues. This observation is explained by the manner in which the lycopene in food is attached to proteins, forming part of a crystalline matrix. Cooking and grinding are more efficient in promoting the release of lycopene from these compounds, so that it becomes more accessible, increasing its bioavailability. However, cooking can reduce the effectiveness of flavonoids, vitamin C and vitamin E.

It is also likely that cooking is responsible for the isomerization that occurs during the absorption process, changing the configuration of lycopene from the trans to the cis geometric isomer, which can alter the biological action of this carotenoid. Other physiological processes may also account for the different ratios of cis and trans isomers that are found in food and tissues [15].

Certain studies on women have shown that the consumption of 0.15 g / kg of body weight of fibers (e.g., pectin, guar gum, alginate, cellulose or wheat), together with an antioxidant supplement containing 0.7 mg / kg of lycopene, 0.4 mg / kg of trans-beta-carotene, 0.2 mg / kg of canthaxanthin, 0.4 mg / kg of lutein and 1.4 mg / kg of alpha-tocopherol resulted not only in an insignificant decrease in the enrichment of carotenoids and alpha-tocopherol in LDLs but also found there to be minimal resistance of these LDLs to oxidation [27]. Other studies showed that fiber intake would affect the formation of micelles, leading to a reduction in the absorption of carotenoids. Thus, the results released thus far do not appear to be consistent.

In tomato fruits, lycopene is present mainly in the pericarp, which is located in the cell compartment of the chloroplasts, in which the crystals are associated with membrane structure [28].

The concentration of lycopene in fruits and vegetables changes according to its type, climate and geographical effect, planting site, season, maturity, post-harvest handling and storage. Thus, higher concentrations are found in vegetable and fruit peelings compared with the pulp of the same vegetable or fruit, and higher concentrations are found in foods produced in warmer regions and in processed food. In general, the redder the food, the greater the concentration of lycopene it contains.

The structure and the physical and chemical properties of lycopene in foods will determine how the human body uses it. The bioavailability of lycopene also appears to be related to its isomeric forms. As already mentioned, despite being primarily present in food in the form of the trans-isomer (80% to 97%), the cis isomer form of lycopene appears to be better absorbed and is found in the human body in greater quantities, due to the shorter length of the molecule and to its greater solubility in the micelles. The trans-isomers can aggregate in the intestine and form crystals, strongly reducing its absorption by the micelles [15]. The pH of the stomach does not appear to significantly influence the transformation of trans to cis-isomers.

The ability of lycopene to capture singlet oxygen has been attributed to its extensive system of conjugated double bonds. The maximum protection is achieved when the carotenoid has nine or more double bonds. It was also found that lycopene, being acyclic, is more efficient than di-cyclic β-carotene, although both have eleven conjugated double bonds.

The action of lycopene in human health has received great attention in recent years [29, 30], with the strongest scientific evidence relating to cancers of the lung, esophagus, breast, uterus, bladder and prostate. Lycopene may also play a role in preventing cardiovascular diseases and other chronic degenerative diseases [14, 19, 31-36].

A multicenter study involving 10 European countries concluded that lycopene, or another correlated substance, could contribute to the protective effect of vegetables against the risk of myocardial infarction, which was not observed with α- and β-carotene [37]. Another prospective study in women, with a duration of approximately seven years, indicated that dietary lycopene or other phytochemicals derived from tomato fruits confer cardiovascular benefits [38].

Yet another 12-year prospective study, conducted in the United States, found a reduced risk of coronary artery disease in women who engaged in a high consumption of foods rich in α-or β-carotene; however, the same effect was not observed when the food consumed was instead rich in lycopene, β-cryptoxanthin or lutein / zeaxanthin [39].

Among the foods containing lycopene, tomato fruits and its derivatives have raised great interest for their appreciable carotene content, which has been referred to as being associated with a reduced risk of various diseases. Specifically, regarding cardiovascular disease, lycopene is known to inhibit the oxidation of the particles of low-density cholesterol (LDL) cholesterol, which are a major risk factor for the advance of atherosclerosis. Lycopene also induces the formation of endothelial enzymes generating nitric oxide (a compound with anti-atherosclerotic properties), which acts as a vasodilator and an antithrombotic and anti-inflammatory agent. Furthermore, lycopene reduces cholesterol levels at the expense of the LDL-cholesterol fraction, which appears to be due to the ability of this carotenoid to inhibit 3-hydroxy-3-methylglutaryl coenzyme A (HMG-CoA) reductase, an enzyme that is involved in cholesterol synthesis and in the increased activity of receptors in the clearance of plasma LDL-cholesterol [19, 40-42].

In a study of hypercholesterolemic adult men who had received 60 mg/day of lycopene for three months, it was found that the subjects experienced an average reduction in the LDL-cholesterol of approximately 14% without any change in the HDL-cholesterol fraction [18].

Additionally, a review of 72 epidemiological studies that investigated the relationship between tomato fruits consumption and cancer demonstrated an inverse relationship between the intake of lycopene-rich foods and cancer incidence. The consumption of foods high in lycopene content increases the lycopene concentration in the blood, which is associated with a lower risk of digestive tract, lung, pancreas, and especially of prostate cancer. Lycopene is found in the human prostate, suggesting the biological possibility of a direct effect of this carotenoid in the prostate function and the reduction of this organ's carcinogenesis [18, 35, 43].

Although there is a large amount of evidence on the protective effects of lycopene regarding prostate cancer, several studies have shown inconsistent results about this effect [44]. However, lycopene is an efficient inhibitor of cellular proliferation, and these inconsistencies can be partially explained by the fact that the lycopene may present different bioavailabilites, depending on the food source [45].

Carotenoids decreased the risk of lung cancer in non-smokers by 32%. They also found that, for a higher intake of alpha-carotene, this risk reduction rose to 63%. However, in smokers and those using other antioxidants, the risk reduction was negligible except when it came to lycopene. Smoking modifies the plasma concentrations of various carotenoids. However, this alteration does not occur in lycopene, and there is even a significant decrease in cancer risk with higher lycopene consumption. Lycopene is more effective in protecting cell membranes against damage caused by the radical NO that exists in smoke and thus appears to play a role in preventing lung cancer [45].

According to a study in Canada that conducted questionnaires regarding the frequency of food consumption, the mean intake of lycopene was 25 mg per day, of which 50% was fresh tomato fruits. Considering that the lycopene bioavailability in fresh tomato fruits is lower than that of processed tomatoes, the authors suggested encouraging people to increase their consumption of commercially widespread processed foods [20].

However, we still cannot be precise about the specific amount, minimum or maximum, of lycopene that should be ingested by a healthy population. Rao and Agarwal, [21] postulated that, to meet the daily needs, to prevent oxidative stress and cell damage, the optimal daily intake of lycopene should be 35 mg/day. According to Rao [19], 5 mg to 10 mg of lycopene per day would be sufficient to obtain the benefits of this nutrient. Other authors [46] reported that supplementation with 15 mg/day of lycopene inhibited the progression of benign prostatic hyperplasia.

Thus, in several pathologies, a higher lycopene intake appears to be necessary. That finding indicates that further studies will be required to assess what amounts are safe to administer to subjects that are either healthy or have a pathological condition, and what its effects are, as there is no agreement regarding the nutritional recommendations for the intake of lycopene [19].

Table 2 presents a brief description of the diseases in which a protective role of lycopene was evidenced.

Table 2. Diseases which may be prevented by lycopene ingestion

Diseases	References
Cancer	
Gastric cancers (esophagus, stomach, colon, pancreatic and rectum)	[47-49]
Breast	[50, 51]
Lung	[52]
Uterus and ovaries	[33]
Prostate	[35, 36, 43-46, 53-58]
Cardiovascular diseases	
(hypercholesterolemia, atherosclerosis, myocardial infarction, stroke)	[16, 18, 19, 21, 31, 38, 40-42, 59]
Chronic degenerative diseases	
Alzheimer and Parkinson	[60-64]
Amyotrophic lateral sclerosis	[65]
Osteoporosis	[66-68]
Other Diseases	
Diabetes	[69]
Skin	[70]

Lycopene Variation in Tomato Fruits after Harvest

The lycopene content of tomato fruits is highly dependent on the maturation stage: more mature tomatoes have higher amounts of lycopene. Significant variation in lycopene content was not observed among the tomato types or production areas [71]. However, because tomato fruits are usually picked before the maturation process is complete and ripening continues after harvest, it is important to understand the variation in lycopene content during the maturation period after harvest.

Thus, the main purpose of this section was to evaluate the lycopene content during maturation after harvest in three of the most widely consumed types of tomato fruits in Portugal (round, elongated and cherry) to obtain an optimal time period between harvest and consumption for maximizing the lycopene intake.

MATERIAL AND METHODS

Material and Reagents

The determination of lycopen concentration was conducted on a Gilson HPLC system (Villiers le Bell, France), consisting of a pump (model 321) and a UV-VIS detector (155 model) set to λ = 450 nm. A 25 cm × 4.6 mm i.d. Hichrom 5 C18 column (Reagente 5, Oporto, Portugal), a Nucleosil C18 guard column (Palo Alto, California, USA) and a Rheodyne injector (Bensheim, Germany) with a 100-µl loop were used for lycopene quantification. An LG computer (Coimbra, Portugal) equipped with UniPoint Gilson software was used to control the chromatographic system and for data acquisition and processing. A KLS Edmund Buehler mixer (Tuebingen, Germany), a BV-401C Fagor domestic homogenizer (Mondragon, Spain), a Mettler Toledo AG285 balance (Greifensee, Switzerland), an Ariston freezer (Coimbra, Portugal), a sample concentrator using nitrogen (Reagente 5) and 0.45 µm filters from Millipore (Bedford, MA, USA) were utilized for sample processing. All of the reagents used were of chromatographic quality and were purchased from Merck (Darmstadt, Germany). Ultrapure water was obtained from a Milli-Q system (Millipore).

Samples

A total of 144 tomato fruits were collected from a greenhouse farm located at Arazede, Montemor-o-Velho, Coimbra (Portugal, Center Region, GPS N 40.30183°/W -8.687932°). All plants from each tomato type were grown under the same soil and climatic conditions and similar agricultural practices. The samples were collected from the same plant at 3 different stages of maturation (green, turning and red) for each of the three types of tomato fruits analyzed (round, elongated and cherry). The ripening stage for all samples was selected according to US Department of Agriculture criteria [72]:

- Green – tomato fruit surface is completely green; the shade of green may vary from light to dark;
- Turning – 10% to 30% of the tomato fruit surface is not green; in the aggregate, shows a definite change from green to tannish-yellow, pink, red, or a combination thereof;
- Red – more than 90% of the tomato fruit surface is not green; in the aggregate, shows red color.

Tomato fruits of the same type with a similar size and weight were chosen. An overview of the types, weights, dimensions, maturation stages and respective number of tomato fruits selected is shown in Table 3.

Table 3. Tomato fruit sampling specifications

Type	Weight (g)	Dimensions (mm)	Maturation stage	Number of samples
			red	6
Cherry	12 – 18	Ø 20 – 30	turning	24
			green	24
		length 80 – 100	red	12
Elongated	90 – 120		turning	12
		width 30 – 50	green	18
			red	9
Round	180 – 220	Ø 75 – 100	turning	15
			green	24

Sample Preparation

After the sample collection, the maturation process was conducted at room temperature (18 – 25°C) in the shade with a light intensity similar to that which is found in retailing conditions (alternate cycles of 12 hours in the dark and indirect light with a 35 lumens per watt incandescent lamp). The ripening process was then interrupted and the samples were frozen, as displayed in Table 4.

Each sample consisted of 3 round or elongated or 6 cherry tomato fruits. After the tomato fruits were cut into small pieces, a homogenization procedure was conducted.

A total of 2.5 ± 0.5 g of tomato fruit was weighed in an amber flask to which 100 mL of extraction solvent (hexane: acetone: ethanol 2:2:1 v: v: v) was added, and the obtained mixture was homogenized for 30 minutes. After the homogenization, 15 mL of water was added to the mixture, and homogenization was performed again. The aqueous layer was then separated from the organic layer, and a 1.0 mL aliquot of the organic layer was evaporated to dryness under a stream of nitrogen at 40°C. The dry residue was dissolved into the mobile phase (acetonitrile:dichloromethane:methanol, 70:20:10) and injected into the chromatographic system at a flow rate of 1.5 mL/min, and an isocratic elution of the mobile phase was conducted to achieve lycopene quantification [73, 74].

RESULTS AND DISCUSSION

The data for lycopene are presented in figure 5 a, b and c (n=3, mean ± SD).

Obtained data show that the lycopene contents in red tomatoes were 2.39 ± 0.17, 3.47 ± 0.42 and 3.75 ± 0.81 mg/100 g, for cherry, round and elongated types, respectively. However, the ripening period after the harvest is an important factor influencing the lycopene

Table 4. Scheme of the tomato fruits sampling during postharvest maturation

Type	Maturation	Day 0	Day 4	Day 8	Day 11	Day 15	Day 18	Day 22	Day 30
	red	Cr0	-	-	-	-	-	-	-
Cherry	turning	Ct0	Ct4	Ct8	-	Ct15	-	-	-
	green	Cg0	Cg4	Cg8	-	Cg15	-	-	-
	red	Er0	Er4	Er8	Er11	-	-	-	-
Elongated	turning	Et0	Et4	Et8	Et11	-	-	-	-
	green	Eg0	Eg4	Eg8	Eg11	Eg15	Eg18	-	-
	red	Rr0	Rr4	Rr8	-	-	-	-	-
Round	turning	Rt0	Rt4	Rt8	Rt11	Rt15	-	-	-
	green	Rg0	Rg4	Rg8	Rg11	Rg15	Rg18	Rg22	Rg30

Cr –Cherry red; Ct – Cherry turning; Cg – Cherry green;

Er – Elongated red; Et – Elongated turning; Eg – Elongated green;

Rr – Round red; Rt – Round turning; Rg – Round green.

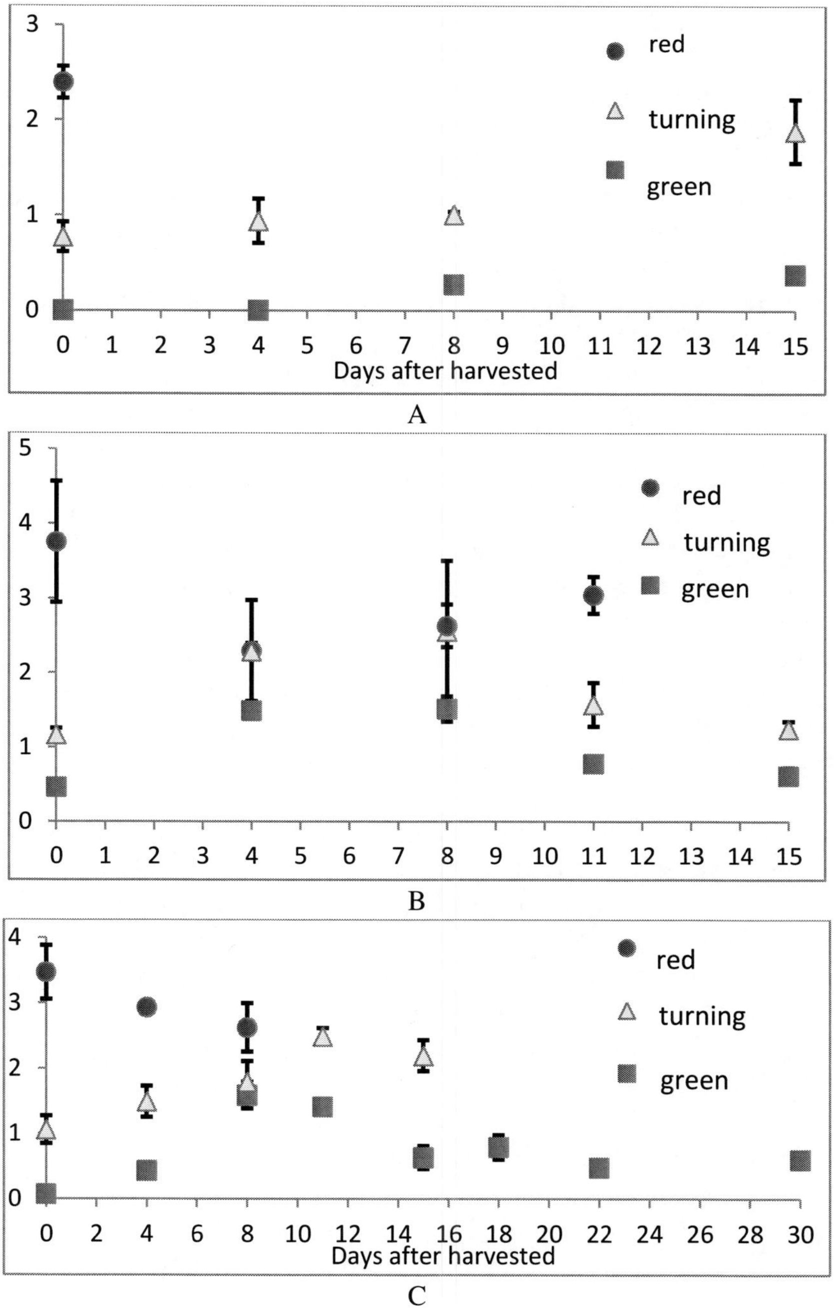

Figure 5. Lycopene content's (mg/100 g) of tomato fruits during postharvest maturation (a – cherry, b – elongated and c – round).

concentration. The obtained results indicated that the lycopene content changes during the ripening process after harvest.

On day 0 (zero), when the tomatoes were picked, the lycopene content was highest in the red tomatoes, followed by the turning and green tomatoes, independent of the type. In the following days, the lycopene content significantly decreased in the red tomatoes, whereas the quantity of lycopene increased in the turning and green tomatoes. However, the significantly increase in turning and green tomatoes never reached the level achieved in the red tomatoes picked at day 0.

As shown in Figure 5a, the lycopene content in the cherry tomatoes illustrates a significant increasing of lycopene content from day 0 to day 15 after harvest for both green and turning tomatoes; however, this trend does not apply to round and elongated tomato fruits (Figures 5 b and c). In both green and turning round tomatoes, a significant increase in the lycopene content was observed between day 0 and day 8 after harvest, followed by either a not significant decrease lycopene content up to day 11. However, a significantly decrease in the lycopene content was observed after the eleventh day. In elongated turning tomatoes, the lycopene content significantly increase from day 0 to day 11 after harvest. A significant increase in the lycopene content was also observed in green elongated tomatoes from day 0 to day 8, followed by a significantly decrease until day 15. From day 15 to day 18, the lycopene content again increased.

In general, the results obtained revealed that the lycopene content varies throughout the ripening process after harvest: the lycopene content in turning tomato fruits increases significantly during maturation after harvest. This find could be due to biochemical changes in tomato fruits during the ripening process, which leads to a loss of chlorophyll and yellow and orange carotenoids and the synthesis of lycopene [75-77].

The results also showed that, although lycopene synthesis occurs in green tomato fruits after harvest, the concentrations in green tomatoes were lower than those in red tomatoes picked at day 0, regardless of the type. An increase in lycopene content in red tomato fruits was not observed after the harvest. Thus, red tomatoes should be consumed immediately after being harvested.

Many factors affect the lycopene concentration in tomato fruits, including cultivation practices and sampling period. Although the lycopene content in tomato fruits has been described in various publications, heterogeneity of the results is frequently observed, even in the same study. For instance, Marković and co-workers [78] reported lycopene concentrations between 1.82 and 11.19 mg/100 g from tomato fruits originating from three regions of Croatia, although the type of tomato fruit used in the study is unclear. In a Hungarian study, Lugasi and co-workers [79] clearly demonstrated that the lycopene contents in the tomatoes picked in the winter and early spring were only 0.85 and 1.10 mg/100 g, respectively. However, a high level of lycopene was observed in the tomatoes picked in the summer, ranging from 7.6 to 13.6 mg/100 g.

CONCLUSION

The literature survey on lycopene makes clear the importance of this carotenoid in the prevention of an increasing number of diseases, namely the different types of cancers and cardiovascular disease. Moreover, the importance of tomato fruits as a food source of lycopene in most diets is also an indisputable fact.

Thus, a better understanding of the formation of lycopene in tomato fruits after harvest is an important contribution that is described and highlighted above.

Overall, a maturation period of more than eleven days after harvest led to a decrease in the total content of lycopene in the three types of tomato fruits studied. Thus, to consume the highest lycopene content in turning tomato fruits, a period of eleven days from harvest should not be exceeded. These results are applicable for the three types of tomato fruits studied and are in agreement with a previous study that evaluated the behavior of mineral contents after harvest [80].

However, future research is needed to identify the ripening period that yields the most benefit from lycopene ingestion by the consumer. This knowledge is important for consumer health and for agricultural and economic development.

ACKNOWLEDGMENTS

The authors are gratefully to Mr. José Marques, manager of VALMARQUES Enterprise, for samples supplying. The authors are also grateful to the Portuguese Foundation for Science and Technology (POCTI-SFA-8-177) for supporting this research.

REFERENCES

[1] M.A. Del Valle, *Utilidade del subproducto obtenido en la elaboración de derivados de tomate*. Tesis Doctoral – Facultad de Farmácia – Universidade Complutense de Madrid, Madrid (2004).

[2] D. Almeida, *Manual de Culturas Hortícolas* – Volume II. 1st edition. Editorial Presença, Lisboa (2006).

[3] S.F. Santos Domingos, *Valorização do Tomate - Extracção supercrítica de compostos bioactivos a partir de repiso de tomate*. Dissertação de Mestrado, Instituto Superior Técnico, Universidade Técnica de Lisboa, Lisboa (2009).

[4] J.L. Carvalho, L.G. Pagliuca, *Hortifruti Brasil*, 58, 6 (2007).

[5] Observatório dos Mercados Agrícolas e Importações Agro-Alimentares, www.observatorioagricola.pt accessed in 2012, January, 31st .

[6] FAO, Food and Agricultural commodities production (FAOSTAT), http://faostat.fao.org/site/339/default.aspx accessed in 2011, November, 9th.

[7] I. Martins, A. Porto, L. Oliveira, *Tabela de Composição de Alimentos Portugueses*, INSA, Lisboa (2006).

[8] D.B. MacDougall, *Color in food: Improving quality*, Woodhead Publishing Limited, Cambridge (2002).

[9] M.L. Gonçalves, *Conceitos Farmacológicos Associados ao Uso de Nutracêuticos*, in Nutracêuticos e Alimentos Funcionais, edited by J. F. Pinto, Lidel-edições Técnicas, Lda, Lisboa (2010).

[10] J. Von Lintig, *Ann. Rev. Nutr.*, 30, 35 (2010).

[11] H. Sies, W. Stahl, *Am. J. Clin. Nutr.* 62, 1315 (1995).

[12] S. Antunes, *Micronização Supercrítica do β-Caroteno*. Tese de Mestrado. Instituto Superior de Técnico. Universidade Técnica de Lisboa. Lisboa (2007).

[13] M.E. Figueira, *Nutracêuticos na Fronteira com os Alimentos Funcionais*, in Nutracêuticos e Alimentos Funcionais, edited by J. F. Pinto, Lidel-edições Técnicas, Lda, Lisboa (2010).

[14] L. Arab, S. Steck, *Am. J. Clin. Nutrit.*, 71, 1691S (2000).
[15] T.W. Boileau, A.M.Boileau, J.W. Erdman, jr., *Experim. Biol. Med.*, 227, 914 (2002).
[16] –C.W. Hadley, K.S. Clinton, J.S. Schwartz, *J. Nutr.* 133, 727 (2003).
[17] http://en.wikipedia.org/wiki/Lycopene accessed in 2012, February, 13th.
[18] B.N. Lombán, J.M.P. Sanchez, *Fitoquímicos de origen vegetal en la prevención de enfermedades degenerativas*, in Nutrición y alimentación en la promoción de la salud, edited by R. M. Ortega, A.M. Requejo, R. M. Martínez, Madrid (2007).
[19] A.V. Rao, *Experim. Biol. Med.*, 227, 908 (2002)
[20] A.V. Rao, Z. Wassem, S. Agarwal, *Food Res. Intern.*, 31, 737 (1998).
[21] A.V. Rao, S. Agarwal, *J. Am. Col. Nutrit.*, 19, 563 (2000).
[22] P.V. Minorsky, *Plant Physiol.*, 130, 1077 (2002).
[23] P.M. Brambley, *Phytochem.*, 54, 233 (2000).
[24] S.K. Clinton, *Nutr Rev.*, 56, 32 (1998).
[25] V. Böhm, R. Bitsch, *Eur. J. Nutr.*, 38, 118 (1999).
[26] C.M. Allen, A.M. Smith, S.K. Clinton, S.J. Schwartz, *J. Am. Diet. Assoc.*, 102, 1257 (2002).
[27] L. Yonekura, A. Nagao, *Mol. Nutr. Food Res.*, 51, 107 (2007).
[28] W. Bian, C. Barsan, I. Egea, E. Purgatto, C. Chervin, M. Zouine, A. Latché, M. Bouzayen, J.-C. Pech, *J. Botany*, Article ID 289859, 13 pages, doi:10.1155/2011/289859, (2011).
[29] A.V. Rao, S. Agarwal, *Nutr. Res.*, 19, 305 (1999).
[30] F. Khachik, L. Carvalho, P.S. Bernstein, G.J. Muir, D.Y. Zhao, N.B. Katz,. *Experim. Biol. Med.*, 227, 845 (2002).
[31] T. Rissanen, S. Voutilainen, K. Nyssonen, J.T. Salonen, *Experim. Biol. Med.*, 227, 900 (2002).
[32] M.L. Nguyen, S.J. Schwartz, *Food Technol.*, 53, 38 (2003).
[33] D.W. Cramer, H. Kuper, B.L. Harlow, L. Titus-Ernstoff, *Intern. J. Cancer*, 94, 128 (2001).
[34] C.W. Hadley, E.C. Miller, S.J. Schwartz, S.K. Clinton, *Experim. Biol. Med.*, 227, 869 (2002).
[35] E. Giovannucci, A. Ascherio, E.B. Rimm, M.J. Stampfer, G.A. Colditz, W.C. Willett, *J. Natl. Cancer Inst.*, 87, 1767 (1995).
[36] M. Stacewicz-Sapuntzakis, P.E. Bowen, *Biochim. Bioph. Acta*, 1740, 202 (2005).
[37] L. Kohlmeier, J.D. Kark, E. Gomez-Gracia, B.C. Martin, S.E. Steck, A.F. Kardinaal, J. Ringstad, M. Thamm, V. Masaev, R. Riemersma, J.M. Martin-Moreno, J.K. Huttunen, F.J. Kok, *Am. J. Epidemiol.*, 146, 618 (1997).
[38] H.D. Sesso, S. Liu, J.M. Gaziano, J.E. Buring, *J. Nutr.*, 133, 2336 (2003).
[39] K.S. Osganian, M.J. Stampfer, E. Rimm, D. Spiegelman, J.E. Manson, W.C. Willett, *Am. J. Clin. Nutr.*, 77, 1390 (2003).
[40] S. Agarwal, A.V. Rao, *Can. Med. Assoc. J.*, 163, 739 (2000).
[41] P. Palozza, N. Parrone, R. Simone, A. Catalano, Arch. Biochem. Bioph., 504, 26 (2010).
[42] P. Palozza, R. Simone, A. Catalano, N. Parrone, G. Monego, F. Ranelletti, *J. Nutr. Biochem.*, 22, 971 (2011).
[43] T.W.M. Boileau, S.K. Clinton; J.W. Erdman, jr., *J. Nutr.*, 130, 1613 (2000).
[44] T. Shirai, M. Asamoto, S. Takahashi, K. Imaida, *Toxicol.*, 181, 89 (2002).

[45] N.J.I.E. Shami, E.A.M. Moreira, *Rev. Nutr.*, 17, 227 (2004).

[46] M.E.S. Bozetto, *Efeito do licopeno e do extrato de tomate sobre os níveis séricos de PSA total e livre, testosterona, IGF-1 e sintomas prostáticos em pacientes com hiperplasia prostática benigna : um ensaio clínico randomizado controlado*. Tese de Doutoramento, Universidade Federal do Rio Grande do Sul. Faculdade de Medicina., Porto Alegre (2009).

[47] S.K. Clinton, *Nutrition Rev.*, 56, 35 (1998).

[48] E. De Stefani, F. Oreggia, P. Boffetta, H. Deneo-Pellegrini, A. Ronco, M. Mendilaharsu, *Oral Oncol.*, 36, 47 (2000).

[49] P. Greenwald, J.A. Milner, D.E. Anderson, S.S. McDonald *Cancer Metastasis Rev.*, 21, 230 (2002).

[50] J. Levy, E. Bosin,B. Feldman, Y. Giat, A. Miinster, M. Danilenko, Y. Sharoni, *Nutr Cancer*, 24, 66 (1995).

[51] H.D. Sesso, J.E. Buring, S.M. Zhang, E.P. Norkus, J.M. Gaziano, *Cancer Epidemiol Biomarkers Prev* ,14,1074 (2005).

[52] Y. Ito, K. Suzuki, S. Suzuki, R. Sasaki, M. Otani, K. Aoki, *Int J Vitam Nutr Res.,* 72, 237 (2002).

[53] E. Giovannucci, *J Natl Cancer Inst*, 91 317 (1999).

[54] O. Kucuk, F.H. Sarkar, W. Sakr *Cancer Epidemiol Biomarkers Prev*, 10, 861 (2001).

[55] L. Chen, M. Stacewicz-Sapuntzakis, C. Duncan, R. Sharifi, L. Ghosh, R. van Breemen, D. Ashton, Ph.E. Bowen, *J Natl Cancer Inst*. ,93, 1872 (2001).

[56] E. Giovannucci, Y. Liu, M.J. Stampfer, W.C.Willett, *J Natl Cancer Inst*, 94, 391 (2002).

[57] O. Kucuk, D.P. Wood Jr., *J. Urol.*, 167, 651 (2002).

[58] K. Wertz, U. Siler, R. Goralczyk, *Arch. Biochem. Bioph.*, 430, 127 (2004).

[59] L.J. Ignarro, M.L. Balestrieri, C. Napoli, *Cardiovasc. Res.,*15, 326 (2007).

[60] D.V.M. Mohsen Meydani, *Nutrit. Rev.,* 59, S75 (2001).

[61] M.P. Mattson, S.L. Chan, W. Duan, *Physiol. Rev.*, 82, 637 (2002).

[62] A.V. Rao, B. Balachandran, *Nutr. Neurosci.*, 5, 291 (2003).

[63] - P. Rinaldi, M.C. Polidori, A. Metastasio, E. Mariani, P. Mattioli, A. Cherubini, M. Catani, R. Cecchetti, U. Senin, P. Mecocci, *Neurobiol. Aging*, 24, 915 (2003).

[64] H. Kaur, S. Chauhan, R. Sandhir, *Neurochem. Res.*, 36, 1435 (2011).

[65] M.P. Longnecker, F. Kamel, D.M. Umbach, T.L. Munsat, J.M. Shefner, L.W. Lansdell, D.P. Sandler, *Neuroepidemiol.*, 19, 210 (2000).

[66] L.G. Rao, *Tomato lycopene and bone health: Preventing osteoporosis*. In Tomatoes, lycopene and human health, edited by A.V. Rao, Scotland: Caledonian Science Press (2006).

[67] L.G. Rao, E.S. Mackinnon, R.G. Josse, T.M. Murray, A. Strauss, A.V. Rao, *Osteoporos. Int.*, 18, 109 (2007).

[68] S. Sahni' M.T. Hannan, J. Blumberg, L.A. Cupples, D.P. Kiel, K.L. Tucker, *J. Bone Miner. Res.*, 24, 1086 (2009).

[69] A. Kuhad, R. Sethi, K. Chopra, *Life Sci.*, 83, 18128(2008).

[70] M.C. Kowalczyk, Z. Walaszek, P. Kowalczyk, T. Kinjo, M. Hanausek, T.J. Slaga, *Carcinogenesis*, 30, 1008 (2009).

[71] N. Kozukue, M. Friedan, *J. Sci. Food Agr.*, 83, 195 (2003).

[72] US Department of Agriculture. *Standards for Grade of Fresh Tomatoes* (7CFR980.212), US Government Printing Office, Washington DC (2009).
[73] C.H. Lin, B.C. Chen, *J. Chromatogr. A*, 1012, 103 (2003).
[74] A.I.O. Barba, C.M. Hurtado, M.C.S. Mata, V.J. Ruiz, M.L.S. Tejada, *Food Chem.* 95, 328 (2006).
[75] B. Piechulla, R.E. Glick, H. Bahl, A. Melis, W. Gruissem, *Plant Physiol.*, 84, 911 (1987).
[76] P.W. Goodenough, T.H. Thomas, Ann. Appl. Biol., 98, 507 (1981).
[77] H. Gautier, V. Diakou-Verdin, C. Bénard, M. Reich, M. Buret, F. Bourgaud, J. L. Poëssel, C. Caris-Veyrat, M. Génard, *J. Agric. Food Chem.*, 56, 1241, (2008).
[78] K. Marković, M. Hruškar, N. Vahčić, *Nutr. Res.*, 26, 556 (2006).
[79] A. Lugasi, L. Bíró, J. Hóvárie, K.V. Sági, S. Brandt, E. Barna, *Nutr. Res.* 23, 1035 (2003).
[80] F. Costa, M.L. Baeta, D. Saraiva, M.T. Verissimo, F. Ramos, *Food Anal. Methods*, 4, 410 (2011).

In: Tomatoes: Cultivation, Varieties and Nutrition
Editor: Tadahisa Higashide

ISBN: 978-1-62417-915-0

Chapter 14

HEALTH PROPERTIES OF TOMATO ANTIOXIDANTS

Giulia Graziani[1,*], Antonello Santini[1] and Alberto Ritieni[2]
[1]Department of Food Science, University of Napoli "Federico II", Portici, Napoli, Italy
[2]Department of Pharmaceutical and Toxicological Chemistry, University of Napoli "Federico II", Napoli, Italy

ABSTRACT

Epidemiologic evidence of a protective role for fruits and vegetables in cancer prevention is substantial. Antioxidants are important in disease prevention in both plants and animals, inhibiting or delaying the oxidation of biomolecules by preventing the initiation or propagation of oxidising chain reactions. Tomato is a versatile vegetable that is consumed fresh as well as in the form of processed products. Lycopene, the primary carotenoid in tomatoes, is frequently hypothesized to be one of the principle phytochemicals in tomato products with anticancer properties. Although lycopene is the most abundant carotenoid in tomato products, the tomato is also an excellent source of other nutrients and phytochemicals that may also contribute to the health benefits of tomatoes. Other tomato components such as phenolics, flavonoids and vitamins C and E are also responsible for the antioxidant capacity of raw tomatoes and processed tomato products. Epidemiological studies regarding intake of tomatoes and tomato-based products and blood lycopene reported an inverse associations between tomato intake or blood lycopene level and the risk of cancer. The evidence for a benefit was strongest for tumors of the prostate, lung, and stomach. Numerous other potentially beneficial compounds are present in tomatoes, and, conceivably, complex interactions among multiple components may contribute to the anticancer properties of tomatoes. Antioxidant nutrients are believed to slow the progression of atherosclerosis because of their ability to inhibit the tissue damage deriving to oxidative process. Several studies reported that thermal processing enhanced the nutritional value of tomatoes by increasing the bioaccessible lycopene content and total antioxidant activity and are against the notion that processed fruits and vegetables have lower nutritional value than fresh

* Corresponding author: Giulia Graziani, Department of Food Science, University of Napoli "Federico II", Via Università 100, 80055 Portici (Napoli), Italy. Tel/Fax.: 0039 (81) 2539317. E-mail address: giulia.graziani@unina.it.

produce. Other studies demonstrated that lycopene is rather thermally stable in tomato-based food systems during mild thermal treatments, while more intense thermal treatments can cause lycopene degradation reactions. Tomato, one of the most important vegetables worldwide, contains a range of flavonoids and phenolic acids in addition to lycopene, which are regarded as potentially useful compounds with respect to health benefits. Composition data in fresh tomatoes vary due to genetic and environmental factors and cultural practices. Breeding programs aim to produce tomatoes with enhanced levels of flavonoids and other phenolics.

INTRODUCTION

It is widely accepted that a healthy diet is an important factor in preventing chronic diseases such as cancer, cardio-vascular and neuro-degenerative diseases, and in improving energy balance and weight management. In the scientific literature, studies have shown strong inverse correlations between tomato consumption and the risk of certain types of cancer, cardiovascular diseases and age-related macular degeneration [1, 37, 2, 14, 3, 16, 17].

From epidemiological studies, clinical trials and experiments on animals as well as in vitro studies, this protective effect has been mainly attributed to provitamin A [18] and other carotenoids. Moreover, carotenoids are a major class of compounds providing precursors to essential vitamins and antioxidants. Because tomato is the second-most important vegetable in the world after potato, with an annual production of around 122.9 million tones of fresh weight (FW), this horticultural crop is the predominant source of carotenoids. Not all the carotenoids present in food are found in plasma due to selective uptake by the digestive tract. Most of them is derived from either fresh or processed tomato; major ones being lycopene, α- and β-carotene, lutein, zeaxanthin and β-cryptoxanthin.

Lycopene, which constitutes about 80–90% of the total carotenoid content of red ripe tomatoes [19] is the most efficient antioxidant among carotenoids through its quenching activity of singlet oxygen and scavenging of peroxyl radicals [20]. On the other hand, β-carotene, a potent dietary precursor of vitamin A, accounts for around 7% of tomato carotenoid content [21]. Ascorbic acid (vitamin C), while being a most effective antioxidant in plants, is also an important phytochemical of tomato fruit. Except for some wild related genotypes (*S. chilense, S. hirsutum, S. cheesmanii, S. lycopersicoides*) or new commercial varieties recently developed [22], tomato (*Solanum lycopersicum*) fruits are not usually reported to contain anthocyanin. The limited caloric supply, relatively high fibre content, and provision of minerals, vitamins, and phenols such as flavonoids make the tomato fruit an excellent ''functional food'' providing additional physiological benefits as well as meeting basic nutritional requirements.

Tomatoes (Lycopersicon esculentum Mill.) have been recognized also as an important source of dietary flavonoids because of a high consumption worldwide. Flavonoids are considered as potentially health-promoting substances, and efforts are made to produce tomatoes with a standardized content of flavonoids and other compounds [23, 24], or to elevate the concentration of especially quercetin 3-rutinoside (rutin). Tomatoes are worldwide among the most consumed vegetables and play an important role in the human diet. In previous literature, however, flavonoids of tomatoes have scarcely been completely characterized. The occurrence of flavonoids in fruits of tomatoes is almost exclusively restricted to their skin, leaving only negligible quantities in the remaining parts of the fruit [25]. The main flavonoids in fruits of tomatoes identified in previous literature have been reported to be rutin [26, 27], naringenin [26, 28, 29], and chalconaringenin [23, 30,31]. Moreover, some minor flavonoids have been identified from tomato fruits as reviewed by Moco and co-workers [32], among them kaempferol 3-rutinoside [33, 29] and naringenin 7-glucoside [30, 34].

Table 1. Concentration of carotenoids in tomatoes, tomato-based food products and several fruits and vegetables with similar carotenoid profiles

Foods	Concentration of carotenoids (mg/100 g)									
	Lycopene	α-Carotene	β-Carotene	γ-Carotene	Neurosporene	ζ-carotene	Phytofluene	Phytoene	Lutein	Zeaxanthin
Tomato paste	55.45	-	1.27	9.98	6.95	0.84	3.63	8.36	0.34	-
Tomato sauce	17.98	-	0.45	3.17	2.48	0.29	1.27	2.95	trace	-
Catsup	17.23	-	0.59	3.03	2.63	0.33	1.54	3.39	-	-
Tomato puree	16.67	-	0.41	2.94	2.11	0.25	1.08	2.40	0.09	-
Spaghetti sauce	15.99	-	0.44	3.02	3.15	0.34	1.56	2.77	0.16	-
Tomato juice	10.77	-	0.27	1.74	1.23	0.18	0.83	1.90	0.06	-
Vegetable juice	9.66	0.21	0.83	-	N.D.	-	0.69	1.71	0.08	-
Tomato	9.27	-	0.23	1.50	1.11	0.21	0.82	1.86	0.08	-
Grapefruit, pink	3.36	-	2.34	1.38	0.38	-	0.01	0.02	-	-
Papaya	2.52	0.02	0.22	0.01	0.05	0.19	0.44	0.68	0.02	0.02
Vegetable soup	1.93	0.41	1.50	-	-	-	0.31	0.60	0.16	-
Minestrone soup	1.48	0.21	0.92	-	-	-	0.17	0.28	0.15	-
Vegetable beef soup	0.31	0.46	1.51	-	-	-	0.19	0.35	0.11	-
Apricots fresh	0.01	0.02	1.59	0.08	-	0.04	0.36	0.59	0.09	-
Apricots canned	0.07	0.14	4.76	0.12	-	0.02	0.82	1.01	0.17	-
Apricots dried	0.86	1.60	8.56	0.13	-	0.14	2.00	2.27	0.36	-
Orange	-	0.02	0.06	0.004	0.01	0.10	0.06	0.08	0.35	0.25
Orange, mandarin	-	0.05	0.08	0.05	0.01	0.12	0.14	0.11	0.07	0.06
Squash, butternut	-	1.85	5.21	0.36	0.14	0.72	0.20	0.55	2.38	0.28

TOMATO BIOACTIVE METABOLITES: CAROTENOIDS

Fruits and vegetables contain in excess of 40 carotenoids that are routinely absorbed and metabolized by humans [35, 36]. Intake of tomato and tomato-based food products contributes to the absorption of a wide range of carotenoids in human serum and tissues. The prominent carotenoid in tomatoes is the red pigment lycopene that is also among the major carotenoids found in human serum. A prospective cohort study by [37] indicated that intake of lycopene-rich foods (e.g., tomato sauce, tomatoes, and pizza) was inversely associated with risk of prostate cancer whereas the overall intake of fruits and vegetables were not related. The nutritional significance of lycopene has also been demonstrated in an *in vivo* study of colon cancer involving rodents [38]. [39] have shown that lycopene is more potent than γ - and β -carotene in inhibiting the cell growth of various human cancer cell lines. One hypothesis for the biological activity of carotenoids in disease prevention is based on the antioxidant ability of these compounds that can quench singlet oxygen and other oxidizing species and protect the cells from oxidative damage. In early 1990, we first reported on the detection of the oxidation products of lycopene and lutein in human plasma [35]. Because these carotenoids were also abundant in fruits and vegetables associated with a reduced risk of cancers in epidemiological studies, we proposed metabolic pathways for the formation of their oxidation products and concluded that lycopene and lutein can serve as potentially useful chemopreventive agents [40]. Although lycopene has no vitamin A activity, it has been shown to possess strong antioxidant activity *in vitro* [41]. In literature, was reported on the synthesis and characterization of two oxidation products of lycopene detected in human plasma and tomato-based foods [42] and proposed possible pathways for the formation of the oxidative metabolites of this carotenoid [43]. This chapter, we present a review of the distribution of lycopene and related tomato carotenoids in fruits, vegetables, tomato-based food products, human serum, and tissues and discuss the chemistry and metabolic transformation of these compounds in humans.

Tomatoes and tomato-based food products are the major source of lycopene and a number of other carotenoids, such as phytoene, phytofluene, ζ-carotene, γ-carotene, β-carotene, and neurosporene [35, 36]. Other commonly consumed fruits in the United States that contain lycopene are pink grapefruit and papaya. Apricots (fresh, canned, dried) also contain low concentrations of lycopene and related carotenoids [35] (table 1). Among these foods that are the major source of hydrocarbon carotenoids, only ζ -carotene, γ -carotene, and β -carotene are precursors of vitamin A (Figure 1). Because the reduced risk of prostate cancer has been specifically correlated with the high consumption of tomato- based food products [37], this protective effect has been largely attributed to lycopene. Although lycopene is the major carotenoid in these foods, the presence of a wide range of other carotenoids in tomato-based food products cannot be overlooked. It is quite likely that lycopene in combination with other related tomato carotenoids mentioned above may be responsible for the observed biological activity. In 1995, [44] reported on the qualitative and quantitative distribution of carotenoids in namebrand and store-brand tomato-based food products purchased in three major U.S. cities. These foods were extracted and analyzed by high-performance liquid chromatography (HPLC) according to the methodology developed by [36].

TOMATO BIOACTIVE METABOLITES: FLAVONOIDS

The flavonoids are a large family of low molecular weight polyphenolic compounds that are found in plant tissues and include the flavones, flavonols, flavonones, isoflavones, flavan-3-ols, and anthocyanins [45]. More than 4000 flavonoids have been described, most are conjugated to sugar molecules and are commonly located in the upper epidermal layer of leaves. The role of flavonoids in plants is unclear, although many functions have been proposed, including protection against UV-B radiation and defense against pathogen attack [46, 47]. High dietary intake of fruits and vegetables is known to reduce the incidence of some cancers [48, 15]. Although it is thought that non-nutritive compounds, such as flavonoids, may play a role in the protective effect of fruit and vegetables, epidemiological studies have drawn conflicting conclusions on the association between flavonoids and the incidence of cancer. The Zutphen study found no correlation between flavonoid intake, in the form of flavonols and smaller amounts of flavones, and lung, colorectal or all-cause cancer [49] and a Spanish study similarly found no significant link between flavonol intake and protection against lung cancer [50]. In contrast, a clear inverse association between flavonol intake and the incidence of lung cancer was observed in a Finnish study [52], and a protective effect of flavonols against gastric cancer was also found in a Spanish study involving patients with gastric adenocarcinoma [51]. Thus, information on the potency of flavonols as anticarcinogens is at present equivocal, and further studies are required before any firm conclusions can be drawn. Indications of a protective effect of dietary flavonols against the development of free radical-derived damage to endothelial cells of the coronary arteries and the development of atherosclerosis [53, 54] have led to epidemiological studies on the relation between dietary flavonol intake and the incidence of coronary heart disease (CHD). The Dutch Zutphen Elderly Study highlighted a positive correlation between the high flavonol/flavone intake and a reduction in CHD [49]. A similar relationship was observed in a Finnish cohort study [52]. Contrasting results on the prevention of CHD by dietary flavonols were found in a study of male health professionals in the United States. Although this investigation did not support a protective effect of high flavonol intake on total incidence of CHD, it could not rule out a possible protective effect on men with established CHD [55]. Daily intake of flavonols and flavones in the Netherlands was estimated at 26 mg, with major sources being identified as tea (18-50 mg/L), onions (284-486 μg/g), and apple (21-72 μg/g) [49]. Recent work [58] has shown that certain types of tomatoes contain much higher levels of flavonols than were measured in the Dutch study [49]. Also, as sizable amounts of the flavonols in tomatoes are recovered after cooking [58], the consumption of tomatoes and tomato-derived products in foods, such as pizza and lasagne, may provide a more significant contribution to the intake of flavonols in the Western diet than is currently realized.

TOMATOES AND THE PREVENTION OF CORONARY HEART DISEASE

Although genetic factors and age are important in determining the risk, other factors, including hypertension, hypercholesterolemia, insulin resistance, and lifestyle factors such as smoking and diet are also major risk factors associated with the disease. Various epidemiological studies over the past decades showed protection against many chronic diseases, including cardiovascular diseases, due to high consumption of plant foods. Recently, oxidative stress induced by reactive oxygen species (ROS) is also considered to play an important part in the etiology of several chronic diseases, including CHD. Oxidation of the circulating low density lipoprotein (LDL) that carry cholesterol into the blood stream to oxidized LDL (LDLox) is thought to play a key role in the pathogenesis of atherosclerosis, which is the underlying disorder leading to heart attack and ischemic strokes (figure 3). Antioxidant nutrients are believed to slow the progression of atherosclerosis because of their ability to inhibit the damaging oxidative processes.

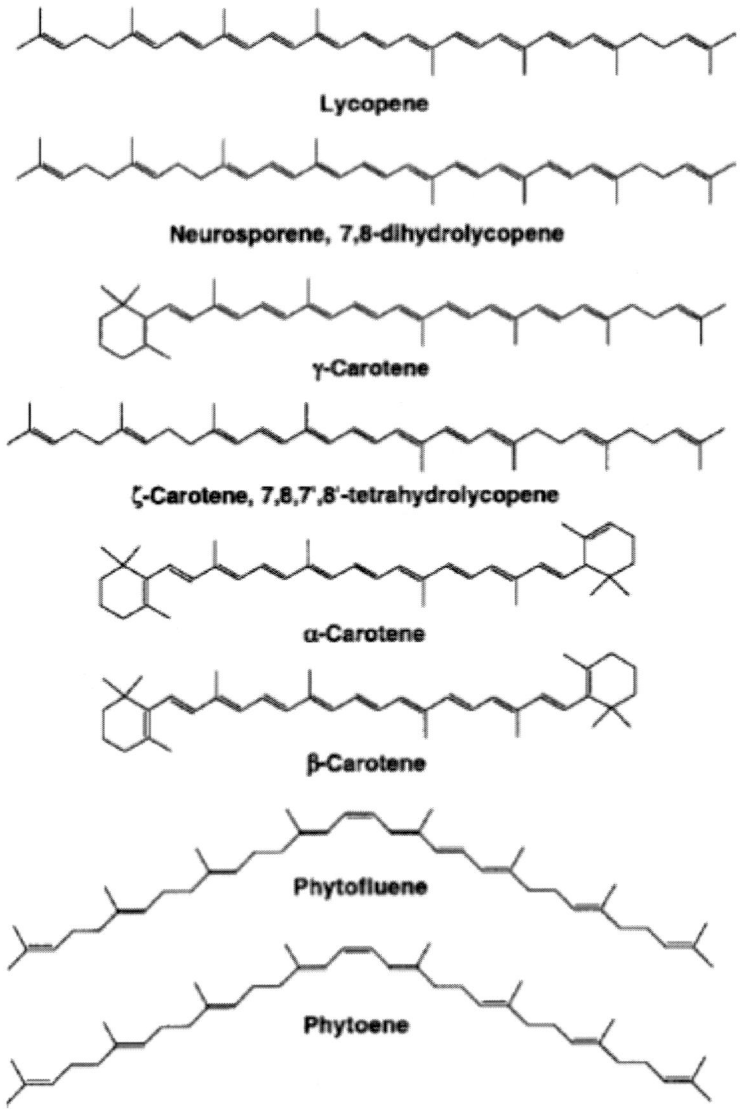

Figure 1. Chemical structures of carotenoids in tomatoes and tomato-based food products.

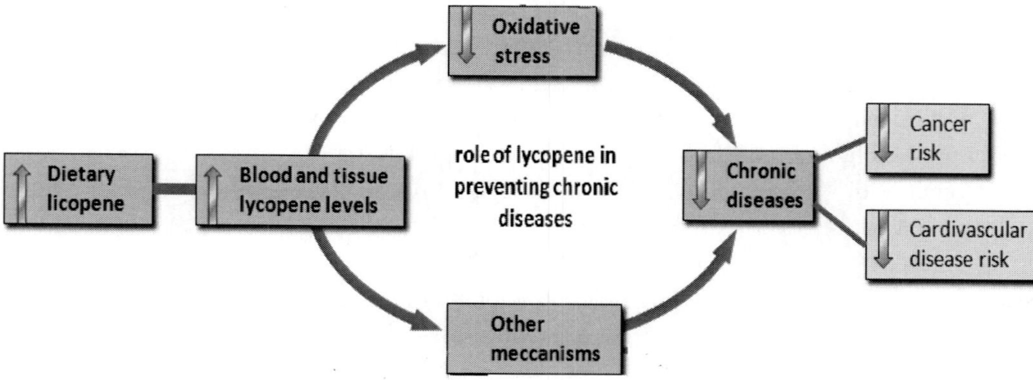

Figure 2. Proposed mechanisms for the role of lycopene in preventing chronic diseases.

Numerous phytochemicals are contained in plant foods, many of which are potent antioxidants, with carotenoids as one group of lipophilic antioxidants [59].

Recently, 139 Cretan (Greece) men aged 79 years and over were compared to men from Zutphen (The Netherlands). The Cretan men had nearly fourfold higher mean levels of lycopene as well as a lower level of oxidative stress and higher levels of antioxidants in plasma than men of the same age from Zutphen [60].

[61] analysed serum samples of 349 subjects on concentrations of conjugated dienes in LDL, being one marker of lipid oxidation. The values were significantly lower in men compared to women. Lycopene content in plasma was significantly negatively associated with content of conjugated dienes. Thus, dietary carotenoids contributed significantly to lowered LDL oxidative modification in vivo. [62] analysed serum samples of 264 healthy Korean women on their contents of lycopene. In addition, arterial stiffness as a possible marker involved in the pathophysiology of CVD was assessed by brachial-ankle pulse wave velocity (baPWV). Serum lipid profile, hs-CRP and contents of oxidized low density lipoprotein (oxLDL) were also analysed. A negative correlation was found between lycopene and oxLDL and also between lycopene and baPWV. Thus, lycopene may be responsible for a reduced oxidative modification of LDL, being possibly one mechanism by which lycopene could reduce arterial stiffness and the risk of CVD. A total of 299 Korean men were investigated on interrelationship between arterial stiffness, antioxidant status and the risk of metabolic syndrome. The authors analysed beside other parameters, baPWV, content of lycopene, lipid profile and oxLDL. baPWV was inversely correlated with lycopene content in serum. A negative correlation was also seen between lycopene and oxLDL. Thus, an interrelationship was shown between circulating lycopene, baPWV and metabolic syndrome [63]. Within a follow-up period of 11.9 years in another study, 3061 participants were asked to fill in a questionnaire and to give serum samples. Lycopene contents in serum tended to be lower for those who had died due to CVD than for those who survived [64]. Two major hypotheses have been proposed to explain the anticarcinogenic and antiatherogenic activities of lycopene: nonoxidative and oxidative mechanisms (figure 2). Among the nonoxidative mechanisms, the anticarcinogenic effects of lycopene have been suggested to be due to regulation of gap-junction communication,, to suppression of phosphorylated regulatory proteins such as p53 and Rb antioncogenes and stop cell division at the G0–G1 cell cycle phase. Lycopene affects the modulation of the liver metabolizing enzyme, cytochrome P450 2E1 and reduces cellular

proliferation induced by insulin-like growth factors, which are potent mitogens, in various cancer cell lines. Lycopene also has been shown to act as a hypocholesterolemic agent by inhibiting HMG–CoA (3-hydroxy-3-methylglutaryl– coenzyme A) reductase. Lycopene has been hypothesized to prevent carcinogenesis and atherogenesis by protecting critical cellular biomolecules, including lipids, lipoproteins, proteins and DNA.

A case-control study with 760 cases and 682 controls showed a decreased risk of acute myocardial infarction with increasing intake of a-carotene, b-carotene and β-cryptoxanthin but no association for lycopene [65]. The CARDIA Study (Coronary Artery Risk Development in Young Adults) with 4580 participants showed that those people with higher lycopene contents tended to have less healthy lifestyles. Serum total and individual carotenoids, with the exception of lycopene, were inversely associated with markers of inflammation, oxidative stress and endothelial dysfunction [66]. The Minnesota Heart Survey Study with 5369 men and 6070 women used a 24-h dietary recall. The authors developed a Heart Disease Prevention Index. This index improved between 1980/1982 and 2000/2002: for men by 2.58 points (8.3%) from 31.14 to 33.72 and for women by 2.44 points (7.9%) from 30.97 to 33.41.

Thus, overall diet quality has moderately improved. However, improvement has plateaued and levelled off during the last 5-year period that was studied. Regarding the carotenoids, only uptake of b-carotene, lutein, zeaxanthin and b-cryptoxanthin significantly increased from 1980/1982 to 2000/2002, but not consumption of lycopene [67]. So, controversial epidemiological data exist regarding the CVD-preventive effects of lycopene.

Oxidation of low-density lipoproteins, which carry cholesterol into the blood stream, may play an important role in the causation of atherosclerosis.

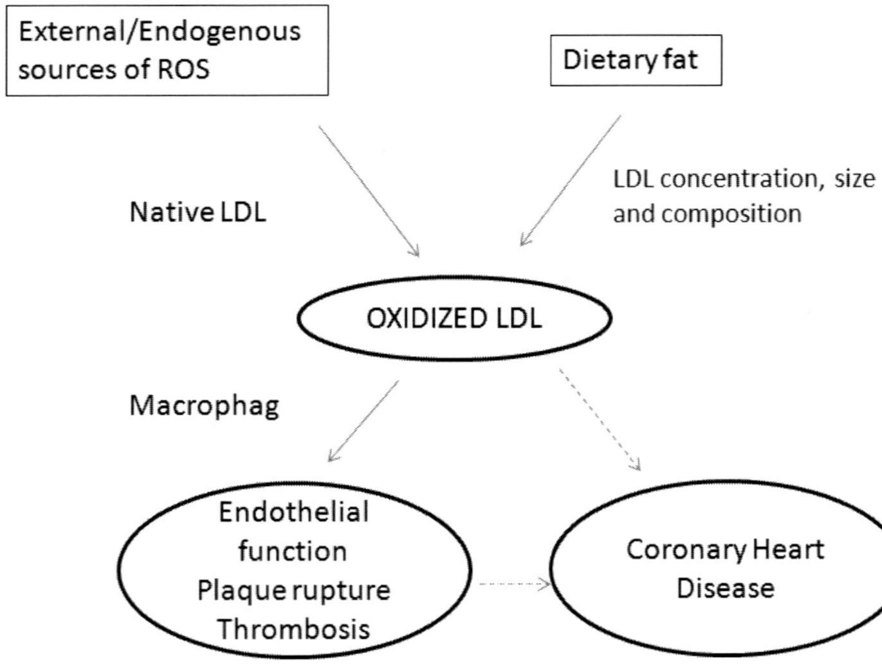

Figure 3. LDL oxidation hypothesis in the etiology of coronary heart.

Antioxidant nutrients are believed to slow the progression of atherosclerosis because of their ability to inhibit damaging oxidative processes. Several controlled clinical trials and epidemiological studies have provided evidence for the protective effect of vitamin E, which has been ascribed to its antioxidant properties. However, in the recently completed Heart Outcomes Prevention Evaluation (HOPE) Study, supplementation with 400 IU/d of vitamin E for 4.5 years did not result in any beneficial effects on cardiovascular events in patients at high risk.

In a multicentre case–control study, the relation between antioxidant status and acute myocardial infarction was evaluated [68]. Subjects were recruited from 10 European countries to maximize the variability in exposure within the study. Adipose tissue antioxidant levels, which are better indicators of long-term exposure than blood antioxidant levels, were used as markers of antioxidant status. Biopsy specimens of adipose tissue were taken directly after the infarction and were analysed for various carotenoids. After adjustment for a range of dietary variables, only lycopene levels, and not β-carotene levels, were found to be protective. A study from Johns Hopkins University, Baltimore, showed that smokers with low levels of circulating carotenoids were at increased risk for subsequent myocardial infarction [69]. Lower blood lycopene levels were also found to be associated with increased risk for and death from coronary artery disease in a population study comparing Lithuanian and Swedish cohorts with different rates of death from coronary artery disease.

Oxidatively modified macromolecules, particularly those derived from lipoproteins deposited in the vessel wall, may be important factors that initially activate the cellular and cytokine networks involved in atherosclerotic lesion formation and progression [70].

It has been suggested that carotenoids may protect lipoproteins and vascular cells from oxidative damage [71]. This is supported by observational studies that report that foods rich in carotenoids and antioxidant vitamins are associated with a reduced risk of cardiovascular disease [72]. Carotenoids have been detected in lipid-rich atherosclerotic plaques; however, the levels of lycopene were very low [73]. A recent study conducted in men with myocardial infarction and matched controls found that lycopene concentration in adipose tissue was a decreased risk factor, with an odds ratio of 0.52 for the contrast between the 10th and the 90th percentiles, after correcting for age, body mass, socioeconomic status, smoking, hypertension, and family history [68]. However, another case-control study examining serum carotenoids and risk of myocardial infarction did not find a relationship [75].

Dietary intake of tomatoes, consumption of tomato extracts, and supplementation with tomato phytonutrients (like lycopene) have all been shown to improve the profile of fats in our bloodstream [76]. Specifically, tomato intake has been shown to result in decreased total cholesterol, decreased LDL cholesterol, and decreased triglyceride levels. It's also been shown to decrease accumulation of cholesterol molecules inside of macrophage cells. Macrophage cells are a type of white blood cell that gets called into action when oxidative stress in the bloodstream gets too high, and the activity of macrophages - including their accumulation of cholesterol - is a prerequisite for development of atherosclerosis. Many phytonutrients in tomatoes are likely to be involved with the improvement of our blood fat levels. Two little-known phytonutrients - one called esculeoside A and the other called 9-oxo-octadecadienoic acid - are currently under active investigation by researchers as tomato phytonutrients especially important in blood fat regulation [77].

A recent study has demonstrated that healthy human subjects who ingested lycopene in the form of tomato juice or tomato sauce for 1 week had a significantly lower level of oxidized LDL than controls [78]. Lycopene was shown to inhibit the activity of an essential enzyme involved in cholesterol synthesis in both an in vitro study and a small clinical study [79], suggesting a hypocholesterolemic effect. Other possible mechanisms include enhanced LDL degradation, LDL particle size and composition, plaque rupture, and altered endothelial functions. Recent epidemiological studies have also shown an inverse relationship between tissue and serum levels of lycopene and mortality from coronary heart disease, cerebrovascular disease, and myocardial infarction [80]. However, the most impressive population-based evidence comes from a multi-center, case-control study where subjects from ten European countries were evaluated with regard to a relationship between antioxidant status and acute myocardial infarction. After adjusting for a range of dietary variables, only lycopene levels were found to be protective [80]. Some 725 middle-aged men, free of coronary heart disease and stroke, took part in the Kuopio Ischemic Heart Disease Risk Factor (KIHD) study. Men in the lowest quartile of serum levels of lycopene had a 3.3 fold increased risk of an acute coronary event or a stroke as compared to other groups (Rissanen et al 2003). In a second study [81], the association between plasma lycopene concentration and intima-media thickness of the common carotid artery wall (CCA-IMT) was examined. This study was part of the Antioxidant Supplementation in the Atherosclerosis Prevention (ASAP) study, in which 520 asymptomatic men and women participated. It was found that low plasma levels of lycopene were associated with an 18% increase in IMT in men compared with men whose plasma levels were higher than the median. In women, the difference did not remain significant after the adjustment [81].

TOMATOES AND CANCER

The Mediterranean diet, which is rich in vegetables and fruits, including tomatoes, has been suggested to be responsible for the lower cancer rates in that region [82]. Dietary intake of tomatoes and tomato products has been found to be associated with a lower risk of a variety of cancers in several epidemiological studies [83]. A high intake of tomatoes was linked to protective effects against digestive tract cancers in a case–control study [84] and a 50% reduction in rates of death from cancers at all sites in an elderly US population [85]. The most impressive results come from the US Health Professionals Follow-up Study, which evaluated the intake of various carotenoids and retinol, from a food-frequency questionnaire, in relation to risk of prostate cancer [86]. The estimated intake of lycopene from various tomato products was inversely related to the risk of prostate cancer. This result was not observed with any other carotenoid. A reduction in risk of almost 35% was observed for a consumption frequency of 10 or more servings of tomato products per week, and the protective effects were even stronger with more advanced or aggressive prostate cancer. In recent studies serum and tissue levels of lycopene were shown to be inversely associated with the risk of breast cancer [87] and prostate cancer [88]; no significant association with other important carotenoids, including β-carotene, was observed. [83] recently reviewed 72 epidemiological studies, including ecological, case–control, dietary and blood-specimen-based investigations of tomatoes, tomato based products, lycopene and cancer. In 57 studies there was an inverse association between tomato intake or circulating lycopene levels and risk of several types of cancer; in 35 cases the association was statistically significant. None of the studies showed adverse effects of high tomato intake or high lycopene levels. Although the epidemiological evidence of the role of lycopene in cancer prevention is persuasive, this role remains to be proven. There are few human intervention trials investigating the effectiveness of lycopene in lowering cancer risk. Most of the workers have investigated the effects of tomato or tomato product (lycopene) supplementation on oxidative damage to lipids, proteins and DNA [89, 90].

A preliminary report has indicated that tomato extract supplementation in the form of oleoresin capsules lowers the levels of prostate-specific antigen in patients with prostate cancer [91].

Tomato consumption showed a consistent inverse relation with the risk of digestive tract neoplasms in an integrated series of studies conducted in Italy in the 1980s [92]. Between 1992 and 1999, a case-control study was conducted in different areas of Italy on patients below age 80 years of age with different kinds of cancers, i.e., oral cavity and pharyngeal, esophageal, colorectal, breast, and ovarian [92]. The control group consisted of 5000 patients, also below the age of 80, with acute, non-neoplastic diseases. The multivariate relative risk of oral, pharyngeal and esophageal cancer decreased with subsequent levels of lycopene intake (0.7 for oral and pharyngeal cancers and 0.7 for esophageal cancer, both with borderline statistical significance). In colorectal, breast, and ovarian cancers, tomato intake was inversely and significantly correlated with cancer risk (RR= 0.8) [92]. Thirty-two patients with localized prostate adenocarcinoma consumed tomato sauce-based pasta dishes for 3 weeks (30 mg lycopene daily) before their scheduled radical prostatectomy. Prostate tissue was obtained as a biopsy at baseline and as a resected tissue at the time of the prostatectomy. Serum and prostate lycopene, serum prostate-specific antigen (PSA) concentrations, and leukocyte DNA 8-OH deoxyguanosine/deoxyguanosine (8OhdG) were measured at baseline and at the end of the intervention. Mean serum PSA concentrations decreased by 17.5% ($P < 0.002$) and leukocyte 8OhdG decreased by 21.3% ($P < 0.005$) after tomato sauce consumption. Resected tissues from tomato sauce-supplemented patients had 28.3% lower prostate 8OhdG compared with the control group ($P < 0.03$). Cancer cell 8OhdG staining in the resected prostate sections was reduced by 40.5% in mean nuclear density ($P < 0.005$) and by 36.4% in mean area ($P < 0.018$). The apoptotic index was higher in hyperplastic and neoplastic cells in the resected tissue after supplementation. These data indicate a significant uptake of lycopene into prostate tissue and a reduction in DNA damage in both leukocyte and prostate tissue [93]. A clinical study investigated the biological and clinical effects of lycopene supplementation among patients with localized prostate cancer [94]. Thirty-six patients with newly diagnosed prostate cancer were randomly assigned to receive tomato extract containing 30 mg of lycopene (15 patients) or no supplementation (11 patients) for 3 weeks before radical prostatectomy. Subjects in the intervention group had smaller tumors (45% vs. 80%), less involvement of surgical margins and/or extraprostate tissues with cancer (73% vs. 18%), and less diffuse involvement of the prostate by high-grade prostatic intraepithelial neoplasia (33% vs. 0%) compared with subjects in the control group. Mean PSA levels were lower in the intervention group than in the control group [94].

As noted, in people of the Mediterranean region with an above-average intake of vegetables and especially of cooked tomatoes, the incidence of prostate cancer is lower, probably the result of inhibition of two kinds of reactions involved in prostate cancer causation. The first reaction relates to the probable genotoxic carcinogens, most likely the heterocyclic amines formed during cooking of meat, especially of beef and pork [95].

Risk for many cancer types starts out with chronic oxidative stress and chronic unwanted inflammation. For this reason, foods that provide us with strong antioxidant and anti-inflammatory support are often foods that show cancer prevention properties.

Prostate cancer is by far the best-researched type of cancer in relationship to tomato intake. The jury verdict here is clear: tomatoes can definitely help lower risk of prostate cancer in men. One key tomato nutrient that has received special focus in prostate cancer prevention is alpha-tomatine. Alpha-tomatine is a saponin phytonutrient and it's shown the ability to alter metabolic activity in developing prostate cancer cells. It's also been shown to trigger programmed cell death (apoptosis) in prostate cancer cells that have already been fully formed. Research on alpha-tomatine has also been conducted for non-small cell lung cancer, with similar findings.

Along with prostate cancer and non-small cell lung cancer, pancreatic cancer and breast cancer are the two best-studied areas involving tomatoes and cancer risk. Research on tomatoes and breast cancer risk has largely focused on the carotenoid lycopene, and there is fairly well documented risk reduction for breast cancer in association with lycopene intake.

ANTIOXIDANT PROPERTIES OF TOMATOES AFTER PROCESSING

Tomatoes are usually stored to some extent and frequently cooked prior to consumption. Processing usually involves heat treatment and/or homogenization, both of which can disrupt the cellular matrix of tomatoes [96]. The intactness of the cellular matrix determines the bioavailability of different nutrients. Published data on the effects of heat treatment are not consistent. [96] found that heat treatment can have a deleterious effect on the micronutrient content of vegetables but at the same time the bioavailability of some nutrients can increase. The production of tomato paste from fresh tomatoes is an example where both homogenization and heat treatment are used and where the bioavailability of carotenoids is enhanced. However, at the same time as bound antioxidants are released by processing [97] other, labile antioxidant compounds are being destroyed [98]. Data on the effects of processing on the carotenoid content of foods are conflicting owing to several factors. These include the type of carotenoid assessed, the processing conditions used and the constituents of the tomato matrix. Furthermore, processing conditions and modes of calculating 'retention' during cooking are frequently not comparable from one study to another [99]. [99] reported that thermal processing increased carotenoid concentration which could be attributed to enzymatic degradation weakening protein– carotenoid aggregates [97, 100] and the effect of reducing the moisture content of the processed product. [101] showed that the antioxidant activity of cultivar Black Persimmon increased during storage at 15 C, while frying in oil or steaming the tomatoes led to a fall in vitamin C content but the overall antioxidant activity increased following these thermal treatments. More than 80% of processing tomatoes produced are consumed in the form of processed products such as tomato juice, paste, puree, ketchup, sauce, and salsa. In general, food processing is thought to decrease the nutritional value of staples due to the loss of certain compounds, such as vitamins [102]. Nevertheless, it has been previously reported that food processing renders lycopene more available in processed tomato products than in raw tomatoes [103]. In this regard, investigations have been generally limited to assessing the effect of heat treatment on carotenoids, especially lycopene, in tomatoes. More recently, studies have also described the effect of thermal processing or high-pressure treatment on some antioxidants in tomato juice [104,105]. However, a scant few studies have considered the joint effect of different industrial treatments commonly used in the food industry not only on carotenoids or some antioxidants, but also on other bioactive compounds present in tomatoes.

Previous studies demonstrated that lycopene is rather thermally stable in tomato-based food systems during mild thermal treatments [36, 106, 7], while more intense thermal treatments can cause lycopene degradation reactions [107,109]. In addition, several studies have investigated isomerization of all-trans-lycopene during food processing.

[106], and [110] only found limited lycopene isomerization during thermal processing of tomato products. However, the results [111] and Shi et al [9] clearly indicated the formation of cis-lycopene during the thermal treatment of tomato puree. On the other hand, it has been suggested that processing of tomatoes increases the bioavailability of lycopene [112]. However, no systematic data are available on the lycopene bioavailability in tomato as a function of processing time and temperature. Since the fraction of an ingested nutrient that can be used by the human body is of major importance, it is a real challenge to identify processing conditions resulting in maximal overall nutrient bioavailability (a balance between lycopene losses and improved bioavailability). In tomatoes, lycopene occurs as carotenoid–protein complexes or membrane bound semicrystalline structures derived from plastids [113] and needs to be transferred into micelles before it is potentially absorbable [114]. Therefore, the matrix in which lycopene is embedded in foods is an important determinant of its bioavailability and release of lycopene from the food matrix is the 1st step in the absorption process [115].

Bioavailability of Tomato Antioxidants

Since humans cannot synthesise carotenoids and must therefore depend on diet to supply these compounds and their beneficial effects, the bioavailability of carotenoids has always been of interest. The concept of bioavailability comes from a fusion of the words biological and availability [116], and its definition has evolved with time. Thus, bioavailability includes accessibility for absorption, absorption, metabolism, transport and tissue distribution, and bioactivity. However the term bioavailability is usually defined as the fraction of an oral dose of a parent compound or active metabolite that reaches the systemic circulation [117], a definition more related with the term bioaccessibility. In general, bioaccessibility can be described as the sum of digestibility and assimilation. Although bioavailability and bioaccessibility are often used indistinctly, it is important to note that definition of bioavailability includes metabolism, transport and tissue distribution, and bioactivity. However, bioactivity generally fades into the concept of bioavailability given that, pharmacologically speaking, once the active ingredient has reached its target organ, it is not inert, but rather carries out some action [112]. Digestibility refers specifically to the fraction of food components that are transformed by digestion into potentially accessible matter through all physical– chemical processes that take place in the lumen. Assimilation refers to the uptake of bioaccessible material through the epithelium by some mechanism of transepithelial absorption. In the field of carotenoids some works that measure digestibility stage refer this term as bioaccessibility while others equate measurement of assimilation with bioavailability [118, 119, 120,121]. Finally, digestibility plus assimilation are often referred to as bioavailability [122]. In many cases, the information published on the bioavailability of carotenoids is obtained by determining the quantity of carotenoids in plasma after the intake of food or supplement either in a single dose (postprandial study) or over a certain period of time (supplementation study). Thus, bioavailability is estimated once the total dose has been administered, and the quantity of carotenoids accumulated in plasma is known. Furthermore, experiments may be designed such that relevant factors that modify either digestibility or assimilation may be determined. For example, the carotenoid composition of the ingested food or the type of food matrix may be changed or other components that may interfere or complement absorption may be determined. Considerable amount of data has been obtained with this strategy [123,124]. However, digestion and assimilation involve several steps, and each one could cause an effect on the nutrient or bioactive compound so that a detailed picture is not obtained with an in vivo approach. Therefore, application of in vitro digestion and assimilation procedures is a suitable process to gain information about factors that modulate different steps of carotenoid bioaccessibility and bioavailability.

It is widely assumed that serum concentrations of carotenoids reflect, at least to some extent, the consumption of carotenoid-containing foods [125]. The influence of diet as a factor of serum carotenoid concentrations has long been known, although both dietary intake and serum concentrations of carotenoids have shown a high variability both within and between subjects in different populations [50, 128]. Seasonal variations in individual carotenoid intake, and serum concentrations, have been reported in some European countries (i.e. Spain) while not in others (i.e. UK, Republic of Ireland, Finland) [126, 127, 128, 129], even when total carotenoid intake may not vary significantly (i.e. Spain) [126]. Although fluctuations between seasons may be observed for several carotenoids both in the diet and serum levels [128, 129] in Spanish diet, these reach statistical significance only for b-cryptoxanthin (higher in winter) and lycopene (higher in summer); these changes are found to be in accordance with the availability and consumption of the major dietary contributors (i.e. citrus fruits and tomato and watermelon, respectively) [129, 130].

A European north – south gradient for the intake of some carotenoids and serum concentrations, both within and between European countries, have been reported [126]. This pattern is consistent with food availability data (i.e. fruits and vegetables) among European countries since southern (Mediterranean) countries (i.e. Greece, Italy, Portugal, Spain) consume greater amounts of fruits and vegetables than northern countries (i.e. UK, Ireland, Scandinavian countries).

In some countries, this geographical trend has been reported for both total and individual carotenoid intake and, overall, it is associated with variations in fruit and vegetables consumption (i.e. in UK, low in the North) and with socioeconomic status and cultural factors. In fact, the specific traditional and cultural factors between the two groups of populations, and in addition the changes in marketing could contribute to the change of life style. Consistently, serum levels also show this distribution trend across north – south axis. Time trends in carotenoid intake have been scarcely assessed in European countries. Nonetheless, changes in major dietary sources of carotenoids (fruits, vegetables, cereals and recently fortified foods) is known to have occurred in European countries during the last decades which is partly explained by changes in socio-economical, demographic and cultural factors. Time variation, on a short-term basis, in carotenoid intake has been assessed in Denmark, where, apparently, intake pattern of carotenoids has not changed from 1995 to 1997 [131]. Similarly, in Spain, using almost the same methodology, a fairly consistent qualitative and quantitative pattern of carotenoid intake from fresh fruits and vegetables was observed on a short-term basis, i.e. between 1996 and 2004, although this pattern was different when data were calculated on a longer time scale, i.e. 1960 – 1980 (it could be due to changes in fruit and vegetables consumption of populations) [130].

The bioavailability of carotenoids from plant foods is highly variable and is influenced by the species and structures of carotenoids present in the food, composition and release of carotenoids from the food matrix, amount consumed and absorption in the intestinal track, transportation within the lipoprotein fractions, biochemical conversions and tissue-specific depositions, as well as by the nutritional status of the ingesting host [43]. Carotenes are entirely lipophilic molecules located in the hydrophobic cores of plant membranes. Similarly, xanthophylls are largely hydrophobic molecules with their polar groups at opposite ends of a non-polar carbon skeleton [129].

Because of their lipophilic nature, biotic or abiotic activities that expose carotenoid molecules to potential oxidation, degradation or isomerization will ultimately have an influence on carotenoid biochemistry and bioavailability.

The first step in carotenoid bioavailability is release from the food matrix. Food processing activities, such as thermal processing, mincing or liquefying result in changes to carotenoid chemistry, probably through isomerization or oxidation reactions [133, 134, 135, 136, 137]. However, freezing or low-temperature storage generally preserves carotenoid concentrations by reducing potential enzymatic oxidation [137]. Processing activities usually increase bioavailability through increased release of bound carotenoids from the food matrix; however, thermal degradations in carotenoid chemistry might adversely affect bioavailability in some food crops. Absorption of carotenoids in humans is passive and follows digestive pathways similar to those of lipids. Protein- or membrane-bound carotenoids must first be released from tissues and dissolved in a hydrophobic domain (oils, fats or bulk lipid emulsions) [138]. Because of their hydrophobic nature, carotenoids in the mostly aqueous environment in plant foods must be transferred to bulk lipids or intestinal micelles in the digesta [138]. Carotenoid absorption requires the presence of dietary fat in the small intestine, which stimulates the release of emulsifying bile acids by the gall-bladder [139]. Recent studies have shown that absorption of carotenoids increases when they are ingested with dietary lipids [140,141]. After release from the food matrix, carotenoids are assimilated and oriented into lipid micelles before uptake by intestinal mucosal cells. After release to the enterocyte, carotenoids are incorporated into chylomicrons, which are eventually delivered to the blood stream and, ultimately, to the liver.

Carotenoid compounds can remain in the liver, or be transferred to low-density or high-density lipoproteins before eventual tissue-specific deposition [139]. Carotenoid bioavailability is usually assessed in blood serum after ingestion in dietary trials or cohort studies. The relatively simple analysis quantifies carotenoid changes in serum at various time intervals following ingestion of whole foods or supplements. Some caveats to interpreting serum carotenoid bioavailability include: (i) serum responses to single oral doses of carotenoids are highly variable; (ii) carotenoids measured in serum signify an equilibrium between intestinal absorption, breakdowns, tissue uptake and tissue release; and (iii) high concentrations of endogenous carotenoids (i.e. a-carotene, b-carotene, lycopene and lutein) are already present in serum [142]. Recent studies have also used in vitro Caco-2 human intestinal cell lines to assess carotenoid bioavailability [143]. In these studies, pure carotenoid compounds and whole food samples are brought through an in vitro digestion and reacted with Caco-2 human intestinal cells. Absorption potential is measured using standard HPLC carotenoid analysis. Current results demonstrate that in vitro Caco-2 cells can be used to predict carotenoid bioavailability from supplements and whole foods.

Increases in serum carotenoid concentrations typically result after ingestion of carotenoids from whole food or mono-molecular supplements [144,145]. Our own research has revealed differences in serum carotenoids and responses to macular pigment optical density (MPOD) evaluations in human subjects administered different doses of lutein from mono-molecular supplements and whole food sources. Serum carotenoid levels were higher than the baseline level after ingestion of 10 mg or 30 mg lutein supplements or spinach fortified with 8 mg or 12 mg lutein per 100 g fresh weight. Increases in MPOD from the baseline to the end of a 12 week intervention occurred in subjects administered 30 mg lutein supplements and spinach with 12 mg lutein per 100 g fresh weight (D.A. Kopsell, unpublished). In vitro and in vivo studies show that carotenoid bioavailability is influenced by source (whole food versus supplement), degree of processing, interactions with other carotenoid compounds, the degree of isomerization before, during and after absorption, transit time in the intestine, and the nutritional status of the human subjects [138].

It is important to realize that the polyphenols that are the most common in the human diet are not necessarily the most active within the body, either because they have a lower intrinsic activity or because they are poorly absorbed from the intestine, highly metabolized, or rapidly eliminated. In addition, the metabolites that are found in blood and target organs and that result from digestive or hepatic activity may differ from the native substances in terms of biological activity. Extensive knowledge of the bioavailability of tomato polyphenols is thus essential if their health effects are to be understood.

Metabolism of polyphenols occurs via a common pathway. The aglycones can be absorbed from the small intestine. However, most polyphenols are present in food in the form of esters, glycosides, or polymers that cannot be absorbed in their native form. These substances must be hydrolyzed by intestinal enzymes or by the colonic microflora before they can be absorbed. When the flora is involved, the efficiency of absorption is often reduced because the flora also degrades the aglycones that it releases and produces various simple aromatic acids in the process. During the course of absorption, polyphenols are conjugated in the small intestine and later in the liver. This process mainly includes methylation, sulfation, and glucuronidation.

This is a metabolic detoxication process common to many xenobiotics that restricts their potential toxic effects and facilitates their biliary and urinary elimination by increasing their hydrophilicity. The conjugation mechanisms are highly efficient, and aglycones are generally either absent in blood or present in low concentrations after consumption of nutritional doses. Circulating polyphenols are conjugated derivatives that are extensively bound to albumin. Polyphenols are able to penetrate tissues, particularly those in which they are metabolized, but their ability to accumulate within specific target tissues needs to be further investigated. Polyphenols and their derivatives are eliminated chiefly in urine and bile. Polyphenols are secreted via the biliary route into the duodenum, where they are subjected to the action of bacterial enzymes, especially -glucuronidase, in the distal segments of the intestine, after which they may be reabsorbed. This enterohepatic recycling may lead to a longer presence of polyphenols within the body.

REFERENCES

[1] S. Franceschi, E. Bidoli, C. La Vecchia, R. Talamini, B. D'Avanzo, E. Negri. *International Journal of Cancer. 59(2)* , 181 (1994).

[2] J. Levy, E. Bosin, B. Feldman, Y. Giat, A. Miinster, M. Danilenko, Y. Sharoni. *Nutrition and Cancer. 24,* 3 (1995).

[3] E. Giovannucci. *J Natl Cancer Inst. 91*, 317 (1999).

[4] G. Graziani, R. Pernice, S. Lanzuise, P. Vitaglione, M. Anese, V. Fogliano. *Eur Food Res Technol. 216*, 116 (2003).

[5] F. Khachik, M.B. Goli, G.R. Beecher, J. Holden, W.R. Lusby, M.D. Tenorio, M.R. Barrera. *J Agric Food Chem. 40*, 390 (1992).

[6] M.L. Nguyen, S.J Schwartz. *Proc Soc Exp Biol Med. 218*, 101 (1998).

[7] K.A. Thompson, M.R. Marshall, C.A. Sims, C.I. Wei, S.A. Sargent, J.W. Scott. *J Food Sci. 65*, 791 (2000).

[8] S.K. Sharma, M. Le Maguer. *Food Res Int. 29*, 309 (1996).

[9] J. Shi, M. Le Maguer, M. Bryan, Y. Kakuda. *J Food Process Eng. 25*, 485 (2003).

[10] J. L. Guil-Guerrero, M.M. Rebolloso-Fuentes. *J. Food Compos. Anal. 22,* 123 (2009).

[11] H.Li, Z. Deng, R. Liu, J.C.Young, H. Zhu, S. Loewen, and R. Tsao. *J. Agric. Food Chem. 59* (21), 11803 (2011).

[12] N. Onogi, M. Okuno, R. Matsushima-Nishiwaki, Y. Fukutomi, H. Moriwaki, Y. Muto & S. Kojima. *Nutr Cancer.32*, 20 (1998).

[13] K. Briviba, K. Schnabele, E. Schwertle, M. Blockhaus & G. Rechkemmer. *Biol Chem 382*, 1663 (2001).

[14] E. De Stefani, F.Oreggia, P. Boffetta, H. Deneo-Pellegrini, A. Ronco, M. Mendilaharsu. *Oral Oncology. 36,* 47 (2000).

[15] E. Giovannucci. *Journal of the National Cancer Institute. 91*, 317 (1999).

[16] F. Khachik, P.S. Garth, J. Muir, D. Zhao, N.B. Katz. *Exp Biol Med. 227*, 10845 (2002).

[17] H.D. Sesso, S. Liu, J.M. Gaziano, J.E. Buring. *J. Nutr.* 1, *133* (7), 2336 (2003).

[18] S.T. Mayne. *The FASEB Journal. 10*, 690 (1996).

[19] J. Shi, M. Le Maguer. *Critical Reviews in Food Science and Nutrition 40*, 1 (2000).

[20] A. Mortensen, L. H. Skibsted. *Free Radical Res. 27*, 229 (1997).

[21] M. L. Nguyen, & S. J. Schwartz. *Food Technology. 53*, 38 (1999).

[22] C. M. Jones, P. Mes, J. R. Myers. *J Hered. 94* (6), 449 (2003).

[23] R. Slimestad, M. J. Verheul. *J. Agric. Food Chem. 53*, 3114 (2005).

[24] J. Stewart, S. Bozonnet, W. Mullen, G. I. Jenkins, M. E. J. Lean, A. Crozier. *J. Agric. Food Chem. 48,* 2663 (2000).

[25] J. N. Davies, G. E. Hobson. *CRC Crit. ReV. Food Sci. Nutr 15*, 205 (1981).

[26] N. Rivas, B. S. Luh. *J. Food Sci 33*, 358 (1968).

[27] M. Wöldecke, K. Herrmann. *Z. Lebens. Unters. Forsch. 155*, 216 (1974).

[28] M. Wu, R. C. Burrell. *Arch. Biochem. Biophys. 74*, 114 (1958).

[29] R. Galensa, K. Herrmann. *J. Chromatogr. A 189*, 217 (1980).

[30] G. M. Hunt, E.A. Baker. *Phytochemistry 19*, 1415 (1980).

[31] M. Krause, R. Galensa. *Z. Lebesm. Unters. Forsch. 194*, 29 (1992).

[32] S. Moco, R.J. Bino, O.Vorst, H.A.Verhoeven, J.de Groot, T.A. van Beek, J.Vervoot, C. H. R de Vos. *Plant Physiol. 141*, 1205 (2006).

[33] J. Tronchet. *Ann. Sci. UniV. Besancon, 3. Ser. Botan.7*, 27 (1971).

[34] K. Herrmann. *Lebensm. Unters. Forsch. 169*, 179 (1979).

[35] F. Khachik, G.R. Beecher, M.B. Goli, W.R. Lusby. *Pure Appl Chem. 63*, 71 (1991).

[36] F. Khachik, G.R. Beecher, M.B. Goli, W.R. Lusby. In: Packer L, Ed. Methods in Enzymology. New York: Academic Press, *213*, Part A. 347 (1992).

[37] E. Giovannucci, A. Ascherio, E.B. Rimm, M.J. Stampfer, G.A. Colditz, W.C. Willett. J *Natl Cancer Inst. 87*, 1767 (1995).

[38] T. Narisawa, Y. Fukaura, M. Hasebe, M. Ito, R. Aizawa, M. Murakoshi, S. Uemura, F. Khachik, H. Nishino. *Cancer Lett. 107*, 137 (1996).

[39] J. Levy, E. Bosin, B. Feldman, Y. Giat, A. Miinster, M. Danilenko, Y. Sharoni. *Nutr Cancer. 24*, 257 (1995).

[40] F. Khachik, G.R. Beecher, J.C. Jr. Smith. *J Cellular Biochem. 22*, 236 (1995).

[41] P. Di Mascio, S. Kaiser, H. Sies. *Arch Biochem Biophys. 274*, 532 (1989).

[42] F. Khachik, A. Steck, U.A. Niggli, H. Pfander. *J Agric Food Chem. 46*, 4874 (1998).a

[43] F. Khachik, H. Pfander, B. Traber. *J Agric Food Chem. 46,* 4885 (1998).b

[44] L.H. Tonucci, J.M. Holden, G.R. Beecher, F. Khachik, C.S. Davis, G. Mulokozi. *J Agric Food Chem. 43*, 579 (1995).

[45] E. Haslam. Cambridge: Cambridge University Press. 1998.

[46] J. Li, T.A. Ou-Lee, R. Raba, R.G. Amundson, and R.L. Last. *Plant Cell. 5*, 171 (1993).

[47] R.A. Dixon, N.L. Paiva, and M.K. Bhattacharyya. In Molecular Methods in Plant Pathology, R.P. Singh and U.S. Singh, eds (Boca Raton: CRC Press), 249 (1995).

[48] G.Block, B. Patterson, & A. Subar. *Nutr. Cancer. 18*, l (1992).

[49] M.G.L. Hertog, P.C.H. Hollman, M.B. Katan, & D. Kromhout. *Nutr. Cancer. 20*, 21 (1993).

[50] R. Garcia-Closas, A. Aguda, C.A. Gonzalez, E. Riboli. *Nutr. Cancer. 32*, 154 (1998).

[51] R. Garcia-Closas, C. A. Gonzalez, A. Agudo, E. Riboli. *Cancer Causes Control. 10*, 71 (1999).

[52] P. Knekt, R. Jarvinen, A. Reunanen, J. Maatela. *Br. Med. J. 312*, 478 (1996).

[53] C.V. De Whalley, S.M. Rankin, R.S. Hoult, W. Jessup, D.S. Leake. *Biochem. Pharmacol. 39*, 1743 (1990).

[54] J.A. Vinson, J. Jang, Y.A. Dabbagh, M.M. Serry, S. Cai. *J. Agric. Food Chem. 43*, 2798 (1995).

[55] E.B. Rimm, M.B. Katan, A. Ascherio, M.J. Stampfer, W.C. Willet. *Ann. Int. Med. 125,* 384 (1996).

[56] A.Crozier, M.E.J. Lean, M.S. McDonald, C. Black. *J. Agric. Food Chem. 43*, 590 (1997). a.

[57] M.G.L. Hertog, P.C.H. Hollman, M.B. Katan. *J. Agric. Food Chem. 40*, 2379 (1992).

[58] A. Crozier, M.E.J. Lean, M.S. McDonald, C. Black. *J. Agric. Food Chem. 43*, 590 (1997). b.

[59] B.J. Willcox, J.D. Curb, B.L. Rodriguez. *Am. J. Cardiol. 101*, 75 (2008).

[60] B. Buijsse, E.J. Feskens, J. Moschandreas. *Eur. J. Cardiovasc. Prev. Rehabil. 14*, 495 (2007).

[61] J. Karppi, T. Nurmi, S. Kurl, T.H Rissanen. *Atherosclerosis. 209*, 565 (2010).

[62] O.Y. Kim, H.Y Yoe, H.J. Kim, J.Y Park. *Atherosclerosis. 208*, 581 (2010).

[63] H.Y. Yeo, O.Y. Kim, H.H Lim, J.Y. Kim. *Metabol. Clin. Exp. 60*, 537 (2011).

[64] Y. Ito, M. Kurata, K. Suzuki, N. Hamajima. *J. Epidemiol. 16,* 154 (2006).

[65] A. Tavani, S. Gallus, E. Negri, M. Parpinel. *Free Radic. Res. 40*, 659 (2006).

[66] Hozawa, D. R. Jacobs, M.W Steffes, M.D Gross. *Clin. Chem. 53*, 447 (2007).

[67] S. Lee, L. Harnack, D. R. Jacobs, L. M. Steffen. *J. Am. Diet. Assoc. 107*, 213 (2007).

[68] L. Kohlmeier, J.D. Kark, E. Gomez-Gracia, B.C. Martin, S.E. Steck, A.F. Kardinaal. *Am J Epidemiol. 146*, 618 (1997).

[69] G.J. Handelman, L. Parker, C.E. Cross. Am J Clin Nutr 63:559-65, 1996.

[70] S.K. Clinton & P. Libby. *Arch Pathol Lab Med. 116*, 1292 (1992).

[71] B. Frei. *Crit Rev Food Sci Nutr. 351*, 83 (1995).

[72] S.T. Mayne. *FASEB J. 10*, 690 (1996).

[73] Suarna, R.T. Dean, J. May, R. Stocker. *Arterioscler Thromb Vasc Biol. 15*, 1616 (1995).

[74] L. Kohlmeier, J.D. Kark, E. Gomez-Gracia. *Am J Epidemiol. 146*, 618 (1997).

[75] D.A. Street, G.W. Constock, R.M. Slkeld. *Circulation. 90*, 1154 (1994).

[76] A.V. Rao . *Exp Biol Med* (Maywood). *227*(10), 908 (2002).

[77] B. Fuhrman, A. Elis, M Aviram. *Biochemical and biophysical research communications. 233*, 658 (1997).

[78] M.L. Silaste, G. Alfthan, A. Aro, Y.A. Kesa¨niemi, S. Horkko. *British Journal of Nutrition. 98*, 1251 (2007).

[79] A. Heber and Q.Y. Lu. *Biol Med. 227*, 10920 (2002).

[80] L. Kohlmeier, J.D. Kark, E. Gomez-Gracia, B.C. Martin, S.E. Steck, A.F.M. Kardinaal, J. Ringstad, M. Thamm, V. Masaev, R. Riemersma, J.M. Martin-Moreno, J.K. Huttunen, F.J. Kok. *Am J Epidemiol. 146*, 618 (1997).

[81] T.H. Rissanen, S. Voutilainen, J.K. Virtanen, B. Venho, M. Vanharanta, J. Mursu, J.T. Salonen. *J. Nutr. 133*, 199 –204 (2003).

[82] C. La Vecchia. *Proc Soc Exp Biol Med. 218*, 125 (1997).

[83] A. Giovannucci. *J Natl Cancer Inst. 91*, 317 (1999).

[84] S. Franceschi, E. Bidoli, C. La Vecchia, R. Talamini, B. D'Avanzo, E. Negri. *Int J Cancer. 59*, 181 (1994).

[85] G.A. Colditz, L.G. Branch, R.J. Lipnick, W.C. Willett, B. Rosner, B.M. Posner. *Am J Clin Nutr. 41*, 32 (1985).

[86] A. Giovannucci, E.B. Ascherio, M.J. Rimm, G.A. Stampfer, W. Colditz, C.Willett. *J Natl Cancer Inst. 87*, 1767 (1995).

[87] J.F. Dorgan, A. Sowell, C.A. Swanson, N. Potischman, R. Miller, N. Schussler. *Cancer Causes Control. 9*, 89 (1998).

[88] P. Gann, J. Ma, E. Giovannucci, W. Willett, F.M. Sacks, C.H. Henneken. *Cancer Res. 59*, 1225 (1999).

[89] S. Agarwal, A.V. Rao. *Lipids. 33*, 981 (1998).

[90] B.L. Pool-Zobel, A. Bub, H. Muller, I. Wollowski, G. Rechkemmer. *Carcinogenesis, 18*, 1847 (1997).

[91] O. Kucuk, F.H. Sakr, Z. Djuric, Y.W. Li, F. Velazquez, M. Banerjee. International Conference on Diet and Prevention of Cancer Tampere, Finland, 1999.

[92] C. La Vecchia. *Experimental Biology and Medicine. 227*, 860 (2002).

[93] P. Bowen, L. Chen, M. Stacewicz-Sapuntzakis, C. Duncan, R. Sharifi, L. Ghosh. Exp *Biol Med. 227*(10), 886 (2002).
[94] O. Kucuk, F.H. Sarkar, Z. Djuric, W. Sakr, M.N. Pollak, F. Khachik. *Exp Biol Med. 227*(10), 881 (2002).
[95] T. Sugimura, K. Wakabayashi, H. Nakagama, M. Nagao. *Cancer Sci. 95*, 290 (2004).
[96] K.H. van het Hof, B.C.J. de Boer, L.B.M. Tijburg, B.R.H.M. Lucius, I. Zijp, C.E. West, J.G.A.J. Hautvast, J.A. Weststrate. *J Nutr. 130*, 1189 (2000).
[97] W. Stahl, H. Sies. *J Nutr. 122*, 2161 (1992).
[98] A.A. Abushita, H. G. Daood, and P. A. Biacs. *J. Agric. Food Chem. 48*, 2075 (2000).
[99] D.B. Rodriguez-Amaya. USAID, OMNI Project, 1997.
[100] E.J. Johnson, B.R. Hammond, K.J. Yeum, J. Qin, X .Dong Wang, C. Castaneda, D.M. Snodderly and Robert M Russell. *American Journal of Clinical Nutrition. 71*(6), 1555 (2000).
[101] R.Y Chen, , J.J. Wu, M.J. Tsai, M.S. Liu. *Taiwanese Journal of Agricultural Chemistry and Food Sciences. 38*, 353 (2000).
[102] Y. Klopotek, K. Otto, and V. Bohm. *J. Agric. Food Chem. 53*, 5640 (2005).
[103] C. Gartner, W. Stahl and H. Sies. *Am J Clin Nutr. 66*(1), 116 (1997).
[104] S. Gahler, K. Otto, & V. Bohm . *Journal of Agricultural and Food Chemistry. 51*, 7962 (2003).
[105] K.C. Hsu. *LWT. 41*, 450 (2008).
[106] M.L. Nguyen, S.J. Schwartz. *Proc Soc Exp Biol Med. 218*, 101 (1998).
[107] S.K. Sharma, & M. Le Maguer. *Food Research International. 29*, 309 (1996).
[108] G.R Takeoka, L. Dao, S. Flessa, D.M. Gillespie, W.T. Jewell, B. Huebner, D. Bertow, S.E. Ebeler. *J Agric Food Chem. 49*, 3713 (2001).
[109] Graziani, R. Pernice, S. Lanzuise, P. Vitaglione, M. Anese, V. Fogliano. *Eur Food Res Technol. 52,* 7005 (2004).
[110] C. Seybold, K. Frohlich, R. Bitsch, K. Otto, & V. Bohm. *Journal of Agriculture and Food Chemistry. 52*, 7005 (2004).
[111] V. Dewanto, X. Wu, K. K. Adom, & R. H. Liu. *Journal of Agriculture and Food Chemistry. 50*, 3010 (2002).
[112] W. Stahl, & H. Sies. *J. Nutr. 121*, 2161 (1992).
[113] J. Shi, & M. Le Maguer. *Food Science and Nutrition. 40*(1), 1 (2000).
[114] R. M., Faulks, & S. Southon. *Biochimica et Biophysica Acta. 1740*, 95 (2005).
[115] T.W.M. Boileau, A.C. Boileau, J.W. Erdman. *Biol Med. 227*, 914 (2002).
[116] C. M. Metzler and D. C. Huang. *Clinical Research and Regulatory Affairs. 1*(2), 109 (1983).
[117] K. Schumann, H.G. Classen, M. Hages, R.P. Langenhol, K. Pietrzik & H.K. Biesalski. *Arzneimittelforschung. 47*, 369 (1997).
[118] F. Granado-Lorencio, B. Olmedilla-Alonso, C. Herrero-Barbudo, I. Blanco-Navarro, B. Perez-Sacristan, & S. Blazquez-Garcıa. *Food Chemistry. 102*, 641 (2007).
[119] E. Hedrén, V. Diaz, & U. Svanberg. *European Journal of Clinical Nutrition. 56*(5), 425 (2002).
[120] D. Hornero-Méndez, & M. I. Mínguez-Mosquera. *Innovative Food Science and Emerging Technologies. 8*(3), 407 (2007).
[121] L. O'Sullivan, M.A. Jiwan, T. Daly, N.M. O'Brien and S. Aisling Aherne. *J. Agric. Food Chem. 58* (9), 5374 (2010).

[122] C. Chitchumroonchokchai, M. L. Failla. *J. Nutr. 136*, 588 (2006).
[123] J.J.M. Castenmiller& C. E. West. *Annu. Rev. Nutr. 18*, 19 (1998).
[124] R.M. Clark, L. Yao, L. She and H.C. Furr. *Lipids. 35*(7), 803 (1997).
[125] A. Brevik, L.F. Andersen, A. Karlsen, K.U. Trygg. *Eur. J. Clin. Nutr. 58*, 1166 (2004).
[126] M.E. O'Neill, Y. Carroll, B. Corridan, B. Olmedilla. A *Br. J. Nutr. 85,* 499 (2001).
[127] K.J. Scott, D.I. Thurnham, D. Hart. *Br. J. Nutr. 75*, 409 (1996).
[128] B. Olmedilla, F. Granado, I. Blanco, E. Rojas-Hidalgo. *Am. J. Clin. Nutr. 60*, 106 (1994).
[129] F. Granado, B.Olmedilla, I. Blanco, E. Rojas-Hidalgo. *Eur J. Clin. Nut. 50,* 246 (1996).
[130] F. Granado,., S. Bl_zquez, B. Olmedilla. *Public Health Nutr. 19*, 1 (2007).
[131] T. Leth, J. Jakobsen, N.L Andersen. *Eur. J. Lipid Sci. Technol. 102*, 128 (2000).
[132] F. Granado, B. Olmedilla, I. Blanco, E. Gil-Martinez, E. Rojas-Hidalgo. *Crit. Rev. Food Sci. Nutr. 37*, 621 (1997).
[133] A.A. Updike, and S.J. Schwartz. *J. Agric. Food Chem. 51*, 6184 (2003).
[134] O. Livny. *Eur. J. Nutr. 42*, 338 (2003).
[135] J.J.M. Castenmiller. *J. Nutr. 129*, 349 (1999).
[136] L.M. Kopas-Lane, and J.J. Warthesen. *J. Food Sci. 60*, 773 (1995).
[137] D.B. Rodriguez-Amaya. *Arch. Latinoam. Nutr. 49*, 38 (1999).
[138] R.M. Faulks, and S. Southon. *Biochim. Biophys. Acta. 1740*, 95 (2005).
[139] S. Zaripheh and J.W. Erdman. *J. Nutr. 132*, 531 (2002).
[140] N.Z. Unlu. *J. Nutr. 135*, 431 (2005).
[141] M.J. Brown. *Am. J. Clin. Nutr. 80*, 396 (2004).
[142] K-J. Yeum, and R.M. Russell. *Annu. Rev. Nutr. 22*, 483 (2002).
[143] C. Chitchumroonchokchai, S.J. Schwartz and Mark L. Failla. *J. Nutr. 134* (9), 2280 (2004).
[144] K.J. Yeum, and R.M. Russell. *Annu. Rev. Nutr. 22*, 483 (2002).
[145] A.V. Rao, and H. Shen. *Nutr. Res. 22*, 1125 (2002).

In: Tomatoes: Cultivation, Varieties and Nutrition
Editor: Tadahisa Higashide

ISBN: 978-1-62417-915-0

Chapter 15

THE PROTECTIVE EFFECT OF TOMATO JUICE ON THE RESPIRATORY SYSTEM

G. S. Maritz and M. Mutemwa
Department of Medical Biosciences, University of the Western Cape, Bellville, South Africa

ABSTRACT

Tomato fruits are rich in various nutrients such as vitamins and minerals. It is also a source of fibre. Tomato fruits are a rich source of anti-oxidants and phytonutrients, such as vitamin C, lycopene, carotenes, vitamin E, polyphenols and anthocyanin. The anti-oxidants present in dietary tomato products, such as tomato juice, are associated with improved human health. There is also evidence of protection against prostate cancer. It was also demonstrated that tomato juice protects the adult human against the damage induced by tobacco smoke. Maternal nicotine intake during pregnancy and lactation prevents the programming of the lungs against premature aging and the development of emphysema later in life. Studies showed that if the mother's diet is supplemented with tomato juice, it prevents the decrease in the anti-oxidant capacity of the pregnant and lactating mother as well as that of the fetus and neonate and thus prevents the nicotine-induced programming. Consequently the protection of the baby by the mother is improved. This ensures that the respiratory system of the baby develop normal even when exposed to high levels of oxidants and nicotine via the placenta and breast milk. As a consequence the susceptibility of the individual to develop respiratory disease after birth and later in life will be reduced compared to those not exposed to the anti-oxidants of tomato fruits and thus not protected against oxidants that are inhaled or present in the food that the mother ate.

INTRODUCTION

The tomato fruit is consumed in many ways, including raw, as an ingredient in various dishes and sauces, as tomato juice, and in drinks. Although it is botanically a fruit, it is considered by most as a vegetable. It is rich in various phytonutrients, of which lycopene is an important one. The latter is a potent anti-oxidant. Tomatoes may therefore be beneficial to our health, especially since we are exposed to many different oxidants in the air, food, and drinks. This is of particular importance for smokers who inhale about 10^{15} oxidants per puff [1].

Tomato fruits are eaten freely throughout the world, and their consumption is believed to benefit various organs such as the heart and the respiratory system. An example is that a short-term treatment with antioxidant-rich tomato extract can slow the progression of atherosclerosis and reduce blood pressure [2]. Some studies also suggest that it can prevent certain cancers [3], because of its anti-oxidant properties. It has indeed been shown that lycopene may inhibit the growth of several cultured lung cancer cells and prevent lung tumorigenesis in animal models through various mechanisms, such as modulation of redox status, cell cycle arrest and/or apoptosis induction, a regulation of growth factor signalling, changes in cell growth-related enzymes, an enhancement of gap junction communication as well as a prevention of smoke-induced inflammation. Furthermore, lycopene also inhibits cell invasion, angiogenesis, and metastasis [4]. It certainly assume the status of a health food if we take into consideration the large body of epidemiological evidence illustrating its role in reducing risk for oxidant associated diseases. It is a reservoir of several different antioxidant molecules, such as ascorbic acid, vitamin E, carotenoids, flavonoids and phenolic acids. The disease-preventing potential of a food, such as tomatoes, is a consequence of a several such constituents which may show some synergistic interactions [5]. This implies that giving only one micronutrient to an individual may not be as effective on its own than when given together with other micronutrients in the correct ratios. It has in fact been shown that pure lycopene is not as effective as an anti-oxidant when used on its own, than when it is in the presence of other phytonutrients such as in tomatoes [6].

Apart from being an important source of anti-oxidants such as lycopene, carotene, anthocyanin, vitamin C and others, it is also a source of fiber which has been shown to reduce cholesterol levels and also play a role in controlling blood glucose levels. A cup of fresh tomatoes provide about 57% of the daily recommended value of vitamin C and about 22% of the daily recommended amount of vitamin A. In addition, the anti-oxidant capacity of a recently discovered phenolic substance that is synthesised by the tomato plant, has the antioxidant power that is 14 times higher than, for example that of resveratrol, a well-known antioxidant, found in red wine, which can delay cellular aging. In addition, it is 4.5 times more potent than vitamin E and 10 times more potent than vitamin C [7]. It is therefore plausible that including tomatoes in the diet will prevent premature aging of cells in the body, such as the lungs, that are exposed to relatively high levels of oxidants. Tomatoes thus appear to be highly protective of the respiratory system that is continuously exposed to relatively high levels of oxygen as well as oxidants in the atmosphere.

Tomatoes also contain significant amounts of niacin, vitamin B6, vitamin K, choline and folate which are associated with reducing the risk of heart disease. Furthermore, tomatoes also have an anti-inflammatory function in that it reduces tumour necrosis factor alpha (TNF-alpha) levels significantly. The primary role of TNF is in the regulation of immune cells. It is also able to induce fever and to induce apoptotic cell death. Since inflammatory markers such as TNF-alpha show a correlation with nearly all degenerative diseases, such as atherosclerosis, cardiovascular diseases, several cancers, osteoporosis and Alzheimer's disease, it is evident that tomatoes are playing an important role in maintenance of health and longevity of individuals.

The protective effect of tomato juice is particularly illustrated by the powerful mix of anti-oxidants, where a glass of tomato juice contains 5.7 mg lycopene, 1 mg beta-carotene, 3.7 mg phytoene, 2.7 mg phytofluene, and 1.8 mg of the alpha-tocopherol fraction of vitamin E. Lycopene has the highest antioxidant activity among all dietary antioxidants. In addition to these properties, lycopene has also been shown to induce cell-to-cell communications and modulate hormones, immune systems and other metabolic pathways [8]. Tomatoes also contain moderate amounts of α- and β-carotene and lutein. β-carotene is known for its pro-vitamin A activity and lutein for reduced risk of lung cancer [9].

Apart from its role in protecting the adult against various oxidant induced degenerative diseases, tomatoes are also important in protecting the unborn baby against the harmful effects of oxidants that enters the fetal circulation from the mother. By virtue of its role in cell-to-cell communication lycopene is conceivably also participating in organ development. Due to the presence of all these micronutrients, tomatoes play an important role in the maintenance of homeostasis in the body. Because of the wide variety of physiological systems that its contents are involved with, tomatoes are certainly not only a healthy component of the diet of the adult, but also in the developing fetus and the growing child by protecting it against oxidants and support systems involved in growth. Thus, the diet of the pregnant and lactating female must also contain amongst others, tomatoes because of its rich supply of micronutrients that are required for growth, development and maintenance of homeostasis before and after birth.

BIO-AVAILABILITY OF PHYTONUTRIENTS FROM TOMATOES

Tomato fruits are a valuable source of various micronutrients and phytochemicals such as the carotenoids, polyphenols, potassium, folate, α-tocopherol, as well as vitamin C [10]. Most of the micronutrients and phytochemicals interact with the host to prevent oxidative stress in humans [11]. Studies consistently associate diets rich in fruits and vegetables, such as tomato fruits, with a lower incidence of several diseases [12]. For these nutrients to be playing a role in maintaining health in humans it must be readily absorbed from tomato products by the gastro-intestinal (GI) tract and transferred into the tissue where they perform their role. In common foods, carotenoids are mostly found as all-trans isomers [13], but significant amounts of cis isomers can be produced during processing thereof [14, 15]. The bio-availabilities of the cis isomers of β-carotene and lycopene are different from those of their corresponding trans isomers [16-19] and may have different but specific functions in the body [20-23]. The efficiency with which all-trans β–carotene is absorbed is generally poor but varies alot, ranging between 3.5 and 90%, depending on the dose, and the matrix in which it is incorporated [24, 25].

Three assumptions are currently made about carotenoid metabolism in the lumen of the human upper GI tract. First, carotenoids cannot be absorbed while they remain embedded in their original vegetable matrices. Secondly, since carotenoids are fat soluble it has to be solubilized in the micelles to be absorbed. Thirdly, carotenoids are absorbed by passive diffusion. Research on the fate of carotenoids in the human upper GI tract showed that; 1) there is no significant cis-trans isomerization of β-carotene and lycopene in the human stomach, 2) the stomach plays a significant role in the bioavailability of carotenoids by initiating their transfer from the vegetable matrix to the fat or micellar, phase of the meal, 3) the proportion of carotenoids recovered in the micellar phase of the duodenum is very low (7%), which probably explains the poor availability of these phytomicronutrients, 4) 13-cis-carotene is more fully solubilized in micelles than the all-trans isomer, explaining its higher bioavailability, and 5) cis isomers of lycopene are sparingly secreted in the chylomicrons, suggesting a post-enterocyte origin for these isomers in the body. Various other factors can also affect the bio-availability of the carotenoids, namely, the amount taken in, interaction with other compounds in the food, food processing, and the nature of the food matrix [26]. It has indeed been shown that, for example, the bio-availability of lycopene, the predominant carotenoid in tomatoes, is greater from tomato paste than from fresh tomatoes. Ingestion of tomato juice cooked in an oil medium result in a two- to 3-fold higher uptake of carotenoids by the gastro-intestinal tract [27].

Despite the various factors affecting the bioavailability of the micronutrients, more than 30 carotenoids, including lycopene and its isomers occur in mother's milk. Nearly all these components are derived from the blood. Consumption of a "standard portion" of tomato products increases the plasma and milk lycopene concentrations in lactating women, and in this way increase the lycopene status of the nursing infant [28]. This means that absorption of the micronutrients from the gastro-intestinal tract and its transfer to the circulation is such that it will supply in the needs of the mother and developing baby in the womb, provided that sufficient quantities occur in the diet.

Distribution of Carotenoids

Processed and fresh tomatoes are the main sources of lycopene. Lycopene supplementation increases the lycopene content of the plasma and all tissues. In studies with rats and ferrets it was shown that lycopene from tomato is primarily stored in the liver compared to lung and kidney [29, 30]. Supplementation with β-carotene also results in a marked increase in the levels in the serum of rats [31]. It has also been shown that peak lycopene concentrations in the blood is reached 24 hours after consumption of heated tomato juice [19].

The six carotenoids found in the highest concentrations in human serum are β-carotene, α-carotene, β-cryptoxanthin, lycopene, lutein, and zeaxanthin [32]. Carotenoid content in tissue samples from infants are lower ranging from 0 to 1.0 nmol/g tissue. Beta-carotene and lycopene were almost always the predominant carotenoids found in liver, kidney and lung tissue [30]. It therefore follows that the diet of the infant must have adequate amounts of all the phytonutrients to ensure that adequate quantities are present for maintenance of homeostasis and to support growth and development.

Are We What Our Mothers Ate?

New research adds to existing evidence that adult health is set to a significant degree by conditions in the womb and suggests the programming may start earlier in pregnancy than previously believed. This means that the mother's diet may influence the future health of the baby because pregnancy is a time of constant change as the fetus develops in the mother's womb. During this phase of rapid change the fetus is very plastic and therefore very sensitive to changes in the *in utero* environment, and after birth, the external environment [33]. The lifestyle of the mother often determines the changes in the environment within which the fetus and new born baby must develop. If the mother smokes, or use alcohol, or her dietary intake is inadequate to maintain her body composition, and to support growth of the baby, it will impact negatively on the developing fetus.

On the other hand, as the nutritional state of many human mothers improved around the world, the characteristics of their offspring, such as body size and metabolism, have also changed. Responsiveness to their mothers' condition before birth may generally prepare the offspring so that they are best prepared for the environment forecasted by cues available in early life. On the other hand, rapid improvements in nutrition and other environmental conditions may have damaging effects on the health of those people whose parents and grandparents lived in impoverished conditions [33]. This implies that the health of the offspring after birth and in adulthood is predetermined by the nutritional status of the mother.

Therefore, nutrition in the womb plays an important role in the development of the baby. Nutritional conditions that exists in the womb while the baby is developing in the mother's womb as well as nutrition in infancy, affects a person's health in the future. This appears to be especially true for heart disease and diabetes mellitus [34] as well as the respiratory system [35]. There are critical windows of time during fetal life and during infancy during which nutritional programming occurs. This means that changes in the environment during these critical windows of time may program the physiology and metabolism of the offspring to render them more prone to diseases in adulthood. The new-born's health, as well as its health in the longer term is determined by the mother's lifestyle.

One of the important sources of nutrients for the fetus is the mother's body during the normal turnover process in her body. The turnover is the continuous process of tissue breakdown and renewal of muscle tissue, fat, and bone during which proteins, fats and calcium and other minerals are released into the blood of the mother. During pregnancy these nutrients that are released into the mother's blood serves as a source of food for the developing fetus. A mother acquires her body composition and turnover throughout her lifetime as fetus, new-born, adolescent and up to adulthood. The mother's body turnover and her diet work in harmony to provide nutrition in the womb through the placenta. It is therefore essential that her nutrient intake is adequate to meet the demands of the growing baby as it plays an important role in ensuring that the baby's metabolism and body functions develop normally [36]. By maintaining a healthy lifestyle the mother will contribute to the maintenance of the body composition of not only her children, but also of her grand-children and in this way contribute to the longer term health of her descendants.

It is therefore clear that good nutrition starts before conception. Poor maternal nutrition leads to alterations in oxidative stress and antioxidant defence capacity of the mother [37]. This implies that the mother will be unable to protect the unborn baby during especially those phases of high plasticity during which they are very sensitive to changes in the environment within which they develops. Thus the lifestyle of the mother, even before conception, is important to ensure that the nutrient intake is adequate to create the environment within which the baby can develop without interference during those phases during which it is sensitive for changes in the environment. This is especially important in that the growing baby in the womb is also dependent on the turnover of the body of the mother for a continuous supply of nutrients [38, 39]. If the supply is inadequate due to lack of nutrients from the mother's body as a consequence of poor eating habits, it will be affecting the development of the baby and its body composition [40].

Apart from the diet, the mother is also exposed to foreign substances such as oxidants in the atmospheric air, food and drink, which may adversely impact on the anti-oxidant capacity of her body and thus also her body composition. In smokers the exposure to oxidants is particularly high [1]. In addition, nicotine, the addictive substance in tobacco, also induces formation of oxidants in the body. These oxidants reduce the anti-oxidant capacity of the mother and in this way her capacity to protect the fetus during those phases of development that they are particularly sensitive to the effect of substances such as nicotine and oxidants. Since the developing fetus is unable to protect itself against these substances it tends to accumulate in its body and in this way impact on fetal development. Therefore, it is important to limit exposure to oxidants, or substances with oxidant properties, or that can induce excessive oxidant release in the body of the fetus. It is also essential to ensure that the mother's diet is adequate to protect not only her health but also that of the fetus and new-born as well as to prevent programming of the offspring since programming may give rise to development of diseases later in life.

Therefore, maternal diet and body composition even before conception plays an important role in the development of the fetus. It is essential to ensure that her diet is adequate and balanced to protect the fetus and to ensure proper nutrient supply to the fetus. Indeed, emerging evidence now suggests that use of micronutrient-containing prenatal vitamins before and during pregnancy is associated with reductions in the risk of congenital defects, preterm delivery, low infant birth-weight, and preeclampsia [41]. It has also been shown that higher maternal total intakes of antioxidants during pregnancy may decrease the risks for wheezing illnesses in early childhood [42, 43]. In this respect tomatoes in the diet can play a major role since it is a rich source of potent oxidants and other phytonutrients which act synergistically to enhance the protective capacity thereof. This is of particular importance for organs that are exposed to relative high levels of oxidants.

Nutrition and the Lung: Oxidants, Antioxidants and Respiratory Health

Adult Lung

The human lung is particularly vulnerable to oxidative damage because of the large volume of oxygen that passes through the lungs per day. The lungs operate under a higher oxygen pressure than any of the other organs. This can result in the conversion of high concentrations of anti-oxidants into pro-oxidants [44]. In addition, the lungs are exposed to the external environment and as such it is also exposed to all foreign substances that we inhale in polluted air, and which is made worse by inhalation of tobacco smoke, whether it is first hand, second hand or even third hand. Smoking in particular affects respiratory health of all that are exposed to it because of the high concentration of inhaled particles, oxidants, carcinogens and other foreign substances. Due to the very high level of oxidants in tobacco smoke, as well as the ability of nicotine to induce oxidant formation in the tissue, the antioxidant/oxidant capacity of the lungs becomes seriously compromised. Oxidative stress results from an oxidant/anti-oxidant imbalance, which may result from an excess of oxidants and/or a depletion of anti-oxidants.

The respiratory tract lining fluid forms an interface between the epithelial cells and the external environment, and, therefore constitutes a first line of defence against inhaled oxidants [1]. The presence of dietary anti-oxidants in the lung alveolar epithelial lining and alveolar lining fluid, therefore, provide protection against oxidative damage. The vitamin C content of this fluid lining is very high [45], but is seriously compromised by smoking [46-48]. The vitamin C in the lining fluid, and the vitamin E in the cells of the alveolar wall, work together in protecting the alveolar wall to prevent damage to the alveolar walls [49] and thus the gradual formation of emphysema-like lesions. Vitamin E is believed to be, by its sheer mass, the first line of defence against damage of the alveolar epithelial membranes. In addition to vitamin C and vitamin E, both lycopene and β-carotene have been measured in lung tissue [44]. This implies that the carotenoids also play a role in the defence of the lung against oxidants. Thus, compromising the anti-oxidant capacity of this this first line of defence against inhaled oxidants not only compromise the mothers' ability to protect herself against the inhaled oxidants, but also expose her unborn baby to much higher levels of oxidants which may then induce programming in the offspring and render them more susceptible to diseases later in life.

Vitamin C is beneficial to the lungs, but when depleted during aging, it may contribute to COPD. When vitamin C levels in the lung drops it may also compromise the synergistic effects between vitamins C and E with the consequent lowering of the protective effect of both vitamins. It is therefore, not surprising that a lack of vitamin C cause emphysema due to oxidative stress; and, in addition, a decrease in the collagen synthesis [50]. An inadequate intake of fruit and vegetables rich in all the anti-oxidants may result in an imbalance in the oxidant/anti-oxidant status of the lungs, as well as an imbalance between those vitamins that act synergistically to protect the lungs and to maintain homeostasis. The latter may enhance the gradual breakdown of the connective tissue framework of the aging lung and the development of the so-called senile lung. A lack of vitamin C and the other anti-oxidants in the diet may therefore induce premature aging of the lungs and thus with the associated diseases such as gradual breakdown of the lung parenchyma. This goes hand in hand with a reduced area for gas-exchange and thus compromised lung function. Ingestion of vitamin supplements instead of fruit and vegetables, such as tomatoes, may also contribute to an imbalance between vitamins that act synergistically.

Furthermore, oxidative stress is thought to play an important role in the pathogenesis of a number of lung diseases, not only through direct injurious effects, but by involvement in the molecular mechanisms that control lung inflammation. The presence of oxidative stress contributes to the development of chronic obstructive pulmonary disease (COPD) by, for example, oxidative inactivation of anti-proteinases and airspace epithelial injury. Chronic obstructive pulmonary disease (COPD) is one of the most common lung diseases. It makes it difficult to breathe. There are two main forms of COPD, namely chronic bronchitis, which involves a long-term cough with mucus, and, emphysema, which involves destruction of the lungs over time. Oxidative stress enhances the inflammation that occurs in smokers and patients with COPD [30], and thereby promote gradual tissue breakdown and development of emphysema [35].

Many researchers has investigated the potential role of anti-oxidant nutrients in prevention of chronic diseases and aging processes in humans [51, 52]. Lycopene is a major carotenoid in tomatoes [53] and is considered to be the most efficient biological carotenoid in quenching singlet oxygen [5]. Since tobacco smoke contains more than 10^{15} oxidant molecules/puff [54], it is conceivable that an inadequate intake of dietary anti-oxidants may render the lungs more susceptible to the development of tobacco smoke-induced emphysema. Furthermore, it was shown that smoke exposure indeed decrease plasma and lung lycopene concentrations by 40% and 90% respectively [55]. Other vitamins with anti-oxidant capacity such as vitamins C and E as well as the carotenoids are also lower after smoke exposure [56] and thereby compromise the ability of the lungs to protect itself against oxidant stress. However, in a study by Kasagi et al, [57] they were able to restore the oxidant/anti-oxidant balance in the lungs of smoke exposed rats by giving them tomato juice. As a consequence the lungs were protected from developing tobacco smoke induced emphysema. They suggested that the lycopene played an important role in this respect. The other anti-oxidants in tomatoes, for example vitamins C and E in all likelihood also contribute in the protection of the lungs against the harmful effects of smoking on lung structure and function.

Experimental and clinical evidences suggest that oxidants play a role in the pathogenesis also of other respiratory disorders than emphysema. One such disorder characterised by chronic airway inflammation, is asthma. The respiratory system is chronically exposed to environmental pollutants and oxidants [58]. Research shows increased oxidative stress in asthma and indicate a potential role for oxidants in the pathogenesis of the disease [59]. Antioxidants may, therefore, have a potential role in the treatment of asthma. Oxidative stress may also reduce glutathione levels and cause inactivation of antioxidant enzymes such as superoxide dismutase, with a consequent increase in apoptosis, shedding of airway epithelial cells and airway remodelling. The oxidant/antioxidant equilibrium in asthmatic patients may be further perturbed by low dietary intakes of the antioxidant vitamins C and E, selenium and flavonoids, with a consequent lowering of the concentrations of these and other non-dietary antioxidants such as bilirubin and albumin in plasma and airway epithelial lining fluid [60].

Fetal and Neonatal Lung

The development of the fetal and neonatal lung depends on proper nutrient intake by the mother. It is well established that on-going malnutrition in infants with chronic lung disease is associated with growth failure and delayed lung repair. On-going malnutrition will also affect the supply of vitamins and minerals that are required for optimal lung growth and development. The growth failure and delayed lung repair is conceivably due to a lack of, for example vitamins A and E, where vitamin A is essential for growth and differentiation of epithelial cells in the lung; and vitamin E, for protection of alveolar epithelial cell membranes from oxidant injury. The vitamin E content of the fetus increases with the fetal fat mass mainly during the last trimester of pregnancy. The placenta is only slightly permeable to the lipid soluble vitamins. Consequently a vitamin E deficiency may rapidly develop typically at about 6-8 weeks of age if the mother's diet is not adequate [49].

All the stages of lung development in utero and after birth requires vitamin A. It has been shown that maternal repletion of vitamin A after undernourishment at dietary recommended levels, before, during, and after pregnancy, improve lung function in the offspring. If this type of intervention is not happening it may have an adverse effect on lung structure and function from early life to adulthood [61].

Since tomatoes contain all these vitamins and other antioxidants such as lycopene, it is, in the juice form, ideal to supply these nutrients to ensure normal lung growth and development. It has indeed been shown that greater consumption of lycopene and tomatoes decreases the risk of oxidant damage in the lungs [44].

Vitamin C should be included in the diet of the mother to ensure that it will support not only the protection mechanisms of the lung against oxidants, but also normal growth and development. This is of particular importance since the human body cannot synthesize vitamin C [50]. Tomatoes contain a mix of anti-oxidants which include vitamin C. Thus by including tomatoes in the diet will contribute to prevent oxidant stress and to support growth and maintenance of the connective tissue framework of the lung.

Studies by Sekhon et al, [62] and Maritz and colleagues [63-65] showed that the lungs of rats that were exposed to nicotine via the placenta and mother's milk are unable to maintain the structure of the lung parenchyma in the longer term resulting in emphysematous lesions and the associated loss of lung function [35]. Based on his findings Maritz [35, 66] suggested that the factors, such as growth factors, that controls lung aging and tissue maintenance was compromised by nicotine, where nicotine itself is genotoxic [67] and have pro-oxidant properties [68]. Furthermore, maternal nicotine exposure also induces oxidant stress in lactating rat offspring which is maintained into puberty, especially in male rats [69-71]. In addition, it also lowers the antioxidant capacity of the lungs of the offspring [48, 72] and therefore the capacity to protect itself against the adverse effects of oxidants. It is plausible that supplementing the mother's diet with a fruit and/or fruit/vegetable juice rich in anti-oxidants will prevent the adverse effects of maternal nicotine exposure during gestation and lactation on the maintenance of the lung parenchyma of the offspring in the long term.

Although certain small molecules might reduce the risk of pulmonary damage induced by oxidants in prematurely born infants [73], clinical trials with single small molecules like vitamin E, or vitamin C, or β-carotene, as well as laboratory experiments have largely been disappointing [66, 74, 75]. In a review by Fardy and Silverman [76] they showed that an imbalance between anti-oxidants can also induce tissue damage. This means that under certain circumstances supplementing single anti-oxidants could be detrimental. It has been shown that anti-oxidant defences are multifaceted and that supplementation with an individual small molecule, unless deficient, will have little effect on the incidence of chronic diseases. On the other hand, a mixture of small molecules, such as those found in fruits and vegetables may provide significant protection [77]. Since tomato products contains all these anti-oxidants such as carotene, anthocyanin, lycopene and vitamin C in relatively high concentrations [27, 78], it is ideal to maintain the anti-oxidant status of the mother and child and in this way ensure the development of a healthy respiratory system in the offspring. It may also prevent programming of the lungs of the offspring during development to develop respiratory diseases later in life.

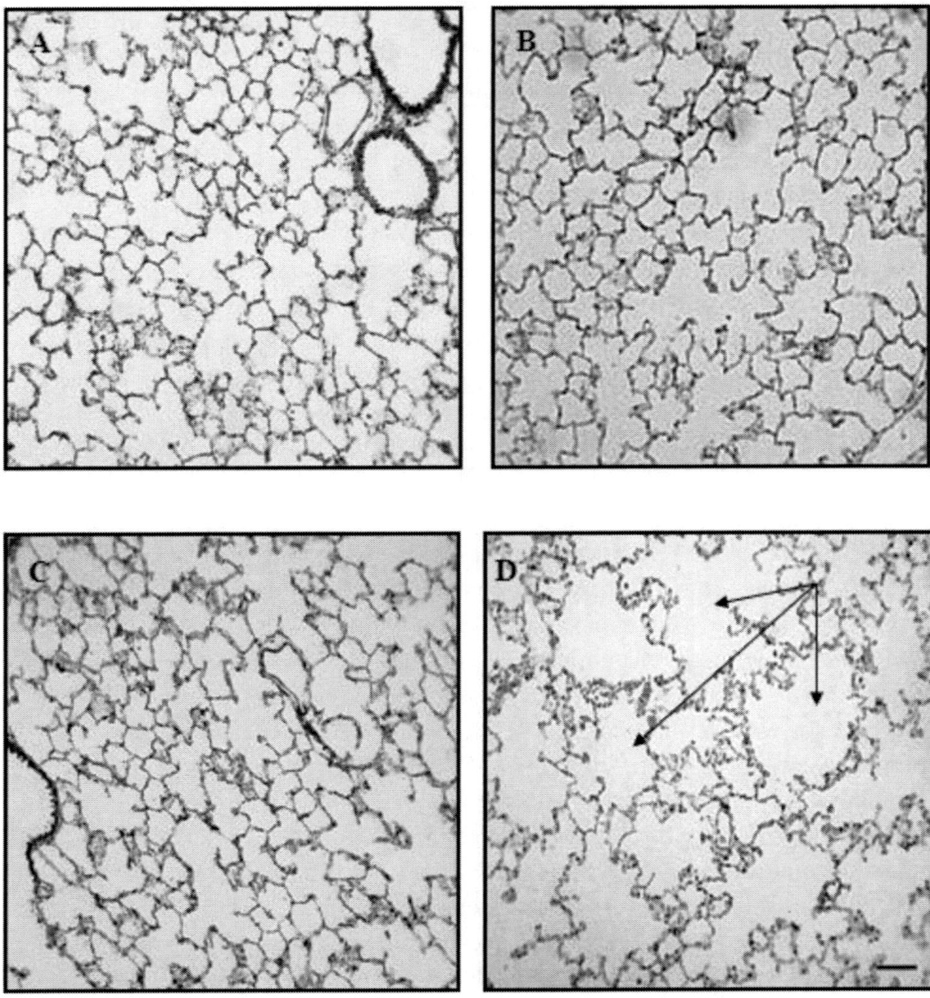

Figure 1. The influence of maternal exposure to nicotine, tomato juice and a combination of nicotine and tomato juice, on lung parenchyma of the 84-day-old offspring. A. = Control; B.= Tomato juice; C. = Nicotine + tomato juice; D. = Nicotine. Arrows = Emphysema. (Bar = 0.5μm).

This suggestion is supported by studies in which we demonstrated that supplementing the diets of female rats during gestation and lactation with tomato juice prevented all the adverse effects of maternal nicotine exposure on the lungs of the offspring [35]. In this study tomato juice supplementation (Figure 1) took place only during gestation and lactation; that is during the phases of lung development during which cell turnover was rapid. It was also when the animals were exposed to nicotine via the placenta and mother's milk and thus when the oxidant release under influence of nicotine [79] was conceivably the highest. The fact that all the adverse effects of maternal nicotine exposure during gestation and lactation was prevented by drinking tomato juice means that tomato juice, and presumably the lycopene and other phytonutrients in tomato juice, protected the lungs during the sensitive periods of lung growth and development. Human studies showed that lycopene prevent DNA damage [80]and cell membrane oxidation in vivo [81]. It is thus plausible that the lycopene in tomato juice improved the protection systems of the mother and fetus to such an extent that it such prevented the adverse effects of maternal nicotine intake on the lungs of the offspring, as programming the lungs of the offspring to age prematurely and the associated decline in lung structure and function in adulthood.

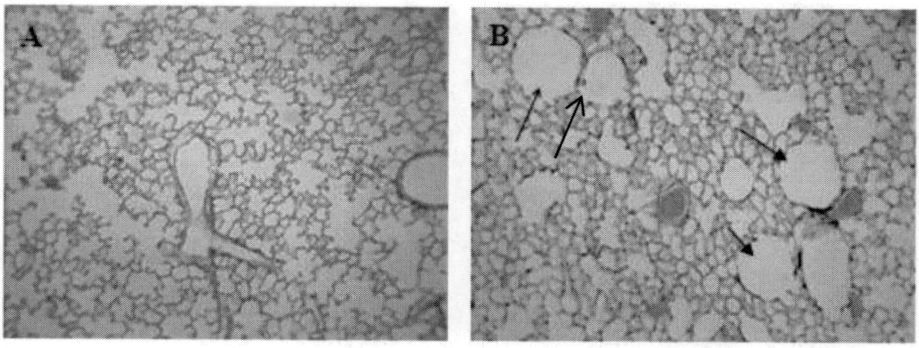

Figure 2. Influence of grand - maternal nicotine exposure during gestation and lactation on the alveolar region of the lungs of the second generation baby. This area where gas exchange occur is damaged (arrows) resulting in a decrease in the area for effective gas exchange. These areas of damage are emphysema-like and appear after lungs reached maturation. This happens even though the baby was never exposed to nicotine which implies these changes were inherited. (A=Control lung, and B=lung with emphysema of animals that originates from grand-mothers that were exposed to nicotine during pregnancy and lactation).

In a recent study it was shown that the structural changes in the lungs of the offspring by grand-maternal nicotine exposure during gestation and lactation can be transferred to at least the next 2 generations (Figure 2). These are programmed changes to induce premature aging and consequently emphysematous changes in the lungs of the offspring after maturation of the animals. By supplementing the diet of the mother with tomato juice prevented the nicotine/oxidant-induced programming. It is therefore, likely that the inclusion of tomatoes, or tomato juice, or other tomato based products in the diet of the mother will be beneficial to her own respiratory health as well as that of her offspring.

Based on the data in our study a daily intake of about one and half glass of tomato juice, where a glass is equal to 250 ml, will supply enough lycopene and other phytonutrients and antioxidants to protect the fetus and neonate in a 60 kg human female against nicotine.

Conclusion

Changes in the external environment or in utero environment may induce changes in the respiratory system of those involved. This involves programming of the developing respiratory system to age prematurely. This programming is induced by maternal nicotine exposure during gestation and lactation. This programming is prevented by supplementing the mothers' diet with tomato juice. Consumption of tomato and tomato products can therefore be considered as a nutritional indicator of good dietary habits and healthy life styles.

REFERENCES

[1] W. MacNee, *Chest*, 117(5 SUPPL. 1), 303S (2000).
[2] Y.N. Engelhard, B. Gazer and E. Paran, *American Heart Journal*, 151, 1 (2006).
[3] B. George et al, *Food Chemistry*, 84, 45 (2004).
[4] P. Palozza et al, *Cancers*, 3, 2333 (2011)..
[5] P. Di Mascio, S. Kaiser and H. Sies, *Archives of Biochemistry and Biophysics,* 274, 532 (1989).
[6] C.M. Jones, P. Mes and J.R. Myers, *Journal of Heredity,* 94, 449 (2003).
[7] Asociación RUVID*, ScienceDaily (*2011).
[8] A.V. Rao and S. Agarwal, *Nutrition Research*, 19, 305 (1999).
[9] H. Sies, *Oxidative stress: oxidants and antioxidants(1991)*1991, London Academic Press..
[10] A. Basu, and V. Imrhan, *Eur J Clin Nutr*, 61, 295 (2007).
[11] A.V.R. Rao and S. Agarwal, *Journal of the American College of Nutrition,* 19, 563 (2000).
[12] H.D. Sesso, et al., *Journal of Nutrition*, 133, 2336 (2003).
[13] W.J. Lessin, G.L. Catigani and S.J. Schwartz, *Journal of Agricultural and Food Chemistry*, 45, 3728 (1997).
[14] D.B. Rodriguez-Amaya, *Archivos Latinoamericanos De Nutricion*, 49, 38s (1999).
[15] J. Shi and M. Le Maguer, *Critical Reviews in Food Science and Nutrition,* 40, 1 (2000).
[16] A.C. Boileau, et al., *The Journal of nutrition,* 129, 1176 (1999).
[17] J. Gaziano, et al., *The American Journal of Clinical Nutrition*, 61, 1248 (1995).
[18] D.E. Holloway, *Free radical research*, 32, 93 (2000).
[19] W. Stahl and H. Sies, *J Nutr,* 122, 2161 (1992).
[20] A.D. Hieber, et al., *Nutrition and Cancer*, 37, 234 (2000).
[21] A. Lavy, A. Ben Amotz and M. Aviram, *European journal of clinical chemistry and clinical biochemistry,* 31, 83 (1993).
[22] G. Levin and S. Mokady, *Free Radical Biology and Medicine,* 17, 77 (1994).
[23] G. Levin, M. Yeshurun and S. Mokady, *Nutrition and Cancer-an International Journal,*. 27, 293 (1997).
[24] P. Borel, et al. *Journal of Lipid Research,* 39, 2250 (1998).
[25] J.W. Erdman, C.L. Poor and J.M. Dietz, *Food Tech,* 42, 221 (1988).
[26] P. Riso, et al., *Pharmacological Research,* 62, 318 (2010).
[27] C.Gartner, W. Stahl and H. Sies, *American Journal of Clinical Nutrition,* 66, 116 (1997).
[28] C.M. Allen, et al., *Journal of the American Dietetic Association,* 102, 1257 (2002).
[29] P.G. Ferreira and A.H. Jaffe, *Monthly Notices of the Royal Astronomical Society*, 312, 89 (2000).
[30] H.H. Schmitz, et al., *J Nutr*, 121, 1613 (1991).
[31] J.D. Ribaya-Mercado, et al., *J Nutr*, 119, 665 (1989).
[32] A. Bendich, J Nutr, 119, 135 (1989).
[33] P. Bateson, et al., *Nature*, 430, 419 (2004).
[34] J.E. Bruin, H.C. Gerstein and A.C. Holloway, *Toxicological Sciences,* 116, 364 (2010).
[35] G.S. Maritz, M. Mutemwa and A.X. Kayigire, *Pediatr Pulmonol*, 46, 976 (2011).

[36] M. Hanson, et al., *Progress in Biophysics and Molecular Biology,* 106, 272 (2011).
[37] J.L. Tarry-Adkins, et al., *Faseb Journal,* 24, 2762 (2010).
[38] W.W. Hay, *J Perinatol*, 26, S13 (2000).
[39] H.W.J. Hay, *Journal of pediatric gastroenterology and nutrition*, 45, S163 (2007).
[40] D.J.P. Barker, *British Medical Bulletin,* 60, 69 (2001).
[41] T.O. Scholl, *Nestle Nutr Workshop Ser Pediatr Program,* 61, 79 (2008).
[42] A.A. Litonjua, et al., *American Journal of Clinical Nutrition,* 84, 903 (2006).
[43] E. Patelarou, et al., *Nutrition Reviews*, 69, 627 (2011).
[44] J. Arab, S. Steck-Scott and A.T. Fleishauer, *Exp Biol Med (Maywood),* 227, 894 (2002).
[45] C.E. Cross, et al., *Environ Health Perspect*, 102, 185 (1994).
[46] G.S. Maritz, *Cell Biol Int*, 17, 579 (1993).
[47] G.S. Maritz, *Pathophysiology,* 4, 135 (1997).
[48] G.S. Maritz, and G. Van Wyk, *Comparative Biochemistry and Physiology - C Pharmacology Toxicology and Endocrinology,* 117, 159 (1997).
[49] H. Bohles, *Int J Vitam Nutr Res,* 1997. 67, 321 (1997).
[50] K. Koike, et al., *American Journal of Physiology - Lung Cellular and Molecular Physiology*, 298, L784 (2010).
[51] A. Bendich, *Annals of the New York Academy of Sciences*, 691, 61 (1993).
[52] S.T. Mayne, *Journal of Nutrition,* 133, 933S (2003).
[53] L.A. Kaplan, J.M. Lau and E.A. Stein, *Clinical physiology and biochemistry,* 8, 1 (1990).
[54] D.F. Church and W.A. Pryor, *Environmental Health Perspectives*, 64, 111 (1985).
[55] X.D. Wang, *Journal of Nutrition*, 135, 2053s (2005).
[56] W. Wei, Y. Kim and N. Boudreau, *American Journal of Public Health,* 91, 258 (2001).
[57] S. Kasagi, et al., *American Journal of Physiology - Lung Cellular and Molecular Physiology,* 290, L396 (2006).
[58] S. Psarras, et al., *Current Pharmaceutical Design*, 11, 2053 (2005).
[59] G. Caramori and A. Papi, *Thorax*, 59, 170 (2004).
[60] N.L. Misso and P.J. Thompson, *Redox Rep*, 10, 247 (2005).
[61] H. Wei, et al., *Journal of Nutritional Science and Vitaminology*, 55, 208 (2009).
[62] H.S. Sekhon, et al., *American Journal of Respiratory Cell and Molecular Biology,* 26, 31 (2002).
[63] G.S. Maritz *Ther Adv Respir Dis*, 3, 193 (2009).
[64] G.S. Maritz and R.A. Thomas, *Cell biology international*, 18, 747 (1994).
[65] G.S. Maritz and S. Windvogel, *Pathophysiology,* 10, 69 (2003).
[66] G.S. Maritz and S.S. Rayise, *Experimental lung research*, 37, 57 (2011).
[67] G. Argentin and R. Cicchetti, *Toxicological Sciences,* 79, 75 (2004).
[68] M.B. Newman, et al., *Life Sciences,* 2002. 71, 2807 (2002).
[69] B.A. Halima, et al., *Hum Exp Toxicol*, 29, 489 (2010).
[70] K. Husain, et al., *Alcohol,* 25, 89 (2001).
[71] J. Hussein, et al., *Toxicology and applied pharmacology*, 218, 1 (2007).
[72] A. Helen, et al., *Toxicology Letters*, 116, 61 (2000).
[73] L.Y.L. Chang, et al., *American Journal of Respiratory and Critical Care Medicine,* 167, 57 (2003).
[74] G. Bjelakovic, et al., *Cochrane Database of Systematic Reviews*, 2, (2008).

[75] G.S. Maritz and G. van Wyk, *Comparative Biochemistry and Physiology Part C: Pharmacology, Toxicology and Endocrinology,* 117, 159 (1997).
[76] C.H. Fardy and M. Silverman, *Archives of disease in childhood. Fetal and neonatal edition*, 73, F112 (1995).
[77] J.W. Erdman, N.A. Ford and B.L. Lindshield, *Archives of Biochemistry and Biophysics,* 483, 229 (2009).
[78] L.X. Zhang, R.V. Cooney and J.S. Bertram, *Carcinogenesis,* 12, 2109 (1991).
[79] J.E. Bruin, *Free Radical Biology and Medicine*, 44, 1919 (2008).
[80] X.F. Zhao, et al., *American Journal of Clinical Nutrition*, 83, 163 (2006).
[81] T.R. Neyestani, et al., *Journal of Endocrinological Investigation,* 30, 833 (2007).

In: Tomatoes: Cultivation, Varieties and Nutrition
Editor: Tadahisa Higashide
ISBN: 978-1-62417-915-0

Chapter 16

Trace Elements: Effect on Tomato Plant and on Human Health After Consumption of Tomato Fruit and Tomato Fruit Food Products

***Charalampos Proestos,*[1,*] *Ioannis N. Pasias,*[2] *Vassiliki Papageorgiou,*[3] *Konstantinos Barmperis*[3] *and Nikolaos S. Thomaidis*[2]**

[1]National and Kapodistrian University of Athens, Department of Chemistry, Food Chemistry Laboratory,
Panepistimiopolis Zografou, Athens, Greece
[2]National and Kapodistrian University of Athens, Department of Chemistry, Laboratory of Analytical Chemistry, Panepistimiopolis Zografou, Athens, Greece
[3]Hellenic Food authority (EFET), Kifisias 124 and Iatridou 2, Athens, Greece

Abstract

The determination of trace elements in tomato fruits and tomato fruit food products has become very important in the last few years since it gives information about the contamination and helps to increase canned food quality and safety. A risk of contamination of the food chain may arise when heavy metals accumulate in plant tissues at concentrations above the threshold levels believed to threaten the health of consumers. Lead, cadmium, arsenic and mercury are the metals and metalloids with the highest toxic potential among those present in foods. Other metals including copper, iron, tin, zinc, antimony, aluminium, manganese, and chromium, which are present in trace amounts in the human diet, may also be the source of toxic hazards to humans under specific conditions. Moreover, the content of these elements in tomato food products varies depending on the type of product and on the procedure followed for the preparation and the type of package material. This chapter deals with the risk

* E-mail: harpro@chem.uoa.gr.

assessment of trace elements found in tomato fruits and tomato fruit food products on human health, the effect of trace elements on tomato plants and the analytical methodologies followed for their accurate and precise determination. The risk assessment of some toxic elements is also discussed.

INTRODUCTION

Lasting recent years, the world tomato trade has grown rapidly, with exports averaging an 8-percent annual increase and topping off at 2.4 million tons for 2007. Indicatively, the Foreign Agricultural Service of the United States Department of Agriculture, in a July 2008 report, presented the world production of tomatoes for processing. Mexico was the leading tomato exporter, accounting for almost half of world exports as shown in Figure 1 [1]. The tomato industry can be described by two different types of culture, namely, glasshouse or greenhouse and open field. All greenhouse-grown tomatoes are marketed fresh, while a high proportion of the tomatoes grown in open fields are processed. Tomatoes may be consumed fresh or, because of their perishable nature, processed to give canned whole peeled tomatoes. Tomatoes may also be processed to give tomato juice and concentrated tomato juice, tomato puree, and tomato paste. Tomato puree and paste may be marketed directly to the consumer or it may be an ingredient in other products, for example, tomato ketchup, soup, and sauces [2,3]. According to Goose et al. [4] and Gould [5], as summarized in a review by Hayes et al. [3], the main categories of tomatoes products are: (i) Tomato pulp, which refers to the crushed tomatoes either before or after the removal of skins and seeds, (ii) Tomato juice, which refers to the juice from whole crushed tomatoes from which the skins and seeds have been removed and which has been subject to fine screening, and is intended for consumption without dilution or concentration, (iii) Tomato paste, which is the product resulting from the concentration of tomato pulp, after the removal of skins and seeds, and contains 24% or more natural tomato soluble solids (NTSS).

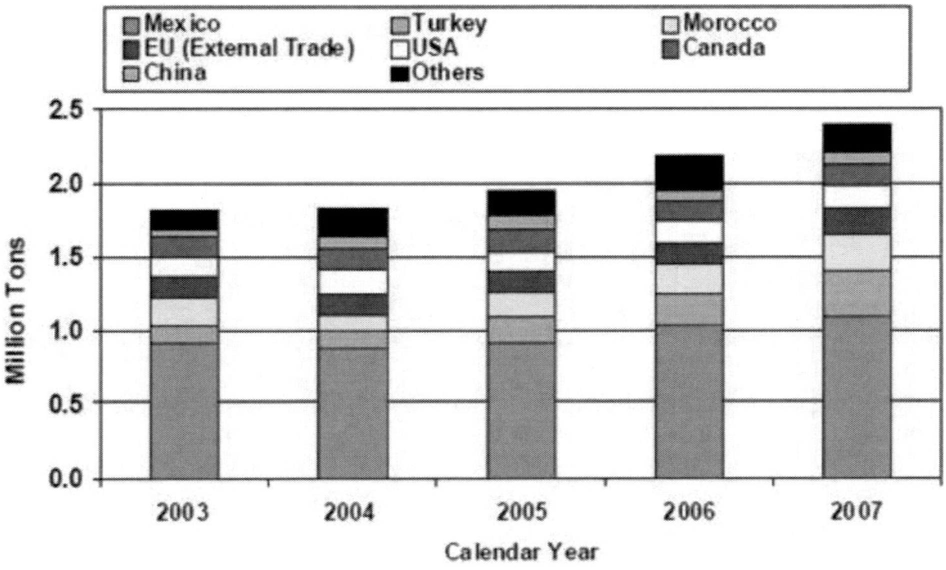

Figure 1. World Tomato exports during the years 2003-2007 [1].

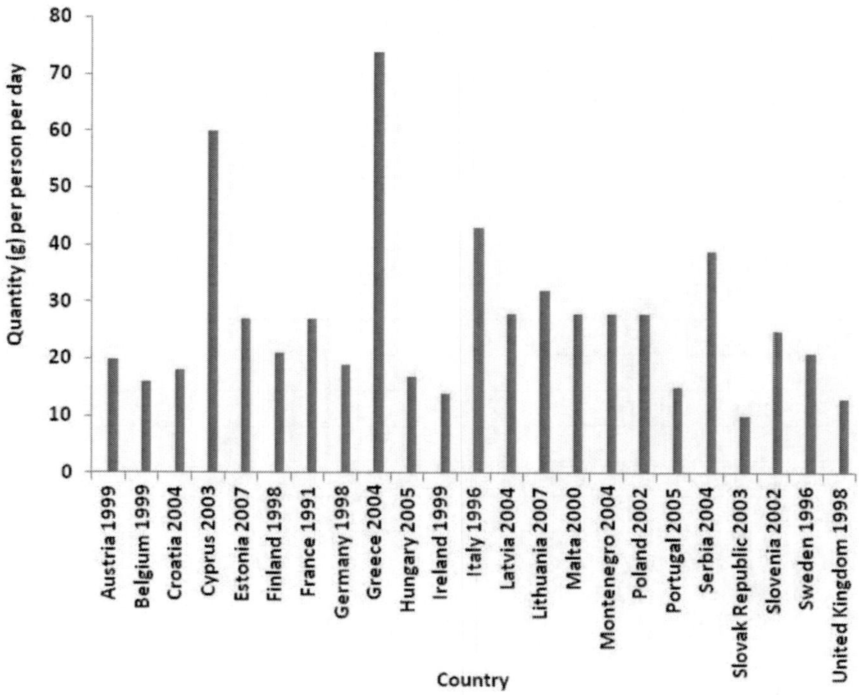

Figure 2. Quantity of tomatoes consumed per person and per day in some European countries at different calendar years.

Tomato paste which is marketed to the consumer in small packs and sold as a condiment may also be described as tomato puree, (iv) Tomato puree, which is the term applied to lower concentrations of tomato paste (containing from 8% to less than 24% NTSS). Rather unfortunately, tomato puree in the U.S. can also be called 'tomato pulp'. A further complication in the U.S. is that if the tomato puree satisfies certain legislative criteria it can also be called 'concentrated tomato juice', (v) Tomato serum, which is tomato juice filtered or centrifuged to completely remove suspended solid material, (vi) Pulp, which is referred to the suspended solid material in tomato juice, puree or paste separated by centrifugation and (vii) Tomato syrup, which is concentrated tomato serum. Data of the average diet from the European program DAFNE (Data Food Network) showed that the consumption of tomatoes was about 27 g per person per day, making tomatoes one of the most highly consumed foods in Europe. More specifically, the quantity consumed per person per day for certain European countries is shown in Figure 2. [6] and the percentage contributed by each tomato product to the total consumption is shown in Table 1 [7]. Tomatoes and their products contain several trace elements. Some of them are essential and others are a result of anthropogenic activities such as mining, industrial processing and application of fertilizers. Furthermore, there are some trace elements, such as tin, that are used as food preservatives and have an effect on food quality. There has been much research into the elemental composition of tomatoes and tomato products, in particular those components considered to be important for their role in human health or indicative of the processes they have undergone, and the aim of this chapter is to present an overview of the accumulation of potentially toxic elements in plants and element transfer to the human food chain, and to describe some analytical methodologies followed for the determination of elemental content in tomatoes and their products.

Table 1. Percentage of consumption of different tomato products to total

Type of tomato product	Percentage contributed by each type to total consumption
Raw tomatoes	36.2
Processed tomatoes	
Soups	8.8
Canned tomatoes	7.0
Grilled	5.4
Fried	3.2
Ketchup	0.4
Tomato-containing products	
Canned food	29.5
Pizza	2.3
Other (e.g. puree)	7.1

1. Trace Elements and Tomato Plant

Soils are the main source for the elemental composition of tomato plants. The concentration of elements in soils varies greatly depending on the composition of parent rocks, mineralization, and soil processes. The exact elemental transport process from soil to plants is still rudimentary in most cases.

Generally speaking, trace elements are transported from the soil into the root and then must cross both cellular and organellar membranes as they are distributed throughout the plant. It seems that some proteins play the role of trace element transporters and are involved not only in normal metal ion homeostasis but also in the overall strategy for heavy metal tolerance [8,9].

The overall accumulation of trace elements in plants can be described as a five-step process: (a) the mobilization and uptake from the soil (b) the compartmentation and sequestration within the root, (c) efficiency of xylem loading and transport, (d) distribution between metal sinks in the aerial parts and (e) sequestration and storage in leaf cells (Figure 3) [10].

Lately, there has been much progress in the field of plant biology and, as noted by Williams et al., “a future challenge will be to determine how these transporters are regulated in plants. There are likely to be many tiers of control and modulation and, although there may be some similar aspects of regulation to those occurring in other organisms, unique regulatory mechanisms may also exist” [9].

The accumulation of trace elements in tomatoes and their transfer to the human food chain is affected by the properties of the soil. Uptake of trace elements by plants depends on many factors, such as pH, clay content, the presence of other ions, the quality of water, the organic matter, the cation exchange capacity, the iron and aluminum oxides, the texture, the aeration, the specific sorption sites and the growth treatment [11, 12, 13].

For example, Wang et al. concluded that the growth of tomato plants was seriously affected by the soil acidity and lowering of uptake of elements was observed for the plants cultivated on acidified soil [14].

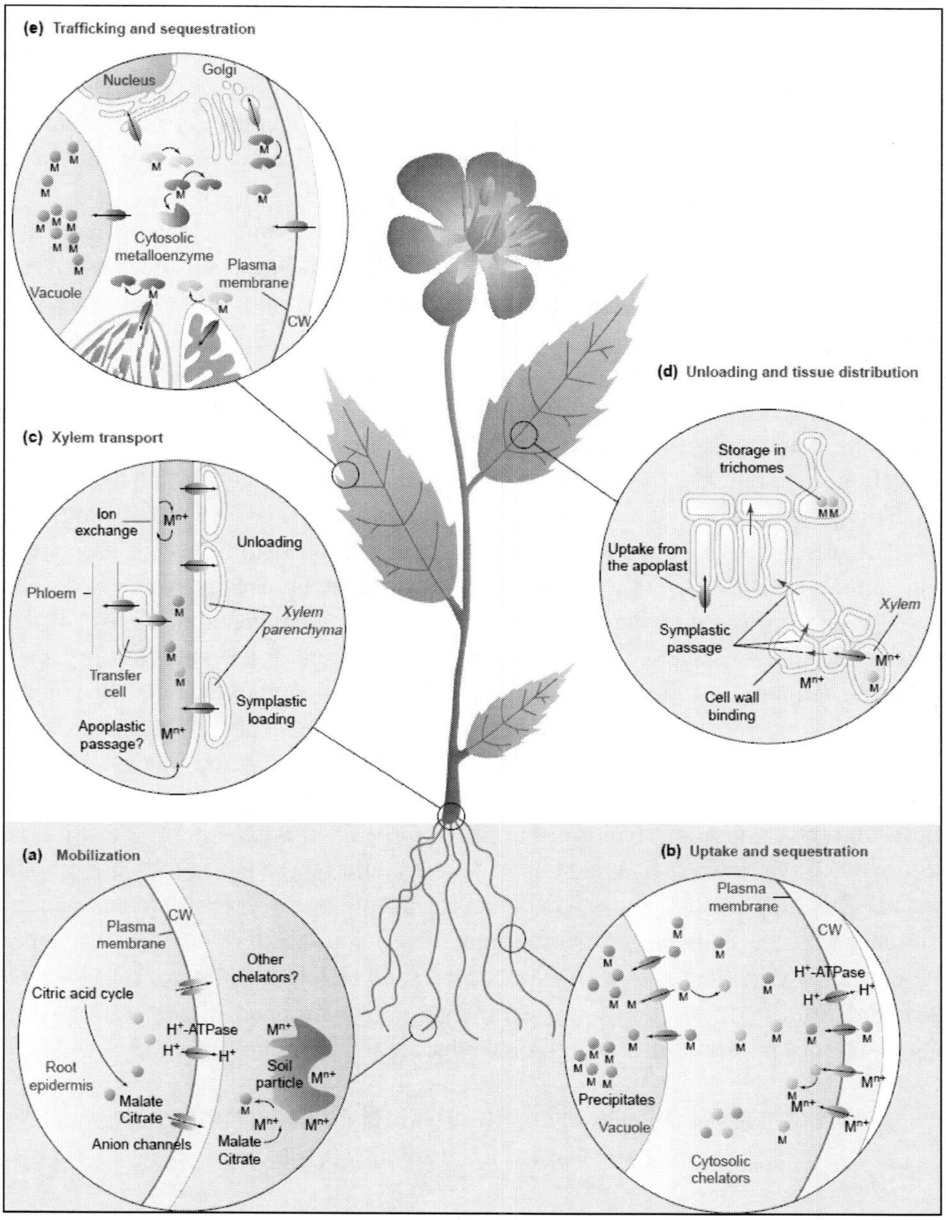

Figure 3. The process for elemental transportation from soils into plants [10].

On the basis of toxicological assessments of a number of substances, the Danish EPA [15, 16] has defined soil quality criteria (no restrictions on land use) and cutoff criteria (any contact with the soil should be omitted) as shown in Table 2. According to Z.L.He et al., the average concentrations (mg kg^{-1}) of As, Cd, Cr, Cu, Hg, Pb, Mo, Ni, Se and Zn in soils worldwide are 9.36, 0.06, 20-200, 20, 0.03, 10–150, 1-5, 40, 0.20 and 10–300, respectively [17]. Trace elements from various anthropogenic and other processes tend to accumulate on soils when they reach the soil surface.

One of the main anthropogenic processes concerning the accumulation trace elements in soils is the use of fertilizers. During the last century many metal-containing pesticides, such as calcium arsenate, copper acetoarsenate, lead arsenate, zinc sulfate and others, have been used for various fruit crops and vegetables.

For example, Burló et al. used four As species at different concentration levels for tomato plants grown under greenhouse conditions and they concluded that the As concentration ratio (shoot As concentration/root As concentration) was significantly affected by both As chemical species and As concentration [12].

Even though the majority of these fertilizers have been banned, the content of these elements in soils is still at high levels. In the last few years there has been an increased interest in elemental content in tomato plants derived from organic rather than conventional growing systems.

The term "organically grown food" denotes products that have been produced in accordance with the principles and practices of organic agriculture. Organic agricultural and food processing practices are wide-ranging and overall seek to foster the development of a food production system that is socially, ecologically, and economically sustainable [18].

The main principle of this kind of cultivation is the use of environmental friendly practices. The differences in trace elements vary from study to study. There seems to be a significant difference in the concentrations of certain elements such as manganese, copper, and zinc between tomatoes cultivated under organic and conventional conditions, but there is no conclusive evidence whether conventional products are more or less safe than organic ones [18,19,20]. Furthermore, a significant difference in elemental composition also occurs among different varieties of tomato plants. For example Fernández-Ruiz et al., after applying multidimensional analysis (principal component and cluster analysis), concluded that wild varieties of tomatoes had higher mineral content than conventional ones [21].

Finally, another important factor influencing the mineral uptake of tomatoes is the cultivation method. Hernández Suárez et al. [22] proved that there is a significant correlation between the cultivation method and trace element concentrations. Trace elements seem to be more influenced by the cultivation than the minerals. Sampling time also had an important influence on the mineral and trace element concentration. Table 3 summarizes the elemental contents found from various studies for several varieties of tomato plants cultivated under different conditions and from different regional areas.

Table 2. Toxicological Soil Quality Criteria and Cut-Off Criteria for Soil (Danish EPA 2000) [15,16]

Trace element	Soil quality criterion (mg/kg dry weight)	Cutoff criterion (mg/kg dry weight)
As	20	20
Pb	40	400
Cd	0.5	5
Cr	500	1000
Cu	500	500
Hg	1	3
Ni	30	30
Zn	500	1000

The effect of some important trace elements on tomato plant is presented below.

1.1. Iron

Iron occurs in the iron-bearing primary minerals of the soil mainly in the ferrous, Fe^{2+} state. Following weathering, the iron is largely converted to the ferric, Fe^{3+}, state and forms various hydrated oxides. Reducing conditions are needed to form the Fe^{2+} state.

There are some crucial factors that affect the content of iron in soils and the possibility of its accumulation in tomato plants. The content of Fe^{3+} in soils mainly depends on pH and is usually associated with water-logged soil. It seems that at pH 6.0 and below it is highly available, while at pH 7.0 availability is already markedly decreased and above this pH it declines dramatically, through precipitation of Fe-compounds, and becomes a limiting factor to plant growth in many cases. Iron deficiency is also associated with the content of other mineral elements in soils, especially Ca, Cu, Zn and P. Fe-deficiency is liable to occur on field-grown crops on calcareous soils. Also, as a result of mismanagement through over-liming, Fe-deficiency is always liable to appear in intensively managed glasshouses and gardens as happens for tomato cultivation [53, 54]. Even under these conditions, iron can be accumulated by tomato plants. Shenker et al. showed that a fungal siderophore produced by *Rhizopus arrhizus* named *raphorin* can act as a Fe carrier to plants growing in nutrient solutions at pH 6.0, 7.0 and 7.3 [55].

1.2. Zinc

Zinc is also an essential element that can become toxic at high concentrations. Zinc is indispensable for the proper growth of tomato plants due to the fact that is has an important role either as a metal component of enzymes or as a functional, structural or regulatory cofactor of a large number of enzymes. It may also play an important role in chlorophyll production [64]. Zinc deficient soils are common all over the world in both tropical and temperate climates and can affect plant growth and yield. Moreover, zinc deficiency results from enhanced P uptake to the shoot. Kaya et al. concluded that Zn stress from exposure to both high and low concentrations have significant effects on plant growth and fruit yield in the tomato cultivars grown in nutrient solution. Reductions in chlorophyll concentration and plant growth in Zn deficient plants appear to be associated mainly with the low Zn concentrations in the leaves. Reduction in fruit yield and quality may be related to both the low Zn and K (which reached deficiency levels) in the leaves. In the high Zn treatment, reduction in plant growth, fruit yield, and leaf chlorophyll concentration corresponds with the excess concentration of Zn in leaves [66].

1.3. Aluminium

Aluminium has a negative effect to plants. Regarding tomato plants, studies showed that aluminum reduced leaf area and leaf stem and root dry weights. Moreover, when the concentration of Al was increased it was found that the length of lateral roots as well as the root system was significantly reduced. The Ca, K Mg, Fe, and Zn content of the roots, stems, and leaves of tomato decreased with increasing amounts of Al in the nutrient solution.

Table 3. Elemental content (mg kg^{-1}) of tomato plants from various areas

	Elemental content (mg kg^{-1})										
Country	Mg	Zn	Se	Fe	Cu	Mn	Cr	Pb	Cd	Ni	Al	As
Portugal [23]	120-200	1.1-2.6	2.0-4.0									
Tenerife, Spain [22]	110-128	0.69-0.86		1.8-2.19	0.24-0.32	0.54-0.66						
Almería, Spain [24]	110-120	1.6-55.0	0.0009-0.014	4.0-20.0	0.45-4.0	0.66-3.0						
Ecuador, Spain, Peru, USA [21]	84	2.6		5.5	4.3	1						
Spain [25]					0.033-0.079	0.022-0.036	0.019-0.072	0.030-0.068	0.013-0.020			
Manisa, Turkey [26]		0.81-1.30		0.56-3.27	0.59-0.84			0.001-0.022	nd			
Greece [27]							0.06-0.13					
Finland [28]	1450	12	nd	29	4	18		0.02	nd	0.02	5	
Poland [29]												0.0003-0.027
Turkey [30]	4.22-6.32	64.8-78.4		1.98-2.41	12.9-18.7			0.38-0.48	0.61-0.71			
Italy [31]					0.46			0.0034	0.0002			
Nigeria [32]		1			0.36							
Mexico [33]	70-440	0.7		2.0-6.0	0.7	0.5		nd-0.13				
okat, Turkey [34]		3.6		4.5	8	0.076	0.037	0.032	0.012	0.06		
Brazil [35]					0.4-0.9							
Slovenia [36]			0.001-0.029									
Thailand [37]			0.012									

nd: not detected.

Aluminum treatment promoted accumulation of P, Mo, and Cu in the roots and decreased the concentration of these nutrients in stems and leaves [79]. Searcy et al. [80] tested pollen from four tomato cultivars differing in sensitivity to aluminum in the sporophyte to determine if A1 sensitivity was also expressed in pollen. The results showed that sensitivity to A1 is expressed in both haploid and diploid generations.

1.4. Manganese

Manganese is also an essential trace element which plays an important role in plants. There are several processes, such as mining and industrial metal production, that are considering responsible for the excess of Mn in soils and plants. Mn is also released in soils, especially in those with low pH values, and plants by water-logging or by high organic matter content. Mn toxicity in tomatoes may also occur after steam sterilization of greenhouse soils, particularly those of high clay content [85]. Excess of Mn in tomatoes can be defined by dark brown/black spots which appear first in the cotyledons and then in the leaves [86].

2. TRACE ELEMENTS OF TOMATO (FRUIT AND TOMATO FRUIT FOOD PRODUCTS) AND HUMAN HEALTH

Consumers are exposed to metals and metalloids mainly via foods. Trace element contamination in tomato fruit and tomato fruit food products can come from different sources, such as agricultural procedures, atmospheric fall-out, pesticide formulations, contamination by chemical fertilizers and irrigation with water of poor quality.

Table 4. Elemental content in ($mg\ kg^{-1}$) in various tomato food products

Food product	Zn	Se	Fe	Cu	Mn	Cr	Pb	Cd	Sn	Ni	Al
Canned tomato paste											
[27]						0.09-0.17					
[42]	0.61			0.40			0.27	0.041	13.3	0.097	
[43]	10.5	0.09	47.7	4.24	6.46	0.52				0.18	1.55
[44]	3.57-7.72		11.4-57.5	1.57-3.37			0.019-0.096	0.023-0.17	0.030-6.04	3.57-7.72	
[45]	6.01		6.03	8.66			0.51	0.31			
Tomato sauce											
[46]	11.24	0.566	37.6	8.46	7.43						0.72
[38]				4.72-9.04			0.20-1.36		0.41-7.32		

Some of them, including copper, iron, tin, zinc, antimony, aluminum, manganese and chromium, are essential to tomatoes but can become hazardous under specific conditions and at levels higher than crucial values. The highest toxic potential among those present in tomato fruits and tomato food products is associated with lead, cadmium, arsenic and mercury and for this reason strict legal restrictions had been adopted for these elements.

As concerns tomato food products especially, there is a variety of processes that are responsible for their elemental content. These include the addition of chemical substances such as food additives, flavoring substances and micronutrients. One of the most familiar food additives is stannous chloride, $SnCl_2$, which is used as a permitted food additive (E512) in small quantities.

Packaging materials are also a source of chemicals in food products and beverages. Canned, dried, frozen or processed foods are normally held from a few weeks to a year or more before being consumed. Several compounds, such as lead, iron, chromium and tin are often added inadvertently during processing, internal rusting and etching of the cans and are possible sources of these heavy metals. Tin is also used in food packaging in combination with steel, as tinplate, because it combines the strength and formability of steel and the corrosion resistance and good appearance of tin [38-41]. The content of trace elements in canned tomato samples depends directly on the package material, the storage time, the temperature and the pH. U. Rafique et al. concluded that the concentration of elements increases with increasing storage time of opened cans and increasing temperature but decreases with increasing pH [42]. Table 4 summarizes the elemental content found in different tomato food products as determined by different researchers.

There are several crucial values for the ingestion of a given nutritional and non-nutritional metals and metalloids, along with the legislation limits as determined by the World Health Organization and the Food and Agriculture Organization (FAO). Such crucial values are: (i) the Lowest Threshold Intake (LTI) corresponding to the quantity of necessary nutrient to assure maintenance, metabolic and physiological function for an individual in good health, (ii) the Provisional Tolerable Weekly Intake (PTWIs) corresponding to the acceptable level of toxic metal that can be ingested on a weekly basis and (iii) the Maximum Tolerable Weekly Intake (MTWI) corresponding to the amount of a substance that can be taken in weekly over a lifetime without appreciable health risk. The limits that have been recommended for the levels of trace elements are usually based on the dry solids content. The risk assessment for each trace element on human health will be presented below.

3. Risk Assessment of the Essential Trace Elements on Human Health

3.1. Copper (Cu)

Copper is one of the essential trace elements and plays an important role in various metabolic processes in humans, animals and plants. It can be found either as soluble and particulate forms in the environment. Due to the antimicrobial properties that it has, Cu is also used as a potential material for food packaging. Copper can be found in the +1 (cuprous) and +2 oxidation states (cupric). Cu functions as a cofactor and is required for structural and catalytic properties of a variety of important enzymes, including cytochrome *c* oxidase, tyrosinase, p-hydroxyphenyl pyruvate hydrolase, dopamine beta hydroxylase,lysyl oxidase,and Cu-zinc superoxidase [48-50]. The effects of copper deficiency in adults include decreased plasma copper levels, decreased caeruloplasmin and erythrocyte copper-zinc superoxide dismutase activities, increased low-density lipoprotein cholesterol, decreased high-density lipoprotein cholesterol, cardiac arrhythmia, electrocardiographic abnormalities and elevated blood pressure [50].

Although normally bound to proteins, Cu may be released and become free to catalyze the formation of highly reactive hydroxyl radicals. Data obtained from in vitro and cell culture studies are largely supportive of Cu's capacity to initiate oxidative damage and interfere with important cellular events. Oxidative damage has been linked to chronic Cu overload and/or exposure to excess Cu. High content of Cu is also responsible for Wilson's disease. Several factors such as the chemical form of the copper and the presence of other dietary components play an important role on the absorption of Cu in the human body. For example, L.M. Gaetke et al. [48] concluded that the presence of a number of nutrients may interact with Cu and alter its toxicity. They also suggested that vitamin E is largely protective against Cu induced oxidative damage. Ascorbic acid is also able to reduce the detrimental properties of Cu. High dietary ascorbic acid and Zn may protect Cu toxicity by reducing its intake. Other components such as Se, beta-carotene, alpha-lipoic acid and polyphenols may also afford some protection against Cu. About 30-50% of ingested Cu, mostly Cu^{+2}, is absorbed in the small intestine, and very small amounts are absorbed in the stomach. Turniund et al. suggested that the minimum dietary copper requirement of young men is > 0.4 mg/d but < 0.8 mg/d [52]. On the other hand World Health Organization (WHO) suggested as an upper limit 12 mg/day for men and 10 mg/day for women. These values take into account the quantity likely to be consumed from the usual diet (< 10 mg/day) and therefore limit both the amount of copper that can be introduced by dietary fortification and the quantity of contaminating copper that can be regarded as tolerable [51].

3.2. Iron (Fe)

The human body contains about 45 g of iron. The majority of the iron, approximately 65%, is bound to haemoglobin, 10% is a constituent of myoglobin, cytochromes, and iron-containing enzymes, and 20 to 30% is bound to the iron storage proteins, ferritin and hemosiderin. There are only trace amounts of unbound iron in the human body [56]. It is well known that iron deficiency result in the widely spread condition, anaemia. As presented in Tables 2 and 3, tomatoes and tomato food products contain a significant amount of iron and for this reason are among the most beneficial foods for iron uptake. On the other hand, high content of iron is responsible for many disorders, such as neurodegenerative disorders and diabetes. Iron is also believed to be responsible for some types of cancer. Kato examined the correlation between iron content and the risk of colorectal cancer and they concluded that higher iron intake (including supplements) does not necessarily lead to iron overload in the body and that luminal exposure to excessive iron may increase the risk of colorectal cancer when combined with a high fat diet [57]. Furthermore, Rasmussen et al. concluded that increased iron stores may play a role in the development of coronary heart disease (CHD) by increasing lipoprotein oxidation [58]. Iron toxicity is mainly related to free radicals. Fe^{+2} has the ability to react with oxygen in the cells and thus the participation of iron is essential in: (a) the production of hydroxyl radical that can subsequently initiate lipid oxidation or oxidize almost every molecule present in biological systems; and (b) the propagation of free radical reactions by decomposing peroxides [56]. A diet plentiful in antioxidants, such as flavonoids or polyphenols and the co-existence of other minerals, like zinc, have a beneficial effect on iron overload. Unfortunately, it is not easy to adopt a minimum and a maximum daily iron intake, due to the fact that iron is essential in many food products, especially red meat. While red meat consumption has been decreasing in recent years, consumption of vegetables, like tomatoes and other iron-rich foods and beverages, has risen. For this reason it difficult to define an exact daily intake and the pathological effects of dietary iron [59]. Finally, as shown in Tables 3 and 4 the canned tomato food products have a higher Fe content than the tomato fruits. An explanation for this phenomenon might be that the use of iron for food and beverage packaging results in some iron dissolving into the food content.

3.3. Tin (Sn)

Tin is an essential trace element and plays an important role in various metabolic processes in humans, animals, and plants. It can be found either as inorganic tin or as organotin compounds. As inorganic tin it can be found in two different oxidation states, in the +2 and +4 states. Tin is one of the most useful elements in package technology, as tinplate in combination with steel, for preservation of tomato food products. Tin compounds combine the strength and formability of steel and the corrosion resistance and good appearance of tin. The use of tin for tomato food preservation is directly based on its ability to minimize the headspace oxygen and to allow the sterilisation of the foodstuff within the hermetically sealed can, giving a long, safe, ambient shelf life with no, or minimal, use of preservatives [60]. Moreover, stannous chloride (E512) is often used in small quantities as an antioxidant and to aid color retention in canned and bottled foods.

Different factors affecting the concentration of tin in canned tomato foods are the lacquer of the can, the pH of the food, the headspace, the presence of nitrates, etc. Pasias et al. investigated the influence of the industrial pretreatment procedure of tomato paste samples and the type of package materials on the total concentration of tin in canned tomato paste samples. They determined tin in different tomato paste canned samples with different pretreatment procedures and the results indicated that the samples where the internal surface of the can was not coated with plastic film layer showed the highest concentration of tin, followed by those where the internal surface was coated with plastic film layer apart from the top and bottom lids. Furthermore, this work proved that the Brix degrees played an important role on the concentration of tin in canned tomato paste samples. In general, samples with lower Brix degrees showed lower concentrations of tin in the analyzed samples [41].

Excessive levels of tin in foodstuffs can cause gastric irritation, so a maximum permissible level of 200 mg/kg of tin in canned food has been set by the European Union [61]. There are some recent studies reporting gastrointestinal perturbations when it is present in concentrations above 200 mg/kg [60,62]. The exact pathological effects of tin overload are manifested as gastric irritation, nausea, vomiting and abdominal discomfort. Inorganic tin salts are poorly absorbed by the gastrointestinal tract and rapidly excreted. Nevertheless, there are several case reports of gastric irritation and vomiting in humans consuming canned foods or beverages, particularly sour fruit products packaged in tinplate cans and containing high levels of tin [63]. Moreover, when the contamination reaches this level the organoleptic properties of the food can be seriously affected.

3.4. Zinc (Zn)

It is important to know the exact concentration of zinc in order to avoid excess amounts of it because negative effects on the nutrient value of the tomato food products can occur [64,65]. Till recently most of the studies concerning zinc had focused on the consequences of zinc deficiency, while little attention has been directed to the potential consequences of excessive zinc intake. Zinc is considered to be relatively nontoxic, particularly if taken orally. Numerous cases of accidental and intentional (self-harm) ingestion of zinc salts have been reported but only a few of these accidental exposures have resulted in death. The majority of the pathological conditions concerning zinc exposure had to do with vomiting, diarrhea, red urine, icterus (yellow mucous membrane), liver and kidney disorders, fatigue and anemia. Zinc is commonly used in pesticides (10 pesticides containing zinc salts including the oxide, sulfate, chloride and phosphide are commercially available in the United States), as food supplements and additives. Zinc oxide and its derivatives confer special fungi-static and chemical properties to the processing and packaging of various tomato products and help to prevent formation of black sulfides which discolor the food. Studies that have been carried out to estimate human requirements, including the metabolic balance studies, showed that growing infants, children, growing adolescents, and pregnant and lactating mothers require more zinc per kilogram of body weight than do mature adults. Furthermore, the recommended value, corresponding to an estimate of the daily exposure to which the human population may be continuously exposed over lifetime without an appreciable risk of adverse health, has been set by the US Environmental Protection Agency (US EPA) as 0.3 mg/kg body weight per day. This value corresponds to 21 mg zinc for a 70-kg male and 18 mg for a 60-kg female and is higher than the recommended acceptable daily intakes recommended by the WHO in some instances. On the other hand, the US Food and Nutrition Board has set the tolerable upper intake level at 40 mg/d for adults older than 19 years [67,68]. Taking into account that the mean consumption from different countries is about 25 g of tomatoes per person per day (Figure 2) and that the mean content of zinc in tomatoes and tomato food products (Tables 3 and 4) is approximately 10 mg/kg, the mean uptake for zinc from tomatoes is equal to 0.25 mg, 100 fold times lower than the crucial value set by EPA for a 70-kg male.

Cells have the ability to protect themselves from zinc toxicity. Palmiter concluded that cells protect themselves from zinc toxicity by inducing proteins such as metallothionein that bind it tightly, by sequestering it in organelles, or by exporting it [69]. Furthermore, foods containing iron, such as red meat, can also help reduce the uptake of zinc.

3.5. Antimony (Sb)

Antimony can be found in soils, waters and air in very small amounts. Antimony will mainly enter into the food chain via food packaging materials. Antimony is used as a catalyst in the production of polyethylene terephthalate (PET) plastics, a material used in the packaging of many foods and beverages, including tomato juices [70-72]. Antimony is a suspected carcinogen with unknown biological function and effects on long term exposure. Low doses of antimony are known to cause headaches, dizziness and depression, but in larger doses can cause violent and frequent vomiting [72,73]. Choe et al. concluded that there is a correlation between migration of antimony (Antimony (III) chloride) from PET and possible oestrogenic effects in humans [74]. There are some values concerning the maximum limits of antimony in drinking. However, these limits are applicable for drinking water itself before it is bottled. In PET packed beverages the specific migration limits according to packaging legislation are 40 mg kg^{-1} in Europe [75] and 50 mg kg $^{-1}$ in Japan [76]. Based on toxicological data, the World Health Organization suggests for humans a tolerable daily intake (TDI) for antimony of 6 mg kg^{-1} body weight [77]. Welle et al. [72] experimentally determined the diffusion coefficients for antimony in PET bottle materials at different temperatures between 105 and 150 °C. It was concluded that antimony levels in beverages due to migration from PET bottles manufactured according to the state of the art can never reach or exceed the European-specific migration limit of 40 mg kg^{-1}. In the overall long-term management of antimony poisoning, chelation therapy can have short-term benefits; however, these benefits must be accompanied by a drastic reduction in exposure to antimony if therapy is to have any long-term benefit [78].

3.6. Aluminium (Al)

Aluminium is the third most common element encountered in nature. Despite this marked presence, there is no evidence that A1 has an essential function in humans. Aluminium has a negative effect to humans. Tomato food products, especially tomato sauces and juices, contain aluminium from packaging materials and cooking utensils during cooking or during storage. For example, Miiller et al. [81] concluded that high amounts of A1 migrated into acidic products such as mashed tomatoes during normal processing in normal, non-coated A1 pans. They showed that after 60 min cooking an A1 content of 10-15 mg/kg was present in tomato sauce. However the average contribution from the daily use of A1 utensils is estimated to be less than 0.1 mg and the studies available so far it is difficult to arrive at a definite conclusion on the extent of leaching of aluminium from cookware and the amount of aluminium made available for ingestion by humans [81,82]. Al has many negative effects on human health. Long term exposure to A1 can cause encephalopathy, osteodystrophy, microcytic anaemia, dialysis arthropathy and it is suspected to be involved in Alzheimer's disease. Jeffery et al. [83] investigated the effect of Al in bones, kidneys and blood and they concluded that there is no single mechanism to explain all the systemic effects of Al. However the percentage of Al accumulated is too low and, therefore, the normal blood level is only 1-3 μg A1/L and the total body burden of A1 in healthy persons is less than 35 mg [81].

3.7. Manganese (Mn)

Manganese is an essential trace element which plays an important role in humans. Mn is mainly related with enzymes and the mitochondrial oxidative processes. For example, Mn is a required co-factor for arginase, which is responsible for urea production in the liver, superoxide dismutase, which is critical to prevent against cellular oxidative stress, and pyruvate carboxylase, an essential enzyme in gluconeogenesis [84]. Mn is also used as a nutritional supplement in many food products [87] and as a material for food packaging, because it has the ability to improve the strength properties of aluminum [88]. This may be the reason why the content of Mn in tomato food products is much higher than the content in tomatoes (Tables 3 and 4).

Although the exact mechanism of Mn toxicity is not well known, it is believed that it is may be associated with the presence of other elements, such as iron, zinc and copper. Regardless of intake, animals generally maintain stable tissue Mn levels as a result of homeostatic mechanisms that tightly regulate the absorption and excretion of this metal. However, high dose exposures are associated with increased tissue Mn levels, resulting in adverse neurological, reproductive and respiratory effects. The brain is the major target organ for Mn toxicity. It retains Mn much longer than other tissues. Following chronic overexposure, Mn can produce a progressive, permanent neurodegenerative disorder with few options for treatment and no cure. To support a risk assessment of manganese exposure in two communities living within a manganese mining district, a cross-sectional study was performed on a sample of the adult population of long term residents by Santos-Burgoa et al. They concluded that an important risk is present which deserves further research and intervention [87]. The mean daily ingestion of tomatoes in different European countries was estimated as 20 g (Figure 2). Taking into account that the mean content of Mn in tomatoes and tomato food products is around 4 mg/kg the mean daily ingestion of Mn from tomatoes is equal to 0.08 mg, which corresponds to 0.8% of the Reference dose (9.8 mg day; EPA, 2001) [89]. The main therapy against Mn toxicity is to remove the patient from the source of Mn. Treatment with levodopa and chelation therapy also has benefits in the improvement of clinical symptoms [84, 90].

3.8. Chromium (Cr)

Chromium is a relatively common element with two main oxidation states of practical importance, III and VI. At physiological pH values, Cr (VI) exists as an oxyanion $(CrO_4)^{2-}$ that resembles sulfate and phosphate oxyanions. Based on this oxyanion molecular mimicry, chromate is therefore transported into and accumulated in cells instead of sulfate or phosphate [91, 92]. Tomato fruits are a good source of chromium. However, it is important to know the exact content of chromium species, because the chemical behavior and biological activity of Cr (III) and Cr (VI) species are very different. Cr (III) seems to have an important role in the metabolism of glucose in the bloodstream together with insulin and benefits diabetic patients by keeping their blood sugar levels under control. One cup of tomato juice provides 7.5 percent of the daily recommended value for chromium. Tomato juice also offers special benefits to diabetes patients who are at increased risk of cardiovascular disease. On the other hand, Cr (VI) has toxic and carcinogenic effects due to its strong oxidizing potential and easy permeation of biological membranes. Thus, hexavalent chromium damages cell macromolecules such as proteins and the DNA [93].

Reported Cr intakes range from 50 to 200 μg/day. However, a comparison of the Cr levels reported by different investigators reveals substantial differences, some of which may be due to the influence of the location where the foods were grown [94]. The estimated daily intake for chromium in tomatoes is equal to 3 μg and is significantly lower than the Reference dose of 105 mg day^{-1} for a 70 kg adult and is also below the daily intake recommended by the US National Research Council (NRC, 1989) for Cr (III) [95]. The therapy of chromium toxicity is also based on the chelation reaction [96].

4. Risk Assessment of Toxic Elements on Tomato Plant and on Human Health

Risk elements are of considerable concern due to their toxicity and cumulative behaviour and their ability to accumulate. This category includes certain metals and metalloids, such as lead, cadmium, arsenic and mercury. Anthropogenic activities, such as mining, industrial processing and application of fertilizers, are the main sources of heavy metal contamination in food chain via plant accumulation.

For example, in the past, indiscriminate application of inorganic arsenicals as pesticides, desiccants, or wood preservatives led to pollution of many agricultural soils and reduction of their productivity [97]. The strategies of heavy metal tolerance in plants are poorly understood but roots seem to be the main accumulation site.

Lead and cadmium are among the most abundant risk elements and are particularly toxic. The excessive content of these metals in food is associated with the appearance of a number of diseases, especially cardiovascular, kidney, nervous as well as bone diseases and in some cases they can cause carcinogenesis, mutagenesis and teratogenesis [98-100].

Moreover, As, in inorganic form, has been classified as a Group 1 carcinogen in humans. Burló et al. investigated the uptake of arsenic (As) species by *Lycopersicum esculentum*, growing under soilless culture and they concluded that the concentrations of As in plants increased significantly with increasing As concentration in solution, which indicates that when tomato plants are exposed to high concentrations of As in nutrient solutions, they may uptake As to concentrations unacceptable for human food [12].

Mercury is also defined as a toxic element. The presence of mercury residues in tomatoes is due to treatment with mercury vaporizing solution. These aerosols usually contain about 0.2% of organically combined mercury, and are applied to growing tomato crops [101]. Methyl mercury poisoning has many effects on human health, such as parestesias and numbness in the hands and feet and at high doses may lead to death. However, due to the low content of methylmercury in tomatoes there is no crucial value corresponding to its toxicity.

The presence of risk elements does not only affect the nutritive values of vegetables but also has detrimental effects on humans consuming these foods. According to the Commission Regulation (EC) No 1881/2006, the maximum permissable levels of Pb and Cd in vegetables are 100 μg/kg and 50 μg/kg, respectively [61].

As Table 3 shows, in the majority of the studies lead and cadmium contents were lower than the crucial values, but there is a recent study from Turkey where the contents of both elements were higher than the crucial values [30]. In general, lead, and cadmium are nowadays the most hazardous among the other toxic elements. On the other hand, the general population does not face a significant health risk from the presence of methylmercury or arsenic in tomatoes due to the low concentrations.

5. Analytical Methodologies for the Determination of Toxic Elements and Trace Elements in Tomato Fruits and Tomato Fruit Food Products

The levels of the toxic elements are routinely monitored in food to protect the consumer. Increasingly, the chemical species of As and Hg are also monitored. The analytical procedures followed for the determination of toxic elements must be accurate and precise with appropriate limits of detection corresponding to the elemental content in tomatoes and the legislation crucial values. The main steps followed for the analysis of tomatoes are (i) the preparation of the samples, (ii) the digestion of the samples for the determination of total elemental content or the appropriate extraction procedure for the determination of specific forms of the elements, (iii) the determination of trace elements using the appropriate technique.

Certain quality assurance criteria must be followed during the procedure. The method followed for trace element determination must be either a standard reference method or an in-house developed method after following all guidelines and protocols for method validation. Several certified reference materials for tomatoes are available and regulatory agencies seeking endorsement and recognition for methods may choose to fully validate a few select analytical procedures by conducting interlaboratory trials under the auspices of organizations. These procedures provide documented intra- and interlaboratory method performance, and are only undertaken after a thorough in-house validation of the method.

Samples of tomato fruits must first be washed, dried, peeled and comminuted after removal of damaged crops. According to the European Directive 333/2007, for sampling and analysis for the official control of the levels of lead, cadmium, mercury and inorganic tin in food samples, "in the course of sampling, precautions shall be taken to avoid any changes which would affect the levels of contaminants, adversely affect the analytical determination or make the aggregate samples unrepresentative. Each sample shall be placed in a clean, inert container offering adequate protection from contamination, from loss of analytes by adsorption to the internal wall of the container and against damage in transit.

All necessary precautions shall be taken to avoid any change in composition of the sample which might arise during transportation or storage. Wherever possible, apparatus and equipment coming into contact with the sample shall not contain those metals to be determined and be made of inert materials e.g. plastics such as polypropylene, polytetrafluoroethylene (PTFE) etc. These should be acid cleaned to minimise the risk of contamination. High quality stainless steel may be used for cutting edges. In the case of inorganic tin, care shall be taken to ensure that all the material is taken into solution as losses are known to occur readily, particularly because of hydrolysis to insoluble hydrated Sn (IV) oxide species" [102,103].

Wet (acid) digestion or dry ashing are among the most widely used procedures for treatment of the tomato samples and retain the analytes of interest in a solution for introduction into the analytical instrument. These techniques are usually used for the determination of total elemental content. Special precautions should be taken to avoid any losses of volatile elements (e.g., As, Cd, Hg) when using these techniques. Method blanks must also be processed through the treatment procedure and analyzed to ensure that the treatment does not bias the accuracy of the food analytical results [104]. For wet digestion many reagents can be used. Nitric acid, hydrochloric acid, sulphuric acid, hydrogen perchlorate, hydrogen peroxide are among the most commonly used and, in particular, the mixture of nitric acid and hydrogen peroxide. Hydrochloric acid is rarely used due to the fact that contains Cl^-, which can react with analytes forming volatile compounds. Hydrogen perchlorate must be used under specific conditions, due to the fact that it is very dangerous. Wet digestion is usually assisted by microwave or ultrasound and other less well-known techniques. For mercury or arsenic speciation a different procedure must be used. This involves the use of H_3PO_4 or HNO_3 at lower concentrations than those used for total elemental content, and the use of other reagents such as EDTA as a cleaning reagent. Microwave assisted or ultrasound extraction techniques are also used but the time and the temperature used are lower than those used for the determination of total trace elemental content [105]. For the majority of toxic elements a wide variety of instrumental techniques are used, such as Electrothermal atomic absorption spectrometry (ETAAS), Inductively coupled plasma–atomic emission spectrometry (ICP–AES), Inductively coupled plasma–mass spectrometry ICP–MS, etc. ETAAS is a method of choice for the routine determination of trace metal content in tomatoes. However, its single-element nature is a severe drawback when the number of metals to be determined and the samples to be analyzed increase. The introduction of multi-elemental ETAAS and High-Resolution Continuum Source Atomic Absorption Spectrometry (H-R CS AAS) system, has increased the analytical potential of this technique. ICP-AES and ICP-MS provide lower detection limits and are multi-element but high cost techniques. For volatile elements, such as As and Hg, hydride generation atomic absorption spectrometry (HGAAS) and cold vapor atomic absorption spectrometry (CVAAS) techniques are used. HGAAS is also used for arsenic speciation. Other techniques for elemental speciation include High-performance liquid chromatography (HPLC) coupled with ICP-MS techniques [104]. For the applicability of the developed "in house" methods, the accepted criteria for analytical method validation must be followed. Accuracy, precision, calibration curves, linearity, methods detection limits, selectivity and robustness must be estimated. For example, for lead and cadmium in tomatoes the limits of detection must be less than 10 μg/kg for lead and 5 μg/kg for cadmium [102]. For inorganic tin in canned tomato food products the limit of detection expressed as total tin content must be less than 5 mg/kg. Precision is estimated by calculating the (%) relative standard deviation under reproducibility conditions and after comparison of the calculated value with the theoretical coming from the Horwitz equation [103]. Method accuracy must be estimated either by the mean of percentage recoveries of the analyte concentrations in spiked tomato sample or by analyzing a certified reference material, such as NIST-SRM 1573a (Tomato leaves). Finally, the expanded uncertainty must be calculated, following, for example, the procedure described in the Eurachem/CITAC Guide [106].

CONCLUSION

Over the last decade the trace element content of food has been of great interest and hence new and innovative approaches to the analysis of food for toxic elements continue to appear in the literature. Humans have traditionally been exposed to low concentrations of metals mainly via water and foods. With time, the range of potentially toxic metal sources has expanded substantially due to increasing agricultural and industrial activities. For the tomato plant, there is a significant correlation between the cultivation method and the trace element concentrations. Lowering of uptake of elements was observed for plants cultivated on acidified soil. As far as human health is concerned, it was shown that lead, cadmium, arsenic and mercury are the metals with the highest toxic potential among those present in tomato fruits and tomato food products and for this reason strict legal restrictions have been adopted. Some other elements such as copper, iron, tin, zinc, antimony, aluminum, manganese and chromium are essential to tomato fruits but can become hazardous under specific conditions and at levels higher than crucial values. Many alternative techniques have been proposed for the determination of toxic elements in tomato fruit and tomato fruit food products. Overall, it seems that the contents of toxic elements in tomato fruits and tomato fruit food products are not at high levels, making them a valuable source for many nutrients.

ABBREVIATIONS

Cold vapour atomic absorption spectrometry (CVAAS)
Coronary heart disease (CHD)
Electrothermal Atomic Absorption Spectrometry (ETAAS)
Food and Agriculture Organization (FAO)
High Performance Liquid Chromatography (HPLC)
High-Resolution Continuum Source Atomic Absorption Spectrometry (H-R CS AAS)
Hydride Generation Atomic Absorption Spectrometry (HGAAS)
Inductively Coupled Plasma–Atomic Emission Spectrometry (ICP–AES)
Inductively Coupled Plasma–Mass Spectrometry ICP–MS
Lowest Threshold Intake (LTI)
Maximum Tolerable Weekly Intake (MTWI)
Natural Tomato Soluble Solids (NTSS)
Polyethylene Terephthalate (PET)
Polytetrafluoroethylene (PTFE)
Provisional Tolerable Weekly Intake (PTWIs)
Tolerable Daily Intake (TDI)
US Environmental Protection Agency (US EPA)
US National Research Council (USNRC)
World Health Organization (WHO)

REFERENCES

[1] FAS, USDA, world markets and trade, office of global analysis, http://www.fas.usda.gov/htp/2007_2008%20Tomato%20Article.pdf
[2] M. Moresi and C. Liverotti, *Int. J. Food Sci. and Tech.* 17, 177 (1982).
[3] W. A. Hayes, P.G Smith and A.E.J. Morris, *Crit. Rev. Food Sci. Nutr.* 38, 537 (1998).

[4] P. Goose and R. Binstead, *Tomato Paste and other Tomato Products,* Food Trade Press Ltd., London (1973).

[5] A. W. Gould, *Tomato Production, Processing and Quality Evaluation*, AVI Publishing, Westport Connecticut (1983).

[6] DAFNE-ANEMOS, Data Food Networking, http://www.hhf-greece.gr/dafnesoftweb/Main.aspx?type=multi (last visited 2-12-2011).

[7] R. Re, G.D. Mishra, C.W. Thane and C.J. Bates, *Europ. J. Clin. Nutr.* 57, 1545 (2003).

[8] M. L. Guerinot, *Biochim. Biophys. Acta Biomembr.* 1465, 190 (2000).

[9] L. E. Williams, J.K. Pittman and J.L. Hall, *Biochim. Biophys. Acta Biomembr.* 1465, 104 (2000).

[10] S. Clemens, M.G. Palmgren and U. Krämer, *Trends Plant Sci.* 7, 309 (2002).

[11] S. Dudka and W.P. Miller, *J. Environ. Sci. Health.* B34, 681 (1999).

[12] F. Burló, I. Guijarro, A.A. Carbonell-Barrachina, D. Valero and F. Martínez-Sánchez, *J. Agric. Food Chem.* 47, 1247 (1999).

[13] M. K. John, C.J. VanLaerhoven and H.H. Chuah, *Environ. Sci. and Technol.* 6, 1005 (1972).

[14] H. F. Wang, N. Takematsu and S. Ambe, *Appl. Rad. Isotopes.* 52, 803 (2000).

[15] Danish Environmental Protection Agency. Guidance Regarding Advice of Inhabitants of Slightly Contaminated Soil (in Danish). Danish EPA, Office of Soil Contamination: Copenhagen, 2000, 75, at http://www.mst.dk/udgiv/publikationer/2000/87-7944-302-8/pdf/87-7944-303-6.pdf.

[16] L. Petersen, E.H. Larsen, P.B. Larsen and P. Bruun, *Environ. Sci. and Technol.* 36, 3057 (2002).

[17] Z. L. He, X.E. Yang and P.J. Stoffella, *J. Trace Elem. in Med. Biol.* 19, 125 (2005).

[18] D. Bourn and J. Prescott, *Crit. Rev. Food Sci. Nutr.* 42, 1 (2002).

[19] L. Malmaurety, D. Parent-Massin, J-L Hardy and P. Vergery, *Food Add. Contam.* 19, 524 (2002).

[20] S. D. Kelly and A.S. Bateman, *Food Chem.* 119, 738 (2010).

[21] V. Fernández-Ruiz, A.I. Olives, M. de Cortes Cámara, M. Sánchez-Mata and M.E. Torija, *Biol. Trace Elem. Res.* 141, 329 (2011).

[22] M. Hernández Suárez, E.M. Rodríguez Rodríguez and C. Díaz Romero, *Food Chem.* 104, 489 (2007).

[23] F. Costa, M. de Lurdes Baeta, D. Saraiva, M.T. Verissimo and F. Ramos, *Food Anal. Meth.* 4, 410 (2011).

[24] J. L. Guil-Guerrero and M.M. Rebolloso-Fuentes, *J. Food Comp. Anal.* 22, 123 (2009).

[25] K. Bakkali, N.R. Martos, B. Souhail and E. Ballesteros, *Food Chem.* 116, 590 (2009).

[26] N. Bagdatlioglu, C. Nergiz and P.G. Ergonul, *J. Verbr. Lebensmittelsich.* 5, 421 (2010).

[27] M. S. Bratakos, E.S. Lazos and S.M. Bratakos, *The Sci. Total Envir.* 290, 47 (2002).

[28] P. Ekholm, H. Reinivuo, P. Mattila, H. Pakkala, J. Koponen, A. Happonen, J. Hellström and M.-L. Ovaskainen, *J. Food Comp. Anal.* 20, 487 (2007).

[29] M. Bronkowska, D. Figurska-Ciura, D. Orzeł, M. Styczyńska, J. Wyka, K. Loźna, A. Zechałko-Czajkowska and J. Biernat, *Food Chem.* 109, 4 (2008).

[30] A. Demirbas, *Food Chem.* 118, 504 (2010).

[31] F. Rossi, F. Godani, T. Bertuzzi, M. Trevisan, F. Ferrari and S. Gatti, *Europ. J. Nutr.* 47, 266 (2008).

[32] P. C. Onianwa, A.O. Adeyemo, O.E. Idowu and E.E. Ogabiela, *Food Chem.* 72, 89 (2001).
[33] C. P. Sanchez-Castillo, P.J.S. Dewey, A. Aguirre, J.J. Lara, R. Vaca, P. L. de la Barra, M. Ortiz, I. Escamilla and W.P.T. James, *J. Food Comp. Anal.* 11, 340 (1998).
[34] M. Tüzen, *Microchem. J.* 74, 289 (2003).
[35] K. S. Ferreira, J.C. Gomes and J.B.P. Chaves, *Food Chem.* 92, 29 (2005).
[36] P. Smrkolj, L. Pograjc, C. Hlastan-Ribi and V. Stibilj, *Food Chem.* 90, 691 (2005).
[37] P. P. Sirichakwal, P. Puwastien, J. Polngam and R. Kongkachuichai, *J. Food Comp. Anal.* 18, 47 (2005).
[38] A, Baysal, M. Ozcan and S. Akman, *Food Chem. Toxicol.* 49, 1399 (2011).
[39] C. F. Oduoza, *Food Chem.* 44, 9 (1992).
[40] L. Jorhem and S. Slorach, *Food Add. Contam.* 4, 309 (1987).
[41] I. N. Pasias, V. Papageorgiou, N.S. Thomaidis, and Ch. Proestos, *Food Anal. Meth.* in press (2011).
[42] U. Rafique, S. Iqbal, S. Faiz and A. Hashmi, *J. Food Proc. Preserv.* 33, 186 (2009).
[43] M. Tuzen and M. Soylak, *Food Chem.* 102, 1089 (2007).
[44] S. Yenisoy-Karakaş, *Food Chem.* in press (2011).
[45] A. Waheed, M. Jaffar and K. Masud, *Nutr. Food Sci.* 33, 261 (2003).
[46] S. Demirel, M. Tuzen, S. Saracoglu and M. Soylak, *J. Hazard. Mater.* 152, 1020 (2008).
[47] J. H. Hotchkiss, *Food Add. Contam.* 14, 601 (1997).
[48] L. M. Gaetke and C.K. Chow, *Toxicol.* 189, 147 (2003).
[49] J. R. Turnlund, *Modern Nutrition in Health and Disease*, Lippincott Williams and Wilkins, Baltimore MD (1999).
[50] R. Uauy, M. Olivares and M. Gonzalez, *The Amer. J. Clin. Nutr.* 67, 952S (1998).
[51] WHO, *Trace elements in human nutrition and health*, Geneva (1996).
[52] J. R. Turniund, K.C. Scott, G.L. Peiffer, A.M. Jang, W.R. Keyes, C.L. Keen, and T.M. Sakanashi, *The Amer. J. Clin. Nutr.* 65, 72 (1997).
[53] P. B. Vose, *J. Plant Nutr.* 5, 233 (1982).
[54] B. N. Smith, *J. Plant Nutr.* 7, 759 (1984).
[55] M. Shenker, I. Oliver, M. Helmann, Y. Hadar and Y. Chen, *J. Plant Nutr.* 15, 2173 (1992).
[56] C. G. Fraga and P.I. Oteiza, *Toxicol.* 180, 23 (2002).
[57] Kato, A.M. Dnistrian, M. Schwartz, P. Toniolo, K. Koenig, R.E. Shore, A. Zeleniuch-Jacquotte, A. Akhmedkhanov and E. Riboli, *Int. J. Canc.* 80, 693 (1999).
[58] M. L. Rasmussen, A.R. Folsom, D.J. Catellier, M.Y. Tsai, U. Garg and J.H. Eckfeldt, *Atheroscler.* 154, 739 (2001).
[59] A-L.M. Heath and S.J. Fairweather-Tait, *Clin. Haematol.* 15, 225 (2002).
[60] S. Blunden and T. Wallace, *Food Chem. Toxicol.* 41, 1651 (2003).
[61] Commission Regulation (EC) No 1881/2006, *Off. J. Europ. Commun.* L364, 5 (2006).
[62] E.S. Boa Morte, M.G.A. Korn, M.L.M.F.S. Saraiva, J.L.F.C. Lima and P.C.A.G. Pinto, *Talanta.* 79, 1100 (2009).
[63] J. Knápek, V. Herman, R. Buchtová and D. Vošmerová, *Cz. J. Food Sci.* 27, S407 (2009).
[64] C. Kaya and D. Higgs, *Sci. Horticult.* 93, 53 (2002).
[65] C. Kaya and D. Higgs, *J. Plant Nutr.* 24, 1491 (2001).

[66] C. Kaya, M.A.S Burton and D.E.B. Higgs, *J. Plant Nutr.* 24, 369 (2001).
[67] J. Nriagu, *Zinc Toxicity in Humans*, School of Public Health, University of Michigan, Elsevier B.V. (2007).
[68] G. J. Fosmire, *The Amer. J. Clin. Nutr.* 1, 225 (1990).
[69] R. D. Palmiter, *PNAS*. 101, 4918, (2004).
[70] A. Leufvh and C. Hermansson, *J. Sci. Food Agric.* 64, 101 (1994).
[71] C. Camelia and T.I. Trască, *Bulletin USAMV-CN.* 63, 497 (2007).
[72] F. Welle and R. Franz, *Food Add. Contam.* 28, 115 (2011).
[73] I.I. Roberts and O.E. Orisakwe, *Sci. Connect.* 2, 1 (2011).
[74] S.-Y. Choe, S.-J. Kim, H.-G. Kim, J.H. Lee, Y. Choi, H. Lee and Y. Kim, *The Sci. Total Envir.* 312, 15 (2003).
[75] European Union, European Commission Directive 2002/72/EC, Last amendment: Commission Directive 2008/39/EC (2008).
[76] Japan External Trade Organisation (JETRO), 2010, http://www.jetro.go.jp/en/reports /regulations/pdf/foodext2010e.pdf.
[77] World Health Organization (WHO). Geneva, Switzerland WHO/SDE/WSH/03.04/74, 2003.
[78] O. Andersen, *Chem. Rev.* 99, 2683 (1999).
[79] L. Simon, T.J. Smalley, J. Benton Jones Jr. and F.T. Lasseigne, *J. Plant Nutr.* 17, 293 (1994).
[80] K. B. Searcy and D.L. Mulcahy, *Theor. Appl. Gen.* 80, 289 (1990).
[81] J. P. Miller, A. Steinegger and C. Schlatter, *Z. Lebensm. Unters. Forsch.* 197, 332 (1993).
[82] P. Rajwanshi, V. Singh, M.K. Gupta and S. Dass, *Envir. Geochem. Health.* 19, 1 (1997).
[83] E. H. Jeffery, K. Abreo, E. Burgess, J. Cannata and J.L. Greger, *J. Toxicol. Envir. Health.* 48, 649 (1996).
[84] J. Crossgrove and W. Zheng, Biomed. 17, 544 (2004).
[85] G. M. Ward, *Can. J. Plant Sci.* 57, 107 (1977).
[86] J. Le Bot, E.A. Kirkby and van M.L. Beusichem, *J. Plant Nutr.* 13, 513 (1990).
[87] C. Santos-Burgoa, C. Rios, L.A. Mercado, R. Arechiga-Serrano, F. Cano-Valle, R.A. Eden-Wynter, J.L. Texcalac-Sangrador, J.P. Villa-Barragan, Y. Rodriguez-Agudelo and S. Montes, *Envir. Res. (A).* 85, 90 (2001).
[88] K. Marsh and B. Bugusu, *J. Food Sci.* 72, R39 (2007).
[89] Environmental Protection Agency, Integrated Risk Information System, US (2001).
[90] M. Aschner, T.R. Guilarte, J.S. Schneider and W. Zheng, *Toxicol. Appl. Pharmacol.* 221, 131 (2007).
[91] M. Costa and B.C. Klein, *Crit. Rev. Toxicol.* 36, 155 (2006).
[92] M. Costa, *Crit. Rev. Toxicol.* 27, 431 (1997).
[93] A. Zhitkovich, *Chem. Res. Toxicol.* 18, 3 (2005).
[94] E. Rojas, L.A. Herrera, L.A. Poirier and P. Ostrosky-Wegman, *Mut. Res.* 443, 157 (1999).
[95] NRC, *Recommended Dietary Allowances,* National Academy Press, Washington, DC (1989).
[96] W. Jr. Banner, M. Koch, D.M. Capin, S.B. Hopf, S. Chang and T.G. Tong, *Toxicol. Appl. Pharmacol.* 83, 142 (1986).

[97] A. R. Marin, S.R. Pezeshki, P.H. Masscheleyn and H.S. Choi, *J. Plant Nutr.* 16, 865 (1993).

[98] WHO, *Environmental Health Criteria,* World Health Organization, Geneva (1992).

[99] WHO, *Environmental Health Criteria,* World Health Organization, Geneva (1995).

[100] IARC, *Beryllium, Cadmium, Mercury and Exposure in the Glass Manufacturing Industry,* International Agency for Research on Cancer, Lyon (1993).

[101] A. S. Beidas and D.J. Higgons, *J. Sci. Food Agric.* 8, 597 (1957).

[102] Commission Regulation (EC) No 333/2007, *Off. J. Europ. Comm.* L88, 29 (2007).

[103] Commission Regulation (EC) No 836/2011, *Off. J. Europ. Comm.* L215, 9 (2011).

[104] S. G. Capar, W.R. Mindak and J. Cheng, *Anal. Bioanal. Chem.* 389, 159 (2007).

[105] K. A. Francesconi and D. Kuehnelt, *Analyst*, 129, 373 (2004).

[106] Eurachem Citac Guide CG4, *Quantifying Uncertainty in Analytical Measurement*, English edition (2000).

In: Tomatoes: Cultivation, Varieties and Nutrition
Editor: Tadahisa Higashide
ISBN: 978-1-62417-915-0

Chapter 17

NUTRACEUTICALS RECOVERY FROM TOMATO PROCESSING WASTE AND BY-PRODUCTS: LYCOPENE

Antonello Santini[1*], Giulia Graziani[1] and Alberto Ritieni[2]
[1]Department of Food Science, University of Napoli "Federico II," Portici (Napoli), Italy
[2]Department of Pharmaceutical and Toxicological Chemistry, University of Napoli "Federico II," Napoli, Italy

ABSTRACT

Nutraceuticals are food or food products that can have health and medical benefits, e.g. the prevention and treatment of diseases. Among these, carotenoids, have a special interest for their high activity as natural antioxidants compounds. These molecules can play a relevant physiological role, and are contained in many fruits and vegetables like tomatoes. There is a growing interest worldwide by many companies in performing the extraction of carotenoids (lycopene, beta-carotene, etc.) from tomato wastes (peels and seeds) in order to meet the requirements and need of the international nutraceuticals, food additives, pharmaceutical, and cosmetic markets. The health benefits of these dietary antioxidants and their use in nutrition to supplement and correct dietary needs, has increased the demand for fruit and vegetable products with recognized and improved antioxidant quality. Tomato (*Solanum Lycopersicon*) is particularly rich in these molecules that are only partially recovered and available after the industrial processing. Lycopene can be an example of the versatility of the carotenoids. It is often included among the major components of pharmaceuticals for the treatment of many pathologies, and it is widely used in cosmetics formulations for its high solubility in fats and oils. It is worth to note that carotenoids like lycopene are still among the major components of waste originating from tomato processing plants, and are dumped with the waste, considered as low value by-product and useless material to discard. The amount of tomato by-products deriving from industrial processes is growing annually. In Europe it is estimated that about ten million tons of tomatoes are processed by the food industry, and the deriving wastes can be quantified in about 0.1 million tons. These residues can be

* Corresponding author: Antonello Santini, Department of Food Science, University of Napoli "Federico II", Via Università 100, 80055 Portici (Napoli), Italy. Tel/Fax.: 0039 (81) 2539317, E-mail address: asantini@unina.it.

used as a source of carotenoids and mainly of lycopene, reincorporated in low-quality tomato products, and also used as an ingredient addressed to animal foodstuff. Countries insisting on the Mediterranean sea, e.g. Italy, are the main producers of tomatoes in Europe, and the waste and residues of the industrial processing of tomatoes represent a low cost and relatively cheap source of molecules with nutraceutical potential use.

INTRODUCTION

Lycopene (*2, 6, 10, 14, 19, 23, 27, 31-Octamethyl-dotriaconta-2, 6, 8, 10, 12, 14, 16, 18, 20, 22, 24, 26, 30-tridecaene*), whose structural formula is shown in Figure 1, is one of the most extensively studied tomato carotenoids [1]. The molecule, contains thirteen double bonds, eleven of which are conjugated, and this structural feature makes it one of the known compounds with the highest molar absorption coefficient value, and determines also its high antioxidant activity. For this reason, lycopene is among the major components of pharmaceuticals for the treatment of prostate cancer and digestive-tract tumours [2-8], it is widely used in cosmetics formulations [9], and, for its high solubility in fats and oils, it is widely used in food and beverages industry and as a colorant in many foodstuffs [10]. As a consequence of an increasing demand, the request for high-purity, low-cost lycopene is growing continuously. The easiest way for obtaining lycopene is its extraction from natural sources. Tomatoes are rich in this compound, and contain about 30-400 mg kg^{-1} in pulp and about 20-30 mg kg^{-1} in peels [11]. These vegetables are usually consumed fresh or cooked. Due to the heating process in the cooking or industrial processing of the fresh product, the final antioxidant lycopene content is different due to a possible heat driven cis-trans isomerism. It has been reported for example that the raw tomato has a content of 2.01 ± 0.04 mg of *trans*-lycopene g^{-1} of tomato, and after 2, 15, and 30 min of heating at 88 °C, the *trans*-lycopene content had increased to 3.11± 0.04, 5.45 ± 0.02, and 5.32 ± 0.05 mg of *trans*-lycopene g^{-1} of tomato ($p < 0.01$) [12]. These data, however, are controversial, and the food processing of the tomatoes can alter the original characteristics and properties [13, 14]. The *trans* form predominates on the *cis* form, more present in the fresh tomatoes, and this form that can be converted to the less bio available all-trans conformation by heating treatment [14]. The industry uses extensively tomatoes for the production of derivatives addressed to the consumers and food industry, e.g. paste, juice, etc., and canned tomatoes, producing a great amount of by products (skin, seeds, etc.) considered as a waste residual to dump.

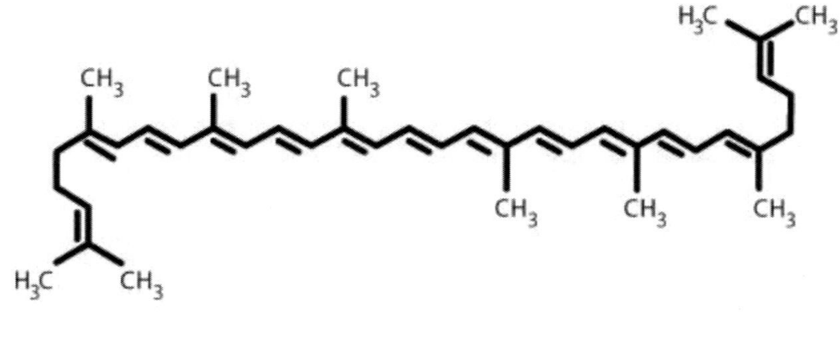

Figure 1. Structure of lycopene.

Considering the increasing amount of tomato by-products deriving from industrial processes that is growing annually, an economic way to recycle these waste and by products into high value nutraceutical value products containing lycopene it would be highly useful.

In Europe it is estimated that about ten million tons of tomatoes are processed by the food industry and the deriving wastes can be quantified in about 0.1 million tons. Their use as a possible source of lycopene could be a safe way to recycle wastes and benefit also a good positive impact on the market for the recovery of an high value nutraceutical value molecule. Moreover, after the extraction and recovery of lycopene, as a possible further use, the remaining residues can be reincorporated in low-quality tomato products, used in animal food as ingredient. This would highly diminish the environmental impact of dumping the waste from tomatoes processing and would produce an high value product recycling residues from industrial processing.

The overall procedures used at present to obtain pure lycopene are based on a preliminary extraction from natural sources mainly tomatoes, but also red vegetables, e.g. red pepper, also a source of capsantin, water melon, and papaya [15], followed by a purification process. All the procedures require at least one organic solvent in the extraction step, which determines some disadvantages: (i) organic solvents could be potentially harmful for health, so they must be completely removed from the exhausted material via a sequence of purification steps before disposal; (ii) after the extraction, tomatoes exhausted matrixes must be dumped as "special residue", and they cannot be re-used; (iii) when using organic solvents molecules different from lycopene, e.g. carotenes, xanthophylls and lipids, may be extracted at the same time.

This last aspect makes lycopene purification usually realized by HPLC, and this procedure is often laborious and time consuming, since ternary or quaternary solvent gradients must be used. In the case of tomato-waste, lycopene recovery can be performed also by supercritical extraction using carbon dioxide, which is quite a safe solvent capable to extract a greater quantity of carotenoids [16]. The maximum total lycopene content of 31.25 g/g of raw tomato was extracted at the highest temperature of 100 °C, and 40 MPa pressure. Unfortunately this method is quite expensive, making the whole process economically not suitable on a large scale, and, moreover, the final product is an oleoresin containing a low percentage of lycopene. Biotechnological production of lycopene using microorganisms has been reported, mainly at the bench scale, and, in most cases, obtaining a low specific yield. Recently a new biotechnology approach that can be scaled-up to an industrial application has been proposed [17]. Considering that the extraction procedures proposed up to now require the use of non aqueous solvents, are time consuming and, consequently seem not to be economically appealing, it is not surprising that has been raised the need of methods for extracting lycopene capable of an high yield and an higher purity level with respect to the commercially available products [18-20].

Lycopene from tomato wastes can be efficiently extracted using a pressurized extraction method that uses the Extractor Naviglio® [21] at a pressure in the 0.7-0.9 MPa range, using as starting material tomatoes skin residues from tomatoes processing in a water dispersion. Lycopene, even if not soluble in water, forms molecular aggregates [22] which can be recovered by the pressure/de-pressure cycles operating into the extractor device. Through a cyclic variation of the pressure gradient between the inner and the outlet of the solid matrix lycopene aggregates are forced to separate. The procedure allows to transfer lycopene aggregates to water as a dispersion, while non polar compounds, insoluble in water, remain in the matrix. Afterwards the aggregates are easily purified in one step by solid phase extraction using small amounts of methanol, and lycopene at more than 98% chromatographic purity is obtained after recovery with a small quantity of solvent acetone that can be easily removed.

LYCOPENE SOURCES: SPECIAL INTEREST OF TOMATO

Lycopene, like other carotenoids, progressively accumulates in the outer skin layer of vegetables during the ripening process. This fat soluble molecule is one of the most extensively studied natural carotenoids, and at the same time is the principal compound responsible for the characteristic red colour of tomatoes. The carotenoids, in fact, are natural pigments that provide the natural yellow, orange, and red colours of fruits, vegetables, and plants. These colours are a result of the presence of conjugated double bonds that also provide the carotenoids with antioxidant properties by the successful delocalization of captured free radical species. Good amount of lycopene is contained in many natural products: tomato (*Solanum Lycopersicon*) in particular has a special interest for its high content in this compound representing an interesting natural source of this molecule. Moreover it is worth to note that tomato is among the most five most important vegetable crops worldwide, second only to potato. Some data regarding lycopene content in tomato common foodstuff are reported in Table 1. Lycopene is a precursor of β-carotene and has a well known activity against the free radicals, reported as at least twice that of the β-carotene itself, it is stable at low temperature and away from light and atmospheric oxygen [23-25]. Due to its pharmacological, anti carcinogenic properties [26-28] and potent antioxidant activity [29] this molecule has been extensively studied, and lycopene has been proved to be effective against some types of cancer, e.g., stomach and prostate cancer, making the demand for this molecule, and for the carotenoids in general, rapidly growing [30-32]. Different synthetic pathways for this molecule have been reported, but in all cases the synthesis of lycopene seems to be a very expensive and economically not convenient procedure [33]. As an example it could be mentioned that the average content of this compound is comprised in the range between 0.00540-1.50 g kg^{-1} in tomato paste, and between 0.020-0.030 g kg^{-1} in tomato peels as raw material, respectively [34-36].

Table 1. Lycopene content reported data from some common food sources

Source	**Quantity mg g^{-1}**
Fresh tomato	8.8-42
Tomato juice	86-100
Tomato sauce	63-131
Tomato ketchup	90-124
Tomato paste	5.4-1500
Tomato peels	20-30

TOMATO WASTE AND BYPRODUCTS RECYCLE TO OBTAIN LYCOPENE

Tomato waste can be considered a special residue of industrial processing, are a bio organic material, and need attention prior to their safely disposal for their potential impact on the environment. The large amount of lycopene and other carotenoids contained by tomato processing by-products make their recycling desirable both from an environmental point of view. The recovery of bioactive compounds like lycopene is in line with a trend of both sustainability and recycle, since this molecule can be recovered from the waste and used as functional ingredient, the formulation of functional food, and as nutraceutical. The extraction of lycopene from natural sources is usually realized using organic solvents, and the extraction procedure is usually followed by a purification step. In the case of lycopene recovery from the tomato-waste, the extraction can also be conducted using supercritical fluid carbon dioxide, and this method allows to recover a greater quantity of carotenoids even if partly mixed in an oleoresin containing around 6% in lycopene [37, 38]. The use of supercritical fluids is quite expensive, making the entire procedure economically not suitable for a large scale process. Production of lycopene using micro-organisms has been proposed too, but always at the bench scale and, in most cases, with a weak specific yield, even if recently a new approach that can be scaled-up to an industrial application has been presented [39, 40]. It is worth to note that Italy is among the main producers of tomatoes in Europe, and that the byproducts and the residues of the industrial processing of tomatoes, now usually reincorporated in low quality tomato products or used as an ingredient in animal food, could be more efficiently used as a source of pure lycopene. The waste quantities deriving from industrial processes are estimated in Europe to be about 0.1 million tons, and represent an interesting low cost recyclable source of lycopene. The availability of the by products from tomato industry and the relevant interest for this molecule and its biological activity, leads to the consideration that better and optimized methods for the extraction and purity characterization and determination of lycopene are needed [35]. New methods, like the high pressure extraction using tap water as extracting solvent of the waste and by products from tomato processing to recover lycopene and valuable compounds have been explored and lead to promising results [21, 42, 43]. Carbon dioxide extraction is in itself a valuable technique and allows to obtain lycopene [44, 45]. Recently an enzyme assisted process to recover lycopene has been proposed with promising results [46]. The starting materials for these extraction procedures can be the solid residue remaining after the industrial processing of tomatoes, that is considered a promising candidate materials for the lycopene recovery. The residue and by products consist mainly of tomato peels and seeds, constituting about 3–5% of total weight of processed tomatoes [47] according to an estimate of the World Processing Tomato Council (WPTC) [48]. Tomato peels and seeds are still rich source of lycopene, since this carotenoid progressively accumulates in the outer skin layer of tomatoes [49]. Lycopene recovery from this material is not straightforward, and in general low yields are achievable with conventional solvent extraction procedures that use organic solvents such as hexane, ethanol, acetone, methanol, tetrahydrofuran, and petroleum ether [50-53].

This could be attributed to the compactness of the plant tissue, preventing the solvent penetration, and the possible degradation of the carotenoid during recovery. In fact, once released from the protective chromoplast environment, lycopene becomes highly susceptible to degradation [54]. *Trans*-to-*cis* isomerization is the first step of degradation, followed by oxidation that forms apo-lycopenals and apo-lycopenones [54, 55]. An extraction method operating under mild conditions and allowing the solvent to overcome mass-transfer limitations in the plant matrix would be definitively advisable. The use of tap water and high pressure is a promising new and safer and fast method of extraction avoiding massive organic solvent use [21]. This procedure is a promising low cost alternative to the solvent extraction and to the other described processes and allows to obtain lycopene in the all *trans* form at a very high grade of purity, not less than 98 % (w/w), with an average recovery of 14 % (w/w) [21]. The use of water and the reduced use of organic solvent represent also an high value to the extraction process, since both the amount and the price of organic solvents directly influence the total cost of producing an acceptable extract/product. Moreover, when the final product is used as a food ingredient, it is absolutely necessary to remove all potentially toxic solvents. As an alternative the use of enzymes has been evaluated. Enzymes catalyzing the hydrolytic cleavage of cell-wall polysaccharides have been successfully utilized to facilitate the release of vegetable oils [56], non-volatile aroma precursors [57], phenols [58] and carotenoids [59-61] from plant materials. Commercial food-grade enzyme preparations with pectinolytic, cellulolytic, and hemicellulolytic activities have been reported to give a good lycopene recovery from tomato skins [62] and tomato paste [63]. The treatment was reported as effective even using short extraction times and room temperature, conditions preventing lycopene degradation [64].

Nutraceutical Interest of Lycopene from Tomato

The term nutraceutical comprises the two words nutrition and pharmaceutical. It can be defined as a substance considered a food or component of a food and provides medical or health benefits including the prevention and treatment of disease. These products range from nutrients to dietary, from supplements to engineered food, herbal products, processed foods e.g. cereals, soups, beverages. The number of nutraceutical substances is very high, and includes different molecules with different chemical nature. The antioxidant substances are among these, and carotenoids are of special interest for their well known positive health effect. Since 1986 it has been shown that retinol and several carotenoids (β-carotene, β-cryptoxanthin, zeaxanthin, lutein, capsorubin, capsanthin, lycopene, and capsanthol) were involved in the cytoprotective injury of gastric mucosa [65]. For this reason these compounds are widely used in foodstuff. Due to the colorant properties, carotenoids that have provitamin A activity, are also widely used as colorants in food. The specific action of lycopene as natural food colorant has been indicated since 1990 [66], and lead to a widespread use of tomato paste as a colorant making that at the same time lycopene a commercially interesting natural pigment.

Antioxidant compounds like lycopene can be added directly to many food products such as butter, popcorn, salad dressings, and beverages or indirectly via animal uptake, for example, chicken and fish [67-69]. Tomato is of special interest for the high lycopene content even if the extraction of this compound is strongly influenced by the extraction procedures [70-72].

The inclusion of lycopene from tomatoes in foodstuff lead to beneficial effects as antioxidant [73, 74] and it would advisable its wide use. The consumption of tomato sauce or juice, fresh tomato and other foods that contain tomato as an ingredient are the major way for assuming this compound with the diet. Other forms of ingestion are represented by the many the synthetic compounds that exist on the market, which can be added to the formulation of different products or consumed in the form of dietary supplements, such as capsules.

The action of lycopene in the risk reduction of diseases like prostate cancer has been documented and values from 1.5 mg day^{-1} to 20 mg day^{-1}, reported suggesting an ingestion of 6.46 mg day^{-1} as effective to reduce the risk of cancer [77, 78]. Tomato enriched snacks can be considered a good option for the assumption of this potent antioxidant [79]. Extruded snacks enriched in lycopene from tomatoes have been already tested and different foodstuff, dairy products e.g. butter, ice cream, and mayonnaise, added with lycopene showed an improved shelf life and bioavailability of lycopene [80, 81] indicating a wide possible use of this molecule as nutraceutical natural ingredient for a wide variety of foodstuff.

CONCLUSION

Nutraceuticals are a novel and interesting category of foodstuff, and are food product that provides health as well as medical benefits including the possible prevention and treatment of health conditions. Phytochemicals and antioxidants are two specific types of nutraceuticals. Foods containing phytochemicals may help to provide protection from diseases such as cancer, diabetes, heart disease, and hypertension. In the last 20 years, many nutraceuticals have become available for self-medication and there is a growing interest for these products due to possible adverse effects of pharmaceuticals and both increased tendency of patients for self-medication.

Lycopene from tomato is an important active molecule that can be recovered from by products and waste of the industrial processing of fresh tomatoes, and since it maintains its potent antioxidant activity that protects cells against free radicals, can be added to foodstuff as nutraceutical ingredient. Its consumption, associated with a balanced diet and a healthy lifestyle, has a protective action and health benefit. The extraction process that uses water as extracting solvent and high pressure, could be an efficient and environmental friendly new approach for lycopene extraction from tomato waste and byproducts, a very low cost starting material. The incorporation of ingredients with functional characteristics like lycopene in food derivatives, e.g. snacks, is an alternative to add nutrients and/or compounds that are beneficial to health, and that are normally found in fresh tomatoes or other vegetables. In general raw fresh tomatoes, retain the high antioxidant original content, and, on the other hand, it can be proven difficult for raw tomatoes to be pleasing to the taste and hence consumed by children or by people having a non vegetarian dietary habit. To avoid this aspect an alternative could be represented by the incorporation in foodstuff of lycopene extracted from by products and waste from tomato industry. These residue can be recycled and represent a rich source of low cost lycopene, and can be considered a great starting material to recover this important molecule that can benefit and add nutraceutical value to foodstuff.

REFERENCES

[1] M. Dorais, D. L. Ehret, A. P. Papadopoulos, *Phytochemistry Reviews* 7, 231 (2008).
[2] G. Lippi and G. Targher, *British Journal of Cancer* 104, 1234 (2011).
[3] K. Hirsch, A. Atzmon, M. Danilenko, J. Levy, Y. Sharoni, *Breast Cancer Res. Treat.* 104, 221 (2007).
[4] S. Franceschi, E. Bidoli, G. La Vecchia, R. Talamini, B. D'Avanzo, E. Negri, *Int. Journal Cancer* 59, 181 (1994).
[5] H. Sies and W. Stahl, *Proc. Exp. Biol. Med.* 218, 121 (1998).
[6] E. Giovannucci, *J. Natl. Cancer Inst.* 91, 317 (1999).
[7] J. Levi, S. Walfisch, A. Atzmon, K. Hirsch, M. Khanin, K. Linnewiel, Y. Morag, H. Salman, A. Veprik, M. Danilenko, *Diet and Cancer* 2, 47 (2011).
[8] M. Stacewicz-Sapuntzakis and P. E. Bowen, *Biochimica et Biophysica Acta* 1740, 202 (2005).
[9] E. Kanno, Patent No. JP 2003212709 (2003).
[10] P. Chattopadhyay, S. Chatterjee, K.S. Sukanta, *African Journal of Biotechnology* 7, 2972 (2008).
[11] A.R. Davis, W.W. Fish, P.A. Perkins-Veazie, *Postharvest Biol. Technol.* 28, 425 (2005).
[12] V. Dewanto, X. Wu, K. K. Adom, R. Hai Liu, *J. Agric. Food Chem. 50*, 3010 (2002).
[13] J. Shi and M. Le Maguer, *Crit. Rev. Food Sci. Nutr.* 40, 1 (2000).
[14] I. Colle, L. Lemmens, S. Van Buggenhout, A. Van Loey, M. Hendrickx, *Journal of Food Science* 75, C753 (2011).
[15] C.Y. Wang and B.H. Chen, *European Food Research and Technology* 222, 347 (2006).
[16] U. Topal, M. Sasaki, M., Goto, K. Hayakawa, *J. Agric. Food Chem.* 54, 5604 (2006).
[17] M.J. López-Nieto, J. Costa, E. Peiro, E. Méndez, M. Rodríguez-Sáiz, J.L. de la Fuente, W. Cabri, J.L. Barredo, *Applied Microbiology and Biotechnology* 66, 153 (2004).
[18] P.M. Choksi and V.Y. Joshi, *Int. J. Food Prop.* 10, 289 (2007).
[19] R.S. Bruno, R. E.C. Wildman, S.J. Schwartz, R.E.C. Wildman (Ed.) Handbook of Nutraceuticals and Functional Foods, Second Edition, CRC Press, Boca Raton, FL, USA, 55-72 (2007).
[20] J.M. Roldàn-Gutiérrez and M.D. Luque de Castro, *Trends in Anal. Chem.* 26, 163 (2007).
[21] D. Naviglio, T. Caruso, P. Iannece, A. Aragon, A. Santini, *J. Agric. Food Chemistry* 56, 6227 (2008).
[22] L. Wang, Z. Du, R. Li, D. Wu, *Dyes and Pigments* 65, 15 (2004).
[23] H. Sies and W. Stahl, *Proc. Exp. Biol. Med.* 218, 121 (1988).
[24] P. Di Mascio, S.P. Kaiser, H. Sies, *Arch. Biochem. Biophys.* 274, 179 (2002).
[25] M.T. Lee and B.H. Chen, *Food Chemistry* 78, 425 (2002).
[26] S. Franceschi, E. Bidoli, C. La Vecchia, R. Talamini, B. D'Avanzo, E. Negri, *Int. J. Cancer* 59, 181 (1994).
[27] E. Giovannucci, *J. Natl. Cancer Inst.* 91, 317 (1999).
[28] O. Livny, I. Kaplan, R. Reifen, S. Polak-Charcon, Z. Madar, B. Schwartz, *J. Nutrition* 132, 3754 (2002).
[29] D. Heber and Q.Y. Lu, *Exp. Biol. Med.* 227, 920 (2002).

[30] M. Stacewicz-Sapuntzakis and P.E. Bowen, *Biochimica et Biophysica Acta* 1740, 202 (2005).

[31] H. Sies and W. Stahl, *Int. Journal Vitam. Nutr. Res.* 73, 95 (2003).

[32] M. Ulrich, Business report: the global market for carotenoids. Business Communication. Norwalk: England, 2000.

[33] L. Zhuo-cai, L. Bo, Y. Hong, C. Zhi-rong, *Hecheng Huaxue* 14, 118 (2006).

[34] D. Montesano, F. Fallarino, L. Cossignani, A. Bosi, M.S. Simonetti, P. Puccetti, P. Damiani, *European Food Research and Technology* 226, 327 (2008).

[35] J.M. Roldàn-Gutiérrez and M.D. Luque de Castro, *Trends in Anal. Chem.* 26, 163 (2007).

[36] N.L. Rozzi, R.K. Singh, R.A. Vierling, B.A. Watkins, *Journal Agric. Food Chem.* 50, 2638 (2002).

[37] A.I. Olives Barba, M. Cámara Hurtado, M.C. Sanchez Mata, V. Fernandez Ruiz, M. Saenz de Tejada, S.D. Lopez, *Food Chemistry* 95, 328 (2005).

[38] E. Vági, B. Simándi, K.P. Vásárhelyine, H. Daood, Á. Kéry, F. Doleschall, B. Nagy, *Journal of Supercritical Fluids* 40, 218 (2007).

[39] G. Sandmann, *Eur. J. Biochem.* 223, 7 (1994).

[40] M.J. López-Nieto, J. Costa, E. Peiro, E. Méndez, M. Rodríguez-Sáiz, J.L: de la Fuente, W. Cabri, J.L. Barredo, *Applied Microbiology and Biotechnology* 66, 153 (2004).

[41] L. Sandei and C. Leoni, Acta Horticulturae 724, 249 (2006).

[42] F. Strati Irini and V. Oreopoulou, *Food Chemistry* 129, 747 (2011).

[43] A. Benakmoum, S. Abbeddou, A. Ammouche, P. Kefalas, D. Gerasopoulos, *Food Chemistry* 110, 684 (2008).

[44] S. Machmudah, Zakaria, S. Winardi, M. Sasaki, M. Goto, N. Kusumoto, K. Hayakawa, *Journal of Food Engineering* 108, 290 (2011).

[45] H. Li, Z. Deng, R. Liu, J.C. Young, Z. H. Zhu, Honghui, S. Loewen, R. Tsao, *Journal of Agricultural and Food Chemistry* 59, 11803 (2011).

[46] A. Zuorro, M. Fidale, R. Lavecchia, *Enzyme and Microbial Technology* 49, 567 (2011).

[47] M. Del Valle, M. Cámara, M.E. Torija, *J. Sci. Food Agric.* 86, 1232 (2006).

[48] WPTC. World production estimate as of 18 October 2010, Release # 33.

[49] S.K. Sharma, L.M. Maguer, *Italian J. Food Science* 2, 107 (1996).

[50] N.L. Rozzi, R.K. Singh, R.A. Vierling, B.A. Watkins, *Journal Agric. Food Chem.* 50, 2638 (2002).

[51] E. Sabio, M. Lozano, V. Montero de Espinosa, R.L. Mendes, A.P. Pereira, A.F. Palavra, *Ind. Eng. Chem. Research* 42, 664 (2003).

[52] M.J. Periago, F. Rincon, M.D. Aguera, G. Ros, *Journal Agric. Food Chem.* 52, 5796 (2004).

[53] I.F. Strati, V. Oreopoulou, *Int. Journal Food Sci. Technol.* 46, 23 (2011).

[54] C. Leoni, *Acta Horti* 613, 357 (2003).

[55] C.C. Caris-Veyrat, A. Schmid, M. Carail, V. Böhm, *Journal Agric. Food. Chem.* 51, 7318 (2003).

[56] S. Xianquan, J. Shi, Y. Kakuda, J. Yueming, *Journal Med. Food* 8, 413 (2005).

[57] P. Hanmoungjai, D.L. Pyle, K. Niranjan, *J. Chem. Technol Biotechnol.* 77, 77 (2002).

[58] A.B. Bautista Ortin, A. Martinez Cutillas, J.M. Ros Garcia, J.M. Lopez Roca, E. Gomez Plaza, *Int. J. Food Science Technol.* 40, 867 (2005).

[59] M. Pinelo, B. Zornoza, A.S. Meyer, *Sep. Purif. Technol.* 63, 620 (2008).

[60] R.I. Santamaria, M.D. Reyes-Duarte, E. Barzana, D. Fernando, F.M. Gama, M. Mota, *Journal Agric. Food Chem.* 48, 3063 (2000).

[61] E. Barzana, D. Rubio, R.I. Santamaria, O. Garcia-Correa, F. Garcia, V.E. Ridaura Sanz, *Journal Agric. Food Chem.* 50, 4491 (2002).

[62] S.M. Choudhari and L. Ananthanarayan, *Food Chem.* 102, 77 (2007).

[63] R. Lavecchia and A. Zuorro, *European Food Res. Technol.* 228, 153 (2008).

[64] A. Zuorro and R. Lavecchia, *Biotechnology and Biotechnological Equipment* 24, 1854 (2011).

[65] J.L. Bureau and R.J. Bushway, *J. Food Sci.* 51, 128 (1986).

[66] G. Sadler, J. Davis, D. Dezman, *J. Food Sci.* 55, 1460 (1990).

[67] G. Britton, Carotenoids. In Natural Food Colorants. Hendry, G. A. F., Houghton, J. D., Eds.; Blackie: Glasgow, U.K., p. 280 (1992).

[68] A. Birtigh, M. Johannsen, G. Brunner, *J. Supercrit. Fluids* 8, 46 (1995).

[69] P.J. Vega, M.O. Balaban, S.F. O'Keefe, J.A. Cornell, *J. Food Science* 61, 757 (1996).

[70] I. F. Strati and V. Oreopoulou, *International Journal of Food Science & Technology* 46, 23 (2011).

[71] A.M. Adalid, S. Rosello, F. Nuez, *Journal of Food Composition and Analysis* 23, 613 (2010).

[72] T. Baysal, S. Ersus, and D. A. J. Starmans, *J. Agric. Food Chem. 48*, 5507 (2000).

[73] E. Rodriguez-Munoz, G. Herrera-Ruiz, G. Pedraza-Aboytes, G. Loarca-Pina, *Plant Foods for Human Nutrition* 64, 46 (2009).

[74] D.S. Sogi, R. Bhatia, S.K. Garg, A.S. Bawa, A. S. *Food Chemistry* 89, 53 (2004).

[75] A.M. Goula, *Journal of Food Engineering* 74, 37 (2006).

[76] R.C. Huang, R. C. *Journal of Food Process Engineering* 29, 1 (2006).

[77] P.H. Gann, *Cancer Research* 59, 1225 (1999).

[78] E. Giovannucci, *Journal of the National Cancer Institute* 87, 1767 (1995).

[79] P. Fernanda Pinto da Costa, M. Bueno Marcondes Ferraz, V. Ros Polski, E. Quast, F. P. Collares Queiroz, C. Joy Steel, *Ciênc. Tecnol. Aliment.* 30, 143 (2010).

[80] Z. Dehghan-Shoar, M. Tafadzwa, A. K. Hardacre, G. W. Reynolds, C. S. Brennan, J. Agric. Food Chem. *59*, 12047 (2011).

[81] K. Devinder, W. Ali Abbas, D. P. Singh, D. S. Sogi, *International Journal of Food Properties* 6, 1217 (2011).

INDEX

A

B

C

D

E

F

G

H

I

J

K

L

M

N

O

P

Q

R

S

T

U

V

W

X

Y

Z